GENETICS, GENOMICS AND BREEDING OF SUNFLOWER

Genetics, Genomics and Breeding of Crop Plants

Series Editor
Chittaranjan Kole
Department of Genetics and Biochemistry
Clemson University
Clemson, SC
USA

Books in this Series:

Published or in Press:
- Jinguo Hu, Gerald Seiler & Chittaranjan Kole: *Sunflower*
- Kristin D. Bilyeu, Milind B. Ratnaparkhe & Chittaranjan Kole: *Soybean*
- Robert Henry & Chittaranjan Kole: *Sugarcane*

Books under preparation:
- Jan Sadowsky & Chittaranjan Kole: *Vegetable Brassicas*
- C.P. Joshi, Stephen DiFazio & Chittaranjan Kole: *Poplar*
- Kevin Folta & Chittaranjan Kole: *Berries*

GENETICS, GENOMICS AND BREEDING OF SUNFLOWER

Editors

Jinguo Hu

USDA-Agricultural Research Service
Western Regional Plant Introduction Station
Pullman, WA
USA

Gerald Seiler

USDA-Agricultural Research Service
Northern Crop Science Laboratory
Fargo, ND
USA

Chittaranjan Kole

Department of Genetics and Biochemistry
Clemson University
Clemson, SC
USA

CRC Press
Taylor & Francis Group
Boca Raton London New York

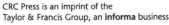

CRC Press is an imprint of the
Taylor & Francis Group, an **informa** business

Science Publishers
Enfield, New Hampshire

CRC Press
Taylor & Francis Group
6000 Broken Sound Parkway NW, Suite 300
Boca Raton, FL 33487-2742

First issued in paperback 2017

ISBN-13: 978-1-57808-676-4 (hbk)
ISBN-13: 978-1-138-11513-2 (pbk)

This book contains information obtained from authentic and highly regarded sources. While all reasonable efforts have been made to publish reliable data and information, neither the author[s] nor the publisher can accept any legal responsibility or liability for any errors or omissions that may be made. The publishers wish to make clear that any views or opinions expressed in this book by individual editors, authors or contributors are personal to them and do not necessarily reflect the views/opinions of the publishers. The information or guidance contained in this book is intended for use by medical, scientific or health-care professionals and is provided strictly as a supplement to the medical or other professional's own judgement, their knowledge of the patient's medical history, relevant manufacturer's instructions and the appropriate best practice guidelines. Because of the rapid advances in medical science, any information or advice on dosages, procedures or diagnoses should be independently verified. The reader is strongly urged to consult the relevant national drug formulary and the drug companies' and device or material manufacturers' printed instructions, and their websites, before administering or utilizing any of the drugs, devices or materials mentioned in this book. This book does not indicate whether a particular treatment is appropriate or suitable for a particular individual. Ultimately it is the sole responsibility of the medical professional to make his or her own professional judgements, so as to advise and treat patients appropriately. The authors and publishers have also attempted to trace the copyright holders of all material reproduced in this publication and apologize to copyright holders if permission to publish in this form has not been obtained. If any copyright material has not been acknowledged please write and let us know so we may rectify in any future reprint.

Cover illustration reproduced by courtesy of Dale Rehder

Library of Congress Cataloging-in-Publication Data

Genetics, genomics and breeding of sunflower / editors, Jinguo Hu, Gerald Seiler.
    p. cm. -- (Genetics, genomics and breeding of crop plants)
  Includes bibliographical references and index.
  ISBN 978-1-57808-676-4 (hardcover)
1. Sunflowers--Genetics. 2. Sunflowers--Genome mapping.
3. Sunflowers--Breeding. I. Hu, Jinguo. II. Seiler, Gerald J., 1949- III. Series.
  QK495.C74G445 2009
  635.9'3399--dc22
                                   2009039713

Visit the Taylor & Francis Web site at
http://www.taylorandfrancis.com

and the CRC Press Web site at
http://www.crcpress.com

Dedication

Professor Loren Rieseberg

This book is dedicated to Professor Loren Rieseberg, Canadian Research Chair in Plant Evolutionary Genomics, University of British Columbia, and Distinguished Professor, Department of Biology, Indiana University for his early and innovative research on plant evolutionary genomics of sunflower (*Helianthus annuus* L.) and its wild relatives. His pioneering research on hybridization and speciation in the genus have raised sunflower to the level of Darwin's finches and stickleback fishes as natural paradigms for the study of adaptive evolution and speciation. This work, in combination with Prof. Rieseberg's studies of the domestication process and habitat diversification, has furthered our understanding of the potential value of the wild sunflower species and their related traits for the expansion of sunflower production in diverse environments. Prof. Rieseberg was one of the first to document the very high levels of gene flow between cultivated and wild sunflowers, implying that transgene escape is likely. He also directs the Compositae Genome Project, a multiple collaborator project, which has developed over 250,000 ESTs for sunflower. More recently, he is leading an effort to sequence the sunflower genome. He has been recognized for his numerous significant accomplishments receiving several awards such as the MacArthur and Guggenheim Fellowships, Class of '54 Endowed Professorship, Fellow of the American Academy of Arts and Sciences, and the American Association for the Advancement of Science, Stebbins Medal, David Starr Jordan Prize, and the Teaching Excellence Recognition Award, Indiana University. During more than 20 years of teaching and research many national and international undergraduate, graduate, and postgraduate students have been beneficiaries of his teaching, expertise, philosophy, and passion for seeking out the how and why of science.

Preface to the Series

Genetics, genomics and breeding has emerged as three overlapping and complimentary disciplines for comprehensive and fine-scale analysis of plant genomes and their precise and rapid improvement. While genetics and plant breeding have contributed enormously towards several new concepts and strategies for elucidation of plant genes and genomes as well as development of a huge number of crop varieties with desirable traits, genomics has depicted the chemical nature of genes, gene products and genomes and also provided additional resources for crop improvement.

In today's world, teaching, research, funding, regulation and utilization of plant genetics, genomics and breeding essentially require thorough understanding of their components including classical, biochemical, cytological and molecular genetics; and traditional, molecular, transgenic and genomics-assisted breeding. There are several book volumes and reviews available that cover individually or in combination of a few of these components for the major plants or plant groups; and also on the concepts and strategies for these individual components with examples drawn mainly from the major plants. Therefore, we planned to fill an existing gap with individual book volumes dedicated to the leading crop and model plants with comprehensive deliberations on all the classical, advanced and modern concepts of depiction and improvement of genomes. The success stories and limitations in the different plant species, crop or model, must vary; however, we have tried to include a more or less general outline of the contents of the chapters of the volumes to maintain uniformity as far as possible.

Often genetics, genomics and plant breeding and particularly their complimentary and supplementary disciplines are studied and practiced by people who do not have, and reasonably so, the basic understanding of biology of the plants for which they are contributing. A general description of the plants and their botany would surely instill more interest among them on the plant species they are working for and therefore we presented lucid details on the economic and/or academic importance of the plant(s); historical information on geographical origin and distribution; botanical origin and evolution; available germplasms and gene pools, and genetic and cytogenetic stocks as genetic, genomic and breeding resources; and basic information on taxonomy, habit, habitat, morphology, karyotype, ploidy level and genome size, etc.

Classical genetics and traditional breeding have contributed enormously even by employing the phenotype-to-genotype approach. We included detailed descriptions on these classical efforts such as genetic mapping using morphological, cytological and isozyme markers; and achievements of conventional breeding for desirable and against undesirable traits. Employment of the in vitro culture techniques such as micro- and megaspore culture, and somatic mutation and hybridization, has also been enumerated. In addition, an assessment of the achievements and limitations of the basic genetics and conventional breeding efforts has been presented.

It is a hard truth that in many instances we depend too much on a few advanced technologies, we are trained in, for creating and using novel or alien genes but forget the infinite wealth of desirable genes in the indigenous cultivars and wild allied species besides the available germplasms in national and international institutes or centers. Exploring as broad as possible natural genetic diversity not only provides information on availability of target donor genes but also on genetically divergent genotypes, botanical varieties, subspecies, species and even genera to be used as potential parents in crosses to realize optimum genetic polymorphism required for mapping and breeding. Genetic divergence has been evaluated using the available tools at a particular point of time. We included discussions on phenotype-based strategies employing morphological markers, genotype-based strategies employing molecular markers; the statistical procedures utilized; their utilities for evaluation of genetic divergence among genotypes, local landraces, species and genera; and also on the effects of breeding pedigrees and geographical locations on the degree of genetic diversity.

Association mapping using molecular markers is a recent strategy to utilize the natural genetic variability to detect marker-trait association and to validate the genomic locations of genes, particularly those controlling the quantitative traits. Association mapping has been employed effectively in genetic studies in human and other animal models and those have inspired the plant scientists to take advantage of this tool. We included examples of its use and implication in some of the volumes that devote to the plants for which this technique has been successfully employed for assessment of the degree of linkage disequilibrium related to a particular gene or genome, and for germplasm enhancement.

Genetic linkage mapping using molecular markers have been discussed in many books, reviews and book series. However, in this series, genetic mapping has been discussed at length with more elaborations and examples on diverse markers including the anonymous type 2 markers such as RFLPs, RAPDs, AFLPs, etc. and the gene-specific type 1 markers such as EST-SSRs,

SNPs, etc.; various mapping populations including F_2, backcross, recombinant inbred, doubled haploid, near-isogenic and pseudotestcross; computer software including MapMaker, JoinMap, etc. used; and different types of genetic maps including preliminary, high-resolution, high-density, saturated, reference, consensus and integrated developed so far.

Mapping of simply inherited traits and quantitative traits controlled by oligogenes and polygenes, respectively has been deliberated in the earlier literature crop-wise or crop group-wise. However, more detailed information on mapping or tagging oligogenes by linkage mapping or bulked segregant analysis, mapping polygenes by QTL analysis, and different computer software employed such as MapMaker, JoinMap, QTL Cartographer, Map Manager, etc. for these purposes have been discussed at more depth in the present volumes.

The strategies and achievements of marker-assisted or molecular breeding have been discussed in a few books and reviews earlier. However, those mostly deliberated on the general aspects with examples drawn mainly from major plants. In this series, we included comprehensive descriptions on the use of molecular markers for germplasm characterization, detection and maintenance of distinctiveness, uniformity and stability of genotypes, introgression and pyramiding of genes. We have also included elucidations on the strategies and achievements of transgenic breeding for developing genotypes particularly with resistance to herbicide, biotic and abiotic stresses; for biofuel production, biopharming, phytoremediation; and also for producing resources for functional genomics.

A number of desirable genes and QTLs have been cloned in plants since 1992 and 2000, respectively using different strategies, mainly positional cloning and transposon tagging. We included enumeration of these and other strategies for isolation of genes and QTLs, testing of their expression and their effective utilization in the relevant volumes.

Physical maps and integrated physical-genetic maps are now available in most of the leading crop and model plants owing mainly to the BAC, YAC, EST and cDNA libraries. Similar libraries and other required genomic resources have also been developed for the remaining crops. We have devoted a section on the library development and sequencing of these resources; detection, validation and utilization of gene-based molecular markers; and impact of new generation sequencing technologies on structural genomics.

As mentioned earlier, whole genome sequencing has been completed in one model plant (Arabidopsis) and seven economic plants (rice, poplar, peach, papaya, grapes, soybean and sorghum) and is progressing in an array of model and economic plants. Advent of massively parallel DNA sequencing using 454-pyrosequencing, Solexa Genome Analyzer, SOLiD system, Heliscope and SMRT have facilitated whole genome sequencing in many other plants more rapidly, cheaply and precisely. We have included extensive coverage on the level (national or international) of collaboration and the strategies and status of whole genome sequencing in plants for which sequencing efforts have been completed or are progressing currently. We have also included critical assessment of the impact of these genome initiatives in the respective volumes.

Comparative genome mapping based on molecular markers and map positions of genes and QTLs practiced during the last two decades of the last century provided answers to many basic questions related to evolution, origin and phylogenetic relationship of close plant taxa. Enrichment of genomic resources has reinforced the study of genome homology and synteny of genes among plants not only in the same family but also of taxonomically distant families. Comparative genomics is not only delivering answers to the questions of academic interest but also providing many candidate genes for plant genetic improvement.

The 'central dogma' enunciated in 1958 provided a simple picture of gene function—gene to mRNA to transcripts to proteins (enzymes) to metabolites. The enormous amount of information generated on characterization of transcripts, proteins and metabolites now have led to the emergence of individual disciplines including functional genomics, transcriptomics, proteomics and metabolomics. Although all of them ultimately strengthen the analysis and improvement of a genome, they deserve individual deliberations for each plant species. For example, microarrays, SAGE, MPSS for transcriptome analysis; and 2D gel electrophoresis, MALDI, NMR, MS for proteomics and metabolomics studies require elaboration. Besides transcriptome, proteome or metabolome QTL mapping and application of transcriptomics, proteomics and metabolomics in genomics-assisted breeding are frontier fields now. We included discussions on them in the relevant volumes.

The databases for storage, search and utilization on the genomes, genes, gene products and their sequences are growing enormously in each second and they require robust bioinformatics tools plant-wise and purpose-wise. We included a section on databases on the gene and genomes, gene expression, comparative genomes, molecular marker and genetic maps, protein and metabolomes, and their integration.

Notwithstanding the progress made so far, each crop or model plant species requires more pragmatic retrospect. For the model plants we need to answer how much they have been utilized to answer the basic questions of genetics and genomics as compared to other wild and domesticated species. For the economic plants we need to answer as to whether they have been genetically tailored perfectly for expanded geographical regions and current requirements for green fuel, plant-based bioproducts and for improvements of ecology and environment. These futuristic explanations have been addressed finally in the volumes.

We are aware of exclusions of some plants for which we have comprehensive compilations on genetics, genomics and breeding in hard copy or digital format and also some other plants which will have enough achievements to claim for individual book volume only in distant future. However, we feel satisfied that we could present comprehensive deliberations on genetics, genomics and breeding of 30 model and economic plants, and their groups in a few cases, in this series. I personally feel also happy that I could work with many internationally celebrated scientists who edited the book volumes on the leading plants and plant groups and included chapters authored by many scientists reputed globally for their contributions on the concerned plant or plant group.

We paid serious attention to reviewing, revising and updating of the manuscripts of all the chapters of this book series, but some technical and formatting mistakes will remain for sure. As the series editor, I take complete responsibility for all these mistakes and will look forward to the readers for corrections of these mistakes and also for their suggestions for further improvement of the volumes and the series so that future editions can serve better the purposes of the students, scientists, industries, and the society of this and future generations.

Science publishers, Inc. has been serving the requirements of science and society for a long time with publications of books devoted to advanced concepts, strategies, tools, methodologies and achievements of various science disciplines. Myself as the editor and also on behalf of the volume editors, chapter authors and the ultimate beneficiaries of the volumes take this opportunity to acknowledge the publisher for presenting these books that could be useful for teaching, research and extension of genetics, genomics and breeding.

Chittaranjan Kole

Preface to the Volume

The sunflower has fascinated mankind for centuries and has served as an inspiration to artists, poets, photographers, and also for business promotions. Being one of the a few agricultural crops that originated from the New World, the conspicuous and showy sunflower is revered worldwide for its magnificence and versatility. The oilseed sunflower contributes approximately 10% of the world's plant-derived edible oil and the confection type sunflower holds a considerable share of the directly consumed snacks market. In addition, sunflower is also grown as an ornamental for cut flowers, as well as in home gardens. Improvement of the sunflower crop has been achieved through traditional breeding in terms of general productivity, profitability to the growers and processors, and nutritional benefits to the consumers. We are now embarking on the age of genomics which will expedite the process of genetic improvement of crops. During the last two decades, there has been an explosion of information on genetic markers, DNA sequences, and genomic resources for most major food crops including sunflower.

This volume is intended to bridge traditional research with modern molecular investigations on sunflower. It begins with basic information about the sunflower plant and germplasm diversity (Chapter 1), followed by classical genetics and traditional breeding (Chapter 2), history and achievement of genome mapping (Chapter 3). Four chapters review activities on mapping single-gene traits (Chapter 4), QTL mapping (Chapter 5), gene cloning (Chapter 6) and marker-assisted breeding (Chapter 7). Since sunflower is one of the most important oil crops, a chapter is devoted to the genetic regulation of seed oil content (Chapter 8). Transgenic sunflower, an interesting topic as old as the technology itself, is also discussed (Chapter 9). The book ends with a chapter on future prospects of the sunflower crop (Chapter 10). We hope that this book is useful to sunflower researchers as well as to people working with other crop species.

Each chapter has been written by one or more experts who have worked diligently in compiling information about their respective areas of expertise. We greatly appreciate their effort and time devoted to this book.

Jinguo Hu
Gerald Seiler
Chittaranjan Kole

Contents

7. Molecular Breeding   221

Begoña Pérez-Vich and *Simon T. Berry*

List of Contributors

Simon T. Berry
Limagrain UK Limited, Woolpit Business Park, Windmill Avenue, Woolpit, Bury St. Edmunds, Suffolk, IP30 9UP, UK.
Email: *simon.berry@nickerson-advanta.co.uk*

André Bervillé
INRA, Department of Genetics and Plant breeding, UMR DIA-PC, 2 place viala, 33, 34060 Montpellier cedex 1, France.
Email: *Andre.Berville@supagro.inra.fr*

Miguel Angel Cantamutto
Departamento de Agronomía, Universidad Nacional del Sur, 8000, Bahía Blanca, Argentina.
Email: *mcantamutto@yahoo.com*

Volker Hahn
University of Hohenheim, Landessaatzuchtanstalt, Fruwirthstr. 21, 70593 Stuttgart, Germany.
Email: *vhahn@uni-hohenheim.de*

Sonia Hamrit
Institute of Biological Sciences, Department of Plant Genetics, University of Rostock, Albert-Einstein-Str.3, 18051, Rostock, Germany.
Email: *sonia.hamrit@uni-rostock.de*

Renate Horn
Institute of Biological Sciences, Department of Plant Genetics, University of Rostock, Albert-Einstein-Str. 3, 18051, Rostock, Germany.
Email: *renate.horn@uni-rostock.de*

Jinguo Hu
USDA-ARS, Western Regional Plant Introduction Station, 59 Johnson Hall, Washington State University, Pullman, WA 99164, USA.
Email: *jinguo.hu@ars.usda.gov*

Chao-Chien Jan
USDA-ARS, Northern Crop Science Laboratory, 1307 18th Street North, Fargo, ND 58105, USA.
Email: chaochien.jan@ars.usda.gov

Begoña Pérez-Vich
Instituto de Agricultura Sostenible (CSIC), Alameda del Obispo s/n. 14004, Córdoba, Spain.
Email: *bperez@ias.csic.es*

Monica Poverene
Departamento de Agronomía, Universidad Nacional del Sur and CERZOS-CONICET, 8000, Bahía Blanca, Argentina.
Email: *poverene@criba.edu.ar*

Seifollah Poormohammad Kiani
INRA, Genetics and Plant Breeding (SGAP UR254), 78026, Versailles, France.
Email: *skiani@versailles.inra.fr*

Ahmad Sarrafi
INP-ENSAT, IFR 40, Laboratoire de Biotechnologie et Amélioration des Plantes (BAP), Castanet Tolosan, 31326, France.
Email: *sarrafi@ensat.fr*

Gerald J. Seiler
USDA-ARS, Northern Crop Science Laboratory, 1307 18th Street North, Fargo, ND 58105, USA.
Email: *gerald.seiler@ars.usda.gov*

Felicity Vear
INRA-UMR ASP, Domaine de Crouelle, 234, Ave du Brezet, 63000 Clermont-Ferrand, France.
Email: *vear@valmont.clermont.inra.fr*

Brady Vick
USDA-ARS, Northern Crop Science Laboratory, Fargo, ND 58105, USA.
Email: *Brady.Vick@ars.usda.gov*

Silke Wieckhorst
Technische Universität München, Wissenschaftszentrum Weihenstephan für Ernährung, Landnutzung und Umwelt, Lehrstuhl für Pflanzenzüchtung, Am Hochanger 4, 85350 Freising, Germany.
Email: *Silke.wieckhorst@wzw.tum.de*

List of Abbreviations

ABA	Abscisic acid
AB-QTL	Advanced backcross QTL
ABRE	ABA-responsive element
ACC	1,Aminocyclopropane-1-carboxylate
ACP	Acid phosphatase
ACP	Acyl carrier protein
AFLP	Amplified fragment length polymorphism
AHAS	Hydroxyacetic acid synthetase
AHAS	Acetohydroxyacid synthase
AP	Arbitrary primer
AS	Asparagine synthetase
ASPE	Allele-specific primer extension
BAC	Bacterial artificial chromosome
BIBAC	Binary-bacterial artificial chromosome
BIO	Biomass
BLAST	Basic local alignment search tool
BP	Base pair
BSA	Bulked segregant analysis
Bt	*Bacillus thuringiensis*
CaMV	Cauliflower mosaic virus
CAPS	Cleaved amplified polymorphic sequence
cDNA	Complementary DNA
CGP	Compositae Genome Program
CGPDB	Compositae Genome Project Database
CMS	Cytoplasmic male sterility
CpT1	Carnitine palmitoyltransferase I
Cry	Cryptochrome (photolyase-like) gene
Cry1Ab	Plant-incorporated protectant *Bt* protein
CSIC	Consejo Superior de Investigaciones Científicas
CUGI	Clemson University Genomics Institute
DDRT-PCR	Differential display reverse transcriptase PCR
DES	Diethyl sulphate
dHPLC	Denaturing high-performance liquid chromatography
DPW	Double petal whirl

DSF	Sowing to flowering
DTF	Emergence to flowering
DUS	Distinctness, uniformity, and stability
ECB	European corn borer
ELIP	Early light induced proteins
EMS	Ethylmethane sulfonate
EPSPS	5-Enolpyruvylshikimate-3-phosphate
EST	Expressed sequence tag
EU	European Union
EXON	Coding DNA sequence
FAD	Fatty acid desaturase
FAS	Fatty acid synthase
FISH	Fluorescence in situ hybridization
FSU	Former Soviet Union
GFP	Green fluorescence protein
GISH	Genomic in situ hybridization
GL	Glyphosate herbicide
GM	Genetically modified
GMO	Genetically modified organism
GPC	Gas chromatography
GS	Glutamine synthetase
GUS	ß-Glucuronidase
GYP	Grain yield per plant
HAP3	Heme-activated protein 3
HAS1	Hyaluronan synthase 1 gene
HD	Head diameter
HD	Homeobox domain
HD-ZIP	Homeodomain-leucine zipper
HISh	Histidin
HO	High-oleic
HOAC	High oleic acid content
HPLC	High-pressure liquid chromatography
HS	High-stearic
HSE	Heat shock element
HSF	Heat shock factor
HSP	Heat shock protein
HT	Herbicide tolerant
HW	Head weight
IDH	Isocitrate dehydrogenase
IFLP	Intron fragment length polymorphism
IMI	Imidazolinone
INDEL	Insertion-deletion
IR	Insect resistant

ISC	International Sunflower Conference
KNOX	*Knotted1*-like homoeobox
LA	Linoleic acid
LAD	Leaf area duration
LAF	Leaf area at flowering
LD	Linkage disequilibrium
LEA	Late-embryogenesis-abundant
LEC1	Leafy cotyledon 1
LEU	Leucine
LG	Linkage group
LIL1	Leafy cotyledon-like 1
LMW	Low molecular weight
LN	Leaf number
LO	Low oleic
LOD	Logarithm of odds
LRR	Leucine-rich repeat
MAS	Marker-assisted selection
MCA	Melting curve analyses
MDH	Malate dehydrogenase
MF	Missing flower
MOD	Microsomal oleate desaturase
MON810	Insect resistant maize line (trade name YieldGard)
MPBQ	2-Methyl-6-phytyl-1,4-benzoquinone
MR	Relative molecular weight
mRNA	Messenger-RNA
MSBQ-MT	2-Methyl-6-solanyl-1,4-benzoquinone methyltransferase
NBS	Nucleotide binding site
NCRPIS	North Central Regional Plant Introduction Station
NILs	Near-isogenic lines
NIRS	Near infrared reflectance spectroscopy
NMR	Nuclear magnetic resonance
NMS	Nuclear male sterile/sterility
NMU	Nitroso-N-methylurea
NPGS	National Plant Germplasm System (of USDA-ARS)
*npt*II	Neomycin phosphotransferase II gene
nsLIP	Non-specific lipid transfer protein
OA	Oleic acid
OA	Oxalic acid
Ol	High oleic mutation
ORF	Open reading frame
OXO	Oxalate oxidase
PAH	Polycyclic aromatic hydrocarbon
PAT	Phosphinothricin acetyltransferase

PC	Phospahatidyl choline
PCR	Polymerase chain reaction
PFU	*Pyrococcus furiosus*
PGD	6-Phosphogluconate dehydrogenase
PGI	Glucose-phosphate isomerase
PGM	Phosphoglucomutase
PH	Plant height
PIC	Polymorphic information content
PIP	Plasma membrane
PK	Protein kinase-like
PP	Photoperiod
PR	Pathogenesis-related
PRX	Peroxidase
PTGS	Post-transcriptional gene silencing
QTL	Quantitative trait loci
QTL x E	QTL x environment
QTN	Quantitative trait nucleotide
RACE	Rapid amplification of cDNA end
RAGE	Rapid amplification of the genomic DNA end
RAPD	Random amplified polymorphic DNA
RFLP	Restriction fragment length polymorphism
RGA	Resistance gene analog
RGC	Resistance gene candidate
RIL	Recombinant inbred line
RR	Roundup Ready® variety
RT	Reverse transcriptase
RT-PCR	Reverse Transcriptase-PCR
S-ACP-DES	Stearoyl-acyl carrier protein desaturase
SAD	Stearoyl-acyl carrier protein desaturase
SBE	Single base extension
SCAR	Sequence characterized amplified region
SD	Seed dormancy
SD	Shoot diameter
SDH	Shikimate dehydrogenase
SDI	Sunflower drought induced
sHSP	Small heat shock protein
SI	Self-incompatibility
SNP	Single nucleotide polymorphism
SP	Self-pollination
SSCP	Single-strand conformation polymorphism
SSLP	Simple sequence length polymorphism
SSR	Simple sequence repeat
STF	Sowing to flowering

STR	Short tandem repeats
STS	Sequence tagged site
SU	Sulfonylurea
SuCMoV	Sunflower chlorotic mottle virus
T_0, T_1, T_2	Succesive generation transformants
TAG	Triacylglycerol
TC	Testcross
TC	Tocopherolcyclase
T-DNA	Transformant-DNA (transgene)
TILLING	Targeting induced local lesions in genomes
TIP	Tonoplast
TIR	Toll interleukin-receptor
TRAP	Target region amplification polymorphism
UPOV	International Union for the Protection of New Varieties of Plants
USDA-ARS	United States Department of Agriculture-Agricultural Research Service
UTR	Untranslated region

Basic Information

Gerald Seiler* and *Chao-Chien Jan*

ABSTRACT

Sunflower is one of a few crops that has its origin in North America. The crop is grown worldwide and performs well in most temperate climates of the world, with significant production occurring in each of the six crop producing continents. It is the second largest hybrid crop and the fifth largest edible oilseed crop grown on 22.9 million hectares in 60 countries with a value of over US$40 billion. Sunflower production continues to face challenges from both abiotic and biotic factors as production is shifting from areas of high productivity to marginal areas with lower yield potential. The challenge for the sunflower community is to breed sunflowers adaptable to these marginal environments while still increasing seed yield. The wild ancestors of the crop have served as a source of many genes for pest resistance, especially for diseases. Molecular biology has added to the scope of plant breeding in sunflower, providing an option to manipulate plant expressions. Researchers will have to strive to combine the best conventional and molecular genetic and genomic approaches to improve sunflower germplasm to keep sunflower as an economically viable global crop. This will require a multidisciplinary team approach and a commitment to a long-term integrated genetic improvement program.

Keywords: Genetic resources; genebanks; botanical description; domestication; wild species utilization; interspecific hybridization; *Helianthus*

USDA Agricultural Research Service, Northern Crop Science Laboratory, Fargo, ND 58105, USA.
*Corresponding author: *Gerald.Seiler@ars.usda.gov*

1.1 Introduction

Sunflower is distinguished from other crops by its single stem and conspicuous large head. Radiant and dazzling, the common sunflower has pleased and fascinated mankind for centuries. *Helianthus*, the genus name of sunflower (*Helianthus annuus* L.) is derived from the Greek words "helios", meaning sun, and "anthus", meaning flower. The Spanish name for sunflower, "girasol", and the French name "tournesol" literally means "turn with the sun", a trait exhibited by sunflower until anthesis, after which the capitula (heads) face the east (Fick 1989). It is a relatively new crop among the field crops and is unique in several aspects. It is one of a few crops that have their origin in North America. Sunflower is further unique in that it has been bred for distinctly different uses: first as a forage crop, later as an oilseed crop, edible confection, birdseed, and to a much lesser extent as an ornamental for home gardens and the cut-flower industry. Cultivated sunflower is a globally important oilseed, food, and ornamental crop. Sunflower derives most of its economic value from the oil extracted from the achenes, with the remaining value from the meal. Sunflower oil is of very high quality and generally sells for a premium price compared to soybean, rapeseed, cottonseed, and groundnut. Sunflower is grown worldwide and performs well in most temperate climates of the world, with significant production occurring in each of the six crop-producing continents.

1.2 Economic Importance

1.2.1 World Production Area

The world production of sunflower in 2005 was estimated to be 22.9 million hectares in 60 countries (Table 1-1) (USDA 2007). It is the second largest hybrid crop, next only to maize, and the fifth largest among the oilseed crops after soybean, rapeseed, cottonseed, and groundnut as a source of edible oil with a value of over US$40 billion (FAO 2008). Cultivated sunflower is primarily grown from single-cross hybrid seed, which was valued at $640 million in 2002, second only to maize. In 2003, sunflower seed accounted for 10% of the world's edible plant derived oil (Kleingartner 2004).

The largest producers of sunflower seed are the Russian Federation with 5.40 million hectares and the Ukraine with 3.69 million hectares, accounting for almost 40% of all production, followed by the European Union, consisting mainly of France, Hungary, Italy, Spain, Bulgaria, and Romania, which produced 3.60 million hectares in 2005 (Table 1-1) (USDA 2007). This was followed by India with 2.34 million hectares and Argentina with 1.89 million hectares. Both China and the United States produced 1.05 million hectares in 2005. Although world sunflower seed production increased by 24% or 5 million metric tons between 1993 and 2003, sunflower seed production declined in terms of market share of the five major oilseeds during the same period (Kleingartner 2004).

Sunflower production in Argentina had the highest yield, 2.03 Mt/ha, in 2005 (Table 1-1). The lowest yield was in India with 0.63 Mt/ha. The largest producer of sunflower, the Russian Federation, averaged 1.19 Mt/ha, while the Ukraine averaged 1.27 Mt/ha. Yield of sunflower from other areas of the world averaged 0.95 Mt/ha, much lower than several of the major sunflower production areas where sunflower is highly managed for maximum yield.

Table 1-1 World sunflower area, production, and yield for 2005 (USDA 2007).

Country	Area harvested (Million Ha)	Production (Million Mt)	Yield (Mt /Ha)
Argentina	1.89	3.84	2.03
China	1.05	1.83	1.74
European Union	3.60	6.02	1.67
Other European	0.41	0.68	1.66
India	2.34	1.49	0.63
Russian Federation	5.40	6.44	1.19
Ukraine	3.69	4.71	1.27
United States	1.05	1.72	1.63
Turkey	0.57	0.78	1.36
Other	2.94	2.79	0.95
Total	22.94	30.30	Mean = 1.32

1.2.2 Nutritional Information

1.2.2.1 Oil and Oil Quality

Sunflower derives most of its economic value from the extracted oil, with the remaining value from the meal. The achenes of oilseed sunflower are usually black. The oil extracted from the achenes accounts for about 80% of the total value of the oilseed sunflower crop (Fick and Miller 1997). Oil content depends on both the percentage of hull and the oil concentration in the kernel. Hull percentage among genotypes may range from 10–60%, while oil content in the kernel may range from 260–720 g/kg. Hull percentages of achenes and oil content in the kernels of present high oil cultivars or hybrids are in the range of 20–25% and 570–670 g/kg, respectively (Miller and Fick 1997). Most oilseed hybrids have a whole achene oil content of 400–500 g/kg. Sunflower oil is premium oil due to its high unsaturated linoleic fatty acid concentration, low linolenic acid content, the absence of significant toxic compounds, and its excellent nutritional properties. Linoleic acid is an essential fatty acid not synthesized by humans, and is the precursor of gamma-linolenic and arachidonic acids (Dorrell and Vick 1997).

The primary use of sunflower oil is as a salad and cooking oil, and as a major ingredient in some margarine and shortening products. Sunflower oil can be used in the manufacture of lacquers, copolymers, polyester films, modified resins, and plasticizers when there is a price advantage to the manufacturer. The high concentration of linoleic acid and very low concentration of linolenic acid means that despite the moderate iodine number of 125–140, it has good drying qualities without the yellowing associated with high-linolenic acid oils (Dorrell and Vick 1997). It can also be used in the manufacture of soap and detergents (Suslov 1968). Emulsifiers and surfactants from fats and oils are also used in formulating pesticides (Pryde and Rothfus 1989).

Traditional sunflower oil is composed of triacylglycerols that exist in the liquid form at room temperature and have a low melting point. The oil is composed primarily of saturated $C16:0$ palmitic (70 g/kg) and $C18:0$ stearic (40 g/kg), monounsaturated $C18:1$ oleic (200 g/kg) and polyunsaturated $C18:2$ linoleic (690 g/kg) fatty acids, with oleic and linoleic accounting for about 89% of the total. Sunflower oil has traditionally been considered as a polyunsaturated oil because of its relatively high linoleic acid concentration of 680–720 g/kg. The highly polyunsaturated nature of sunflower oil makes it an attractive product for health-conscious consumers who desire a diet that maximizes the polyunsaturated-to-saturated fat ratio, which is considered beneficial in reducing the risk of cardiovascular disease (Mensink et al. 1994; Willett 1994).

Traditional sunflower oil (high linoleic acid) is not commonly used for industrial purposes because of its generally higher value, compared to other oilseeds. However, it is used to some extent in paints, varnishes, and plastics because of its good semidrying properties. Along with other vegetable oils, it is has potential value for the production of adhesives, agrochemicals, particularly hydrophobic pesticides, surfactants, additional plastics and plastic additives, fabric softeners, synthetic lubricants, and coatings. Actual use will depend to a large extent on its price relative to that of petroleum-based chemicals.

In the United States, sunflower breeders have developed a mid-oleic (550–700 g/kg) sunflower oil called NuSun® (National Sunflower Association, Bismarck, North Dakota, USA), which possesses a significant advantage over several other popular oils, such as soybean and canola. NuSun® does not have to be hydrogenated prior to its use as frying oil, and therefore has negligible trans fatty acids (Kleingartner 2002; Vick and Miller 2002). This oil offers desirable frying and flavor characteristics, increases the lifespan of the heated oil, and confers a healthful fatty acid composition containing an adequate level of "heart healthy" polyunsaturated fatty acids, yet is free of hydrogenated *trans* fatty acids. Furthermore, the increased oleic acid content has the added benefit of slightly lowering saturated palmitic and stearic fatty acid concentrations to 90 g/kg.

A third type of sunflower oil is high-oleic. High-oleic oil contains 800 g/kg or more oleic acid and 60 g/kg palmitic and 30 g/kg of stearic fatty acids. This type of oil is preferred in the frying industry since it does not require hydrogenation and thus contains no *trans* fats. It is well recognized in the food industry that high oleic and saturated fatty acids provide frying functionality. Increasing saturated fatty acids in frying oil was never an objective because highly saturated oils had already been largely eliminated from the domestic market due to health concerns. High-oleic sunflower oil has only a moderate level of saturated fatty acids at approximately 90 g/kg. Currently, the production of high oleic sunflower oil is limited. The high oxidative stability of this oil makes it a potentially good choice for biofuels, but actual use will depend to a large extent on its price relative to that of petroleum-based chemicals. Several reports have been published evaluating sunflower oil and its blends with diesel as a fuel (Morrison et al. 1995). In addition, with the development of high-oleic sunflower hybrids, sunflower oil has become a more important feedstock for the oleochemical industry, in which the cosmetics industry is a major user (Luhs and Friedt 1994).

Non-oilseed or confectionery sunflower usually has very large black-with-white-striped achenes; i.e., those achenes that pass over a 7.9-mm round hole sieve are used as a confection or snack food, usually roasted and salted. Sunflower kernels are also used in the baking industry, as a condiment for salads and other foods. Non dehulled or partially dehulled sunflower meal can be substituted successfully for soybean meal of equal protein percentage in feeding ruminant animals. Partially or completely dehulled sunflower meal is desirable for feeding swine and poultry. Achenes are also used for feeding birds and in small animal feed.

1.2.2.2 *Protein*

Protein concentration of achenes is of interest for human and livestock consumption, but is also usually related to the production of oil in the achenes. Commercial sunflower meal has a protein concentration of approximately 440 g/kg (dehulled) and 280 g/kg (whole achene) (Dorrell and Vick 1997). Variability among lines or genotypes for protein concentration is sufficient for selection to increase protein concentration of sunflower achenes in a breeding program. However, selection for high protein usually results in lower oil concentration because of a negative correlation between the two traits. Breeding to improve protein concentration of sunflower kernels from about 240 to near 400 g/kg, while maintaining acceptable oil concentration, appears to be a realistic objective (Ivanov and Stoyanova 1978).

Sunflower meal is most efficiently utilized when blended with meal of soybean for use in rations for dairy and beef cattle, sheep, swine, and poultry. Nearly all sunflower meal produced in the USA is used by the livestock feeding industry. However, sunflower protein ingredients derived from dehulled kernels have been evaluated extensively for use in human food, both in Europe and in the USA (Lusas 1982). Sunflower flour and protein concentrates and isolates show promise and are being used to a limited extent in bakery products, infant formula, meat, and meat extenders. Sunflower meal develops a greenish color due to the high concentration of chlorogenic acid that must be removed prior to use in developing products for human consumption.

1.2.2.3 Tocopherol Composition

Tocopherols are powerful natural fat soluble antioxidants that inhibit lipid oxidation in foods and biological systems. Since commercial interest in tocopherols and their antioxidative properties has increased in recent years, enhancing this component of sunflower oil has become a breeding objective. α-tocopherol (vitamin E) is the principal tocopherol in sunflower oil, usually representing over 90% of the total tocopherols. High α-tocopherol concentrations make sunflower oil less stable for frying. A partial substitution of α-tocopherol by a tocopherol derivative with greater antioxidant action would improve the oxidative stability of sunflower oil. Although sunflower oil stability can be enhanced by decreasing the linoleic acid concentration, studies of the tocopherol content of sunflower oil showed that it could be further improved by increasing the proportions of γ- and δ - tocopherols (Warner 2005). Since α-tocopherol is the weakest antioxidant in vitro, its partial replacement by β, γ, and δ or other tocopherol forms, is an important breeding objective.

The tocopherol content and composition of 36 wild *Helianthus* species was examined by Velasco et al. (2004a). Tocopherol concentration averaged 328 mg/kg in achenes with an average profile of 99.0% α-tocopherol, 0.7% β-tocopherol, and 0.3% γ- tocopherol. Cultivated sunflower has an average tocopherol concentration of 669 mg/kg of achenes with 92.4% α-tocopherol, 5.6% β-tocopherol, and 2.0% γ-tocopherol. The maximum tocopherol content in any wild population was observed in *H. maximiliani* with 673 mg/kg. Unusually high levels of β-tocopherol were observed in one population of *H. praecox* with 11.2% of the total tocopherols and one population of *H. debilis* with 11.8%. Increased γ-tocopherol levels were identified in one population of *H. exilis* with 7.4% of the total tocopherols and two populations of *H. nuttallii* with 11.0 and 14.6%.

Dolde et al. (1999) screened 12 modified sunflower oils for total tocopherols with values ranging from 534 to 1858 µg/g, with α-tocopherol as the predominant form, and β-tocopherol found in only two samples at very low levels. Tocopherol concentration does not appear to be related to saturated fatty acid concentration, so breeding for both can be accomplished at the same time.

Demurin (1993) developed the first high β-tocopherol line, LG-15, with β-tocopherol comprising 500 g/kg of the total tocopherols, and the first high γ lines, LG-17 and LG-24, with tocopherol contents of 950 and 840 g/kg of γ-tocopherol, respectively. Velasco et al. (2004b) have released two germplasm lines, T589 with β-tocopherol compositions of 340 to 542 g/kg, and T2100 with 850 g/kg of γ-tocopherol.

New variations of the tocopherol profile have been created by the use of the chemical mutagen ethyl methanesulfate (EMS) creating lines IAST-540 and IAST-1 with a γ-tocopherol content of 940 g/kg (Velasco et al. 2004c). Crosses between lines IAST-1 and T589 led to the production of IAST-5, with 700 g/kg α-tocopherol and 300 g/kg β-tocopherol, and IAST-4 with respective α, β-, γ- and δ-tocopherol contents of 40, 30, 340, and 580 g/kg of the total tocopherol composition.

Warner et al. (2008) found that when the γ-tocopherol content of mid-oleic NuSun® oil was increased from its regular level of 20 to 300–700 ppm, the oxidative stability of the oil increased significantly compared to NuSun® oil with its normal low γ-tocopherol level. The modified oils had α-tocopherol contents up to 300 ppm without negatively affecting the stability of the oil. An oil with one of the best oxidative stabilities had a tocopherol profile of 479 ppm γ, 100 ppm δ and 300 ppm α, indicating that the oxidative stability of modified NuSun® could be improved and still be a good source of vitamin E.

1.3 Crop History

1.3.1 Origin and Domestication

During the past 50 years, Heiser (1951, 1954, 1965, 1978, 1985, 1998) developed the following scenario for the origin and development of the domesticated sunflower from its progenitor, wild *H. annuus*. Prior to the arrival of mankind in the New World, *H. annuus* was restricted to the southwestern USA. Wild *H. annuus* was used by Native Americans for food, and due to its association with humans, it became a camp-following weed and was transported eastward. This weedy sunflower was subsequently domesticated in central USA and carried to the east and southwest.

There is strong archaeological support for the origin of the domesticated sunflower from the central and eastern states in the USA. Large achenes (> 7 mm in length) have been found at several archaeological sites in the central and eastern states, whereas only wild sunflower achenes have been found in sites from the Southwest and Mexico. Until recently, sunflower achenes from the Higgs site in eastern Tennessee (2850 BP) and the Marble Bluff Rock shelter in northwest Arkansas (2843 BP) were the earliest evidence of domesticated sunflower (Brewer 1973; Ford 1985; Crites 1991). However, carbonized sunflower achenes more recently recovered from the Hayes site in Middle Tennessee have yielded a radiometric date of 4625 BP, extending the earliest record of the domesticated sunflower by 1,400 years (Crites 1993).

The possibility that domesticated sunflower may have independent origins in Mexico and the southwestern USA has also been considered (Heiser 1985, 1998). However, until recently, the presence of domesticated achenes in archaeological records outside of the USA was lacking. The morphological similarity among Native American varieties of the domesticated sunflower, and the virtual monomorphism for isozymes and chloroplast DNA (cpDNA) in cultivated lines did not support the multiple origin hypotheses (Rieseberg and Seiler 1990). Recent evidence from an archeological site near San Andres, Tabasco, Mexico, indicates that a second or concurrent center of origin for domestication of sunflower may exist based on a single achene (Lentz et al. 2001). A carbonized achene from that site was dated at 4130 BP. Smith (2006) did not accept the identification of either the achene or seed as sunflower from San Andres. Recently, Heiser (2008) suggested that the achene examined by Lentz et al. (2001) was not *Helianthus*, but bottle gourd (*Lagenaria siceraria*), challenging the timing and possible domestication of sunflower in Mexico. Heiser (2008) felt that there is no convincing evidence for the sunflower in the archaeological record of Mexico, and the historical record provides no support for the domestication or pre-Columbian presence of the sunflower anywhere south of northern Mexico. Lentz et al. (2008) reported the discovery of three more Mexican achenes, this time from Cueva del Gallo in Morales, which date to around 2600 BC. They felt that this discovery strongly supports that the sunflower was a pre-Columbian domesticate in Mesoamerica, but does not preclude a second domestication event in eastern North America. Based on a molecular study, Harter et al. (2004) indicated that Mexican domesticates did not contribute to extant domesticated sunflower germplasm, supporting an independent origin in eastern USA.

Molecular techniques have been used to search for the origin of cultivated sunflower. Morphological, geographical and archeological evidence has led to the hypothesis that the domesticated sunflower was derived from a wild/weedy form of *H. annuus*, possibly in the Midwest. Molecular evidence was concordant with this hypothesis with high degrees of enzymatic and cpDNA sequence similarity observed between wild and domesticated *H. annuus*, and domesticated *H. annuus* contained a subset of the alleles and cpDNA found in wild *H. annuus* (Rieseberg and Seiler 1990). The extensive polymorphism in the wild plants and the virtual monomorphism in the cultivated lines for both isozyme and cpDNA phenotypes further suggest a single origin of the domesticated sunflower from a very limited gene pool. In addition, Native American varieties of the domesticated sunflower were genetically more variable than other cultivated lines, possibly indicating that they gave rise to the other cultivated stocks. Molecular evidence did not, however, allow conclusions as to the exact geographic origin of the domesticated sunflower (Rieseberg and Seiler 1990).

Arias and Rieseberg (1995) used randomly amplified polymorphic DNA (RAPD) loci to investigate the origin and genetic relationships of domesticated sunflower and its wild relatives. RAPD data supported the origin of the domesticated sunflower from wild *H. annuus*; however, because of the high level of identity between the two species, little information was provided regarding the geographic origin of the domesticated sunflower. Cronn et al. (1997), using allozyme variation, concluded that domesticated sunflowers form a genetically coherent group and that wild sunflowers from the Great Plains may include the most likely progenitor of domesticated sunflower. Systematic data does support the southwestern USA as a site of origin for annual sunflowers, including *H. annuus* (Heiser et al. 1969; Rieseberg et al. 1991).

Quantitative trait loci (QTL) controlling phenotypic differences between cultivated sunflower and its wild progenitor were investigated by Burke et al. (2002b). They concluded on the basis of the directionality of QTLs that strong directional selection for increased achene size appears to have played a central role in sunflower domestication. None of the other traits showed similar evidence of selection. The occurrence of the numerous wild alleles with cultivated-like effects, combined with the lack of major QTLs, suggest that sunflower was readily domesticated.

1.3.2 Dispersion

The original geographic range of wild *H. annuus* or the identity of the form from which *H. annuus* is derived is not clear (Asch 1993). Asch (1993) suggests that *H. annuus* originated as a colonizer of natural disturbances, and that bison (*Bison bison* Skinner and Kaiser) created extensively disturbed habitats suitable for colonization by sunflower. Bison may also have served as a dispersal agent for sunflower, by transporting sunflower achenes trapped in matted hair. He also suggested that pre-domestication events generated a wide distribution for wild *H. annuus* throughout the Midwest prior to the arrival of mankind on the scene, and that it was this midwestern form of *H. annuus* that actually gave rise to the domesticated sunflower. Unfortunately, no archaeological evidence exists to either refute or support such hypotheses and archaeological records tell us nothing about the pre-human geographic distribution of wild *H. annuus*.

The achene was originally used directly as food and crudely extracted oil. Native Americans had selected tall, single-headed landraces by the time European explorers reached North America in the 16th century. While sunflower was not a staple of the Native American's diet as were maize, beans, and squash, it was nonetheless cultivated by many tribes from eastern North America throughout the Midwest and as far south as northern Mexico (Putt 1997). The Native Americans also used sunflower hulls as a source of dye, leaves for herbal medicine, and pollen in religious ceremonies.

The domesticated sunflower of North America was introduced to Europe in the early 16th century, perhaps initially by a Spanish expedition in 1510, and later by English and French explorers (Putt 1997). The first published description of sunflower appeared in 1568. Sunflower quickly spread throughout Europe where it was initially cultivated as an ornamental plant or as a novelty. Although it spread rapidly throughout Europe, the sunflower did not have a spectacular success as a crop until it reached Russia, where it was probably introduced by Peter the Great after a trip to western Europe in 1697. Only upon the introduction of the domesticated sunflower was its potential as an oilseed crop recognized. In Russia, the sunflower was quickly adopted as a source of oil, largely because of religious laws that prohibit the consumption of foods rich in oil during the days preceding Christmas and during the season of Lent preceding Easter. Because of its recent introduction, it was not on the list of forbidden foods and could be consumed without infringement of religious laws. By the early 1700s sunflower achenes were eaten as a snack, and in 1716, the first patent for the use of sunflower oil (for industrial uses) was filed in England (Putt 1997). By the 1850s, dozens of sunflower crushing mills were operating in central Europe. However, the oil content of the sunflower achenes was only 250 g/kg. Active selection for high achene oil content began in the 1860s (Heiser 1976). Sunflower historians generally concur that the present cultivated sunflower in North America comes from breeding material with high oil content developed by V.S. Pustovoit, located at the All-Union Research Institute for Oil Crops (VNIIMK), Krasnodar, USSR starting in the 1920s through the mid-1960s.

Most references indicate the latter part of the 19th century was the date for the introduction of sunflower as a non-oilseed crop. By 1880, the "Mammoth Russian" cultivar was available from seed companies in the USA (Beard 1981). Another highly likely route of reintroduction of sunflower from the FSU to North America was via immigrants bringing small quantities of achenes with them. Mennonites from the FSU who immigrated to Canada about 1875 brought sunflower achenes with them for roasting and eating whole (Putt 1997).

Early cultivation of sunflower was primarily for livestock silage and seed for poultry. By the second half of the 20th century, improved Russian varieties with oil contents of 450 to 550 g/kg were available. The discovery of cytoplasmic male sterility (PET1 CMS) by the French scientist Leclercq in the early 1970s laid the foundation for the development of modern day sunflower hybrid. Hybrid sunflower, with higher yields and oil content and more uniformity compared to open-pollinated varieties, provided the last great impetus in establishing sunflower as a global crop.

The dispersion of wild species achenes has been by man and animals. Wild achenes do not have any special mechanisms for effective achene dispersal. Man, of course, has proved to be an important agent and certainly the wide distribution of many species can be largely explained as a result of human activity. Long-distance achene dispersal within North America was probably facilitated by the fact that many species grow along roadsides and railroad yards. Moreover, the perennials with tubers or rhizomes are also spread by man as the results of his development activities. Also, some of the species that are intentionally cultivated have often escaped and become established. However, before man, birds and small mammals were important agents in dispersal for sunflower achenes because they are an attractive food for both. With the exception of being spread by man at times, it seems likely that sunflower achene dispersal is probably highly localized and it is unlikely that long distance dispersal has been involved in extending the distribution of most of the species (Heiser et al. 1969).

1.4 Botanical Description

1.4.1 Taxonomic Position

The Compositae (Asteraceae) is the largest and most diverse family of flowering plants, comprising one-tenth of all known Angiosperm species (Heywood 1978; Funk et al. 2005). Species within this family are characterized by a compound inflorescence that has the appearance of a single "composite" flower. The Compositae family is divided into three major subfamilies and one minor subfamily, with 1,100 to 2,000 genera and over 20,000 species (Cronquist 1977; Heywood 1978; Jansen et al. 1991; Funk et al. 2005). The family has undergone extensive diversifications producing a cosmopolitan array of taxa encompassing ephemeral herbs, vines, and trees that thrive in some of the world's most inhospitable habitats (e.g., vertisols, deserts, and salt marshes). In contrast to some other large Angiosperm families, much of the biodiversity of the Compositae is in extreme environments rather than within the tropics. Representatives of this family are present on every continent and in nearly all habitats except Antarctica (Funk et al. 2005).

Helianthus belongs to the Asteraceae subfamily Asteroideae, tribe Heliantheae, subtribe Helianthineae (Panero and Funk 2002). The classification of Heliantheae is currently supported by data suggesting that *Phoebanthus* is sister to *Helianthus* (Schilling 2001; Schilling and Panero 2002) and that these two genera are sister to another clade containing *Pappobolus* (previously considered South American *Helianthus*), *Simsia, Tithonia,* and some *Viguiera* species (Schilling and Panero, 1996, 2002; Schilling 2001). The genus is native to temperate North America and contains 14 annual and 37 perennial species (Schilling 2006) (Tables 1-2 and 1-3).

The identification of sunflower species has long been problematic. Heiser et al. (1969) concluded that the greatest contribution of sustained efforts to understand sunflower taxonomy was not to provide an easy way to identify sunflower species, but rather to provide explanations for why

Table 1-2 Infrageneric classification of annual *Helianthus* species (Schilling and Heiser 1981; Schilling 2006).

Section	Species	Common Name	Chromosome Number (n)
Helianthus	*H. annuus* L.	Prairie	17
	H. anomalus Blake	Anomalous	17
	H. argophyllus T.&G.	Silver-leaf	17
	H. bolanderi A. Gray	Bolander's, Serpentine	17
	H. debilis		
	ssp. *debilis* Nutt.	Beach	17
	ssp. *cucumerifolius* (T.&G.) Heiser	Cucumber- leaf	17
	ssp. *silvestris* Heiser	Forest	17
	ssp. *tardiflorus* Heiser	Slow-Flowering	17
	ssp. *vestitus* (Watson) Heiser	Clothed	17
	H. deserticola Heiser	Desert	17
	H. exilis A. Gray	Serpentine	17
	H. neglectus Heiser	Neglected	17
	H. niveus		
	ssp. *niveus* (Benth.) Brandegee	Snowy	17
	ssp. *tephrodes* (Gray) Heiser	Ash-Colored, Dune	17
	H. paradoxus Heiser	Pecos, Puzzle, Paradox	17
	H. petiolaris		
	ssp. *canescens* (A. Gray) E.E. Schilling	Gray	17
	ssp. *fallax* Heiser	Deceptive	17
	ssp. *petiolaris* Nutt.	Prairie	17
	H. praecox		
	ssp. *hirtus* Heiser	Texas	17
	ssp. *praecox* Engelm. & A. Gray	Texas	17
	ssp. *runyonii* Heiser	Runyon's	17
Agrestes	*H. agrestis* Pollard	Rural, Southeastern	17
Porteri	*H. porteri* (A. Gray) J. F. Pruski	Confederate Daisy, Porter's	17

Table 1-3 Infrageneric classification of perennial *Helianthus* species.

Section	Series	Species	Common Name	Chromosome Number (n)
Ciliares	*Ciliares*	*H. arizonensis* R. Jackson	Arizona	17
		H. ciliaris DC.	Texas blueweed	34, 51
		H. laciniatus A. Gray	Alkali	17
Ciliares	*Pumili*	*H. cusickii* A. Gray	Cusick's	17
		H. gracilentus A. Gray	Slender	17
		H. pumilus Nutt.	Dwarfish	17
Atrorubens	*Corona-solis*	*H. californicus* DC.	California	51
		H. decapetalus L.	Ten-petal	17, 34
		H. divaricatus L.	Divergent	17
		H. eggertii Small	Eggert's	51
		H. giganteus L.	Giant	17
		H. grosseserratus Martens	Sawtooth	17
		H. hirsutus Raf.	Hairy	34
		H. maximiliani Schrader	Maximilian	17
		H. mollis Lam.	Soft, Ashy	17
		H. nuttallii ssp. *nuttallii* T.&G.	Nuttall's	17
		H. nuttallii ssp. *rydbergii* (Brit.) Long	Rydberg's	17
		H. resinosus Small	Resinous	51
		H. salicifolius Dietr.	Willowleaf	17
		H. schweinitzii T.&G.	Schweinitz's	51
		H. strumosus L.	Swollen, Woodland	34, 51
		H. tuberosus L.	Jerusalem artichoke	51
Atrorubens	*Microcephali*	*H. glaucophyllus* Smith	Whiteleaf	17
		H. laevigatus T.&G.	Smooth	34
		H. microcephalus T.&G.	Small-headed	17
		H. smithii Heiser	Smith's	17, 34
Atrorubens	*Atrorubentes*	*H. atrorubens* L.	Purple-disk	17
		H. occidentalis ssp. *occidentalis* Riddell	Fewleaf, Western	17
		H. occidentalis ssp. *plantagineus* (T.&G.) Heiser	Fewleaf, Western	17
		H. pauciflorus ssp. *pauciflorus*	Stiff	51
		H. pauciflorus ssp. *subrhomboides* (Rydb.) O. Spring	Stiff	51
		H. silphioides Nutt.	Odorous	17
Atrorubens	*Angustifolii*	*H. angustifolius* L.	Narrowleaf, Swamp	17

Table 1-3 contd....

Table 1-3 contd....

Section	Series	Species	Common Name	Chromosome Number (n)
		H. carnosus Small	Fleshy	17
		H. floridanus A. Gray ex Chapman	Florida	17
		H. heterophyllus Nutt.	Variableleaf	17
		H. longifolius Pursh	Longleaf	17
		H. radula (Pursh) T.&G.	Scraper, Rayless	17
		H. simulans E. E. Wats.	Muck, Imitative	17
		H. verticillatus Small	Whorled	17

distinguishing species is so difficult. The taxonomic complexity of the genus *Helianthus* stems from many different factors.

Although many botanists over the past two centuries have worked on the systematics of *Helianthus* (de Candolle 1836; Torrey and Gray 1842; Gray 1884; Dewer 1893; Cockerell 1919; Watson 1929), the most comprehensive knowledge of the genus comes from Heiser's extensive morphological work and crossing studies (Heiser et al. 1969). During the first half of the 20th century, many other botanists focused on smaller taxonomic studies within the genus, publishing some 50 papers characterizing *Helianthus* species and hybrids (Timme et al. 2007).

The genus *Helianthus* has been deemed to be comprised of as few as 10 species to more than 200. Linnaeus (1753) originally described nine species in the genus. Gray (1884) recognized 42 species in North America. In the early 20th century, Watson (1929) accepted 108 species, 15 of them from South America. Heiser et al. (1969) recognized 14 annual species and 36 perennial species from North America in three sections and seven series, as well as 17 species from South America. Subsequently, Robinson (1979) transferred 20 perennial species of South American *Helianthus* to the genus *Helianthopsis*. The taxonomic classification of *Helianthus* by Anashchenko (1974, 1979) was a radical departure from all previous schemes. He recognized only one annual species, *H. annuus* (with three subspecies and six varieties), and only nine perennial species with 13 subspecies. Schilling and Heiser (1981) proposed an infrageneric classification of *Helianthus*, using phenetics, cladistics, and biosystematic procedures that places 49 species of *Helianthus* in four sections and six series (Tables 1-2 and 1-3). The classification of Schilling and Heiser (1981) is presented herein with the following six modifications. First, the sectional name *Atrorubens* used by Anashchenko (1974) has taxonomic priority, thus the section *Divaricati* E. Schilling and Heiser is replaced by section *Atrorubens* Anashchenko. Second, *Helianthus exilis* is recognized as a species, as opposed to an ecotype of *H.*

bolanderi due to recent information which has shown it to be morphologically and genetically distinct (Oliveri and Jain 1977; Rieseberg et al. 1988; Jain et al. 1992). Third, the species name *H. pauciflorus* has priority over *H. rigidus* and is treated accordingly herein. Fourth, *Viguiera porteri* has been transferred to *Helianthus porteri* (Pruski 1998; Schilling et al. 1998). Fifth, *Helianthus verticillatus* has recently been rediscovered and redescribed, and is now recognized as a species (Matthews et al. 2002). Sixth, *Helianthus niveus* ssp. *canescens* has been transferred to *Helianthus petiolaris* ssp. *canescens* (Schilling 2006).

In the past, phylogenetic relationships of the perennial species and polyploidy hybrids have been particularly difficult to resolve. Using the external transcribed spacer region of the nucleus 18S–26S rDNA region, Timme et al. (2007) revealed a highly resolved gene tree for *Helianthus*. Phylogenetic analysis allowed for the determination of monophyletic annual *H.* section *Helianthus*, a two-lineage polyphyletic *H.* section *Ciliares,* and the monotypic *H.* section *Agrestis*, all of which were nested within the large perennial and polyphyletic *H.* section *Divaricati*. The distribution of perennial polyploids and known annual diploid hybrids on this phylogeny suggested that multiple independent hybrid speciation events gave rise to at least four polyploids and three diploid hybrids. Also provided by this phylogeny was evidence for homoploid hybrid speciation outside of *H.* section *Helianthus*.

Anashchenko (1979) proposed the first phylogeny for *Helianthus* on the basis of different chromosome sets: the protogenome A for the perennial forms (*Atrorubentes*) of the west coast of North America, the genome B for the annual forms, C for the *Ciliares*, and the protogenome S for perennial shrubs of South America (*Viguiera*). The *Ciliares* group was therefore regarded as a separate group in *Helianthus*. The genome organization in *Helianthus* is therefore still questionable. Chandler (1991) reviewed sunflower genomic relationships and came to the conclusion that there is little evidence of the existence of distinct genomes in *Helianthus*.

Sossey-Alaoui et al. (1998) using RAPD technology proposed a phylogeny of 36 *Helianthus* species based on 33 fragments common to all the *Helianthus* species. Fifty-six were unique to perennial species of sections *Atrorubentes* and *Ciliares*, 24 were unique to section *Atrorubentes*, and 29 were unique to section *Helianthus*, whereas none were unique to section *Ciliares*. Each set of common or specific fragments was assumed to belong to a genome: (1) the **C** genome carrying the fragments common to all species of the three sections, (2) the **H** genome unique to section *Helianthus*, (3) the **P** genome common to perennial species sections *Atrorubentes* and *Ciliares*, (4) the **A** genome unique to section Atrorubentes. The genomic structure was therefore **HC** for section *Helianthus*, **CPA** for section Atrorubentes, and **CP?** for section *Ciliares*. They also concluded that molecular hybridization with amplification products revealed homologies between *Helianthus* genomes and several other genera in the Helianthinae sub-tribe.

1.4.2 Plant Structure and Growth Habit

Sunflower is an annual crop distinguished from other cultivated crops by its single stem and its large conspicuous inflorescence. When sunflower is in full bloom, it is one of the most photogenic crops because of its large inflorescence with showy yellow-orange ray flowers. Flowering sunflower has served as an inspiration for artists, poets, and business promotions.

The inflorescence is a capitulum or head consisting of an outer whorl of showy and generally yellow ray flowers and from 700 to 3,000 disk flowers in oilseed hybrids up to 8,000 disk flowers in non-oilseed hybrids (Pustovoit 1975). The disk flowers are arranged in arcs radiating from the center of the head and are perfect flowers that produce achenes. Involucral bracts or phyllaries, which vary in form and size, surround the head.

The showy flowers on the outer whorl of the head have five elongated petals united to form strap-like structures, which give them the name ray or ligulate flowers. Ray flowers are usually golden yellow, but may be pale yellow, orange-yellow, or reddish. Variation in petal color has been described by Cockerell (1912, 1918) and Fick (1976). Ray flowers are normally sterile, having a rudimentary pistil and vestigial style and stigma, but no anther. Mutants occur producing more than the normal numbers of ray flowers, or occasionally none.

The remaining flowers covering the large discoidal head are called disk flowers. A single disk flower is often referred to as a floret. Each floret is subtended by a sharp-pointed, chaffy bract, a basal ovary, and two pappus scales (often considered being a modified sepal). The disk flower is perfect (contains both a stamen and pistil).

At anthesis the outer whorl of disk flowers opens first, at about the time that the ray flowers open from their folded position against the buds of the disk flowers. Immediately after this stage is reached, the anther locules dehisce, releasing pollen inside the anther tube. An elongation of the lower portion of the style pushes the two-lobed pubescent stigma up the anther tube. The stigma is not receptive at this stage because the two lobes are held together covering the inner receptive surface. The stigma appears at about the 1700 hour of the same day, and by the following morning it is fully emerged, with receptive surfaces exposed. At this time, the staminal filaments lose turgidity and the anther tube begins to recede into the corolla. The beginning of flowering of disk flowers has been described as the R-5.1 stage of sunflower development (Schneiter and Miller 1981). One to four rows of disk flowers open successively daily for 5 to 10 days. The flowering period is prolonged if heads are larger, or if the weather is cool and cloudy. The stigmas remain receptive for up to 4 or 5 days.

The achene, or nut-like fruit of the sunflower, consists of a seed, often called the kernel, and the adhering pericarp, usually called the hull. Achenes mature from the periphery of the whorl to the center. As achenes mature, the withered calyx, corolla tube, anther, stigma, and style drop off at the point of their attachment. The achenes usually are largest on the periphery of the head and smallest at the center. Disk flowers in the center of the head in some breeding lines fail to produce filled achenes. The lack of achene filling in the center of some heads is influenced both by genotype and environment.

The dimensions of achenes range from 7 to 25 mm long and from 4 to 13 mm wide. The achenes of oilseed sunflower are usually black. Achenes are much smaller in the wild species where they are from 2 to 7 mm long and generally from 1 to 2 mm wide. The weight of 100 achenes of cultivated sunflower ranges from 4 to 20 g. The weight of an individual achene ranges from 40 to 400 mg. Lengthwise, the achenes may be linear, ovoid, or almost round, and in cross-section they may be flat to almost round. Large achenes usually have thick hulls and relatively small kernels. Small achenes, in contrast, usually have thin hulls tightly fitting around the kernel. Thinner-hulled achenes usually have higher oil content than the thicker-hulled achenes.

Cultivated hybrid sunflower plants are annual with non-branched stems, while the wild species have dominant branching. Branched types that possess recessive genes for branching are frequently used as parental male breeding lines or pollinator lines in commercial seed production. Stem dimension and development are influenced by the environment and by plant population. Branching can also be influenced by the environment, especially when the terminal buds are injured early in phenological development.

Many degrees of branching occur in sunflower, ranging from a single stem with a large solitary inflorescence in cultivated types to multiple branching from axils of most leaves on the main stem in the wild species. Branch length varies from a few centimeters to a distance longer than the main stem. Branching may be concentrated at the base or top of the stem, or spread over the entire plant. Generally, heads on branches are smaller than heads on the main stem. Occasionally, some first-order branches have a terminal head almost as large as the main head. In most wild species, the head on the main stem blooms first, but generally is no larger than those on the branches.

Stem (plant) height of commercial sunflower cultivars ranges from 50 to over 500 cm, and stem diameter from 1 to 10 cm. Rare plant types 12 m tall were reported by Dodonaeus (cited by Cockerell 1915). Cultivars which are used for forage are usually tall and late flowering. Long-season oilseed hybrids of the southern hemisphere are taller than hybrids of the northern hemisphere. A generally accepted ideotype of productive sunflower is a medium plant height of 160 to 180 cm (Skoric 1988). Commercial hybrids with a height of 120 to 150 cm have been developed and are referred to as semi-dwarf hybrids, while dwarf sunflower are 80 to 120 cm tall (Schneiter 1992). Stem length is determined by the number and length of internodes. Both tall and short plants with many internodes will have thick stems because of the positive association between numbers of internodes and stem thickness. Stems that are short because of fewer internodes will be thinner (Knowles 1978).

Information on development of sunflower roots is limited. Early studies by Weaver (1926) on the cultivar "Russian" at Lincoln, Nebraska, found that a strong central taproot penetrates to a depth of 150 to 270 cm. Pustovoit (1967) reported root depths of 4 to 5 m in cultivars with a growth period of 100 to 110 days. Root depths of 2 m or more have been frequently reported for sunflower (Jones 1984; Gimenez and Fereres 1986; Sadras et al. 1989; Angadi and Entz 2002a, b). The rooting system of sunflower can be considered "explorative", i.e., a large volume of soil is explored with a combination of thick and thin roots, low specific root length, and low root-length density, as opposed to "exploitative", root systems characterized by predominantly fine roots, high specific root length, and large root-length density (Boot 1990). Adventitious roots may be complementary to primary root systems and may function in plant anchorage and in water absorption and conductance.

The leaf consists of the blade (lamina) and a stalk-like part, the petiole, which connects it to the stem. Leaves in sunflower are rarely sessile (without a petiole), except in some wild species. The leaf is highly variable in both structure and function. In sunflower, as seedlings emerge from the soil, cotyledons unfold and reveal the first pair of true leaves at the top of the shoot. Leaves are produced in opposite alternating pairs, and after five opposite pairs appear a shorter form of alternate phyllotaxy develops (Palmer and Phillips 1963). The number of leaves on single-stemmed plants may vary from as few as eight to as many as 70 (Knowles 1978). There appears to be some association between the number of leaves and time to maturity for plants with numerous leaves (Knowles 1978). Plants with numerous leaves are also usually late maturing. The number of leaves on a plant and their expansion, size, shape, and duration can be greatly influenced by environmental factors. There is considerable variation in leaf size, shape of the entire leaf, shape of the tip, base, margin, surface pubescence, and petiole characteristics. The length and width of leaves vary with the height of the stem. The ratio of length:width may be a useful criterion for cultivar description.

1.4.3 Habitat

The North American species of *Helianthus* are found in virtually all parts of the United States with several species extending into Canada and a few into Mexico. With such a diversity of species, their habitats are highly variable, ranging from disturbed areas to tall grass prairies and to climax forests. They occupy a variety of habitats, usually in fully open areas, but a few will grow in rather dense shade. A number of the species could be classified as weeds. Wild *Helianthus annuus*, which has the most extensive distribution of any species, grows mostly in disturbed areas. A few of the species are endemic to specific habitats. *Helianthus exilis* is restricted to the serpentine soils of the California coastal range and Sierra Nevada Mountains in California while *H. californicus*, which is indigenous to central and southern California, is restricted to riparian habitats. Other species such as *H. niveus* ssp. *tephrodes* are restricted to the Algodones Sand Dunes in California, and *H. porteri* is endemic to granite outcrops in the Piedmont region of the southeastern USA. *Helianthus carnosus* is restricted to boggy wetlands in Florida.

1.4.4 Karyotype

Karyotype analyses of sunflower are reviewed by Chandler (1991), although the genus is of economic and evolutionary interest. Classical karyotype describes each of the haploid chromosome sets of an organism based on features such as length of chromosome, ratio of arm lengths, position of centromere and secondary constrictions, and size and position of heterochromatic knobs. Karyotype analysis has provided useful markers in chromosome identification and the designation of chromosomes in many plant species. Most karyotypes of *Helianthus* species have been conducted with mitotic metaphase chromosome preparations in the 1970s and 1980s (Jan and Seiler 2007). The classification of chromosomes as median (m), sub-median (sm), sub-terminal (st), or terminal (t) when the long:short arm ratio was in the range of 1.0 to 1.7, 1.7 to 3.0, 3.0 to 7.0, and 7.0 µ, respectively, following the convention of Levan et al. (1964) and is used in almost all the sunflower literature.

Karyotype studies have been conducted on *H. annuus* and *H. debilis* (Raicu et al. 1976), *H. mollis* (Georgieva-Todorova et al. 1974), *H. salicifolius* (Georgieva-Todorova and Lakova 1978), hybrids of *H. annuus* x *H. hirsutus* (Georgieva-Todorova and Bohorova 1980), *H. hirsutus* and *H. decapetalus* (Georgieva-Todorova and Bohorova 1979), cultivated *H. annuus* (Al-Allaf and Godward 1977), and 12 *Helianthus* species (Kulshreshtha and Gupta 1981). In general, Georgieva-Todorova and Bohorova (1979, 1980) reported large chromosome sizes, with more chromosomes classified as m and sm, and 2 to 4 satellited chromosomes. In contrast, Kulshreshtha and Gupta

(1981) reported smaller chromosome sizes, chromosome types equally distributed among m, sm and st, and only 1 or 2 satellited chromosomes. The large variation in chromosome condensation among studies of different authors was probably due to differences in chromosome preparation. Perhaps the use of common cultivated *H. annuus* as a check would have helped to resolve some of the problems when comparing the total chromosome length of different genomes.

According to Georgieva-Todorova and Lakova (1978), closely related tetraploid species *H. decapetalus* and *H. hirsutus* had remarkable similarity regarding most features of their karyotypes. The $2n = 30$ *H. mollis* studied by Kulshreshtha and Gupta (1981) was much lower than the basic set of $2n = 34$ and was presumably from roots of a nonviable plant. Also, *H. californicus* is expected to be hexaploid with $2n = 102$. The *H. californicus* with $2n = 34$ may have been a misidentified species. Speciation of *Helianthus* was shown to have involved chromosome exchanges such as translocation and inversion (Chandler et al. 1986), as well as the obvious polyploidization for the tetraploid and hexaploid species. One would expect a continuous variation in DNA and total chromosome length in diploid species, and multiples of single or combinations of the diploids for tetraploids and hexaploids.

The karyotype of *Helianthus annuus* was analyzed using Giemsa banding and fluorescence in situ hybridization (FISH) and computer-aided image processing with respect to the chromosome length, arm ratio, occurrence and chromosomal position of intercalary heterochromatin and the position of 18S/25S and 5S ribosomal RNA genes were located using Giemsa banding and fluorescence in situ hybridization (Schrader et al. 1997). In situ hybridization techniques involving GISH, FISH, and BAC-FISH are being optimized for diversity and evolutionary studies between species of the genus *Helianthus* and development of a physical sunflower map allowing a cross reference to the genetic map (Paniego et al. 2007). Measurements of chromosomes in combination with Giemsa banding and FISH should allow for the discrimination of most chromosome pairs of the sunflower.

1.4.5 Polyploidy

The genus *Helianthus* has a basic chromosome number of $n = 17$ and contains diploid ($2n = 2x = 34$), tetraploid ($2n = 4x = 68$), and hexaploid ($2n = 6x = 102$) species (Tables 1-2 and 1-3). The 14 annual species are all diploid, and the 37 perennial species include 27 diploid, 4 tetraploid, 6 hexaploid, and 4 mixi-ploid species. *Helianthus ciliaris* and *H. strumosus* have both tetraploid and hexaploid forms, while *H. decapetalus* and *H. smithii* contain diploid and tetraploid forms.

The origins and relationships of the polyploid hybrid species in *Helianthus* have long been of interest and remain largely unresolved, particularly for perennials (Timme et al. 2007). It is currently unknown which polyploid species are autopolyploids and which are allopolyploids. The hexaploid *H. resinosus* is currently being investigated using molecular cytogenetics and markers in order to elucidate species origin through polyploidy (Carrera et al. 2004).

1.4.6 Aneuploidy

Aneuploids can be the result of interspecific hybridization. Leclercq et al. (1970) obtained trisomic plants ($2n+1$) in backcross progeny of *H. tuberosus* x *H. annuus* hybrids. The progeny were resistant to downy mildew [*Plasmopara halstedii* (Farl.) Berl and deToni] and they postulated that the extra chromosome came from the *H. tuberosus* genome.

Trisomic plants were obtained after backcrossing interspecific hybrids *H. petiolaris* x *H. annuus* (Whelan 1979) and *H. maximiliani* x *H. annuus* (Whelan and Dorrell 1980; Whelan 1982) with cultivated *H. annuus*. Multivalents were frequently observed in these F_1 hybrids indicating reciprocal translocation heterozygosity. The trisomics presumably originated from unequal disjunction of multivalents. Most trisomics appeared normal, but some had distinct morphological features.

Jan et al. (1988) produced tetraploids, triploids, and aneuploids using the inbred line P21. Meiotic chromosome pairing of the autotetraploid P21 was reasonably normal with 28.16 bivalents, 0.85 univalent, and a small number of multivalents. Triploids were obtained by crossing tetraploids with diploid P21, and aneuploids by crossing triploids with the diploids. The chromosome number of the 137 plants from reciprocal triploid x diploid crosses ranged from $2n = 34$ to $2n = 47+t$. In general, plants with a lower chromosome number were more prevalent when triploids were used as the pollen parent, while the frequency of plants with a higher chromosome number increased when triploids were used as the seed-bearing parent. These results suggest the effective transmission of extra chromosomes through both male and female gametes, with the female gametes being better than the male gametes. However, for rapid trisomic production and fast chromosome reduction, triploids are often used as the pollen parent. These results suggest that diploid sunflower tolerates extra chromosomes well and a set of trisomic genetic stocks is possible. Pollen stainability, an indicator of viability, was above 90% for plants with $2n = 34$ to $2n = 37$. As chromosome numbers increased from $2n = 38$ to $2n = 45+t$, pollen stainability decreased to about 40%, but acceptable seed set was still obtained (Jan et al. 1988).

In the past, the phylogenetic relationships of the perennial species and polyploid hybrids have been particularly difficult to resolve. Two main factors have contributed to this lack of resolution. First is the lack of markers that evolved fast enough to record speciation events during the rapid evolution in the group, and the other is the difficulty in detecting and reconstructing the extensive hybrid speciation, both diploid and polyploid, which strongly impacted the phylogenetic history of the genus (Timme et al. 2007). Natural hybridization and introgression among many of the species result in morphological intergradation between otherwise distinct forms. Polyploidy in the perennial species also contributes to the complexity of species classification in *Helianthus*.

Thirteen species have some form of polyploidy, either tetraploidy or hexaploidy. It is not known which polyploid species are autopolyploids and which are allopolyploids. This has led to various taxonomic treatments of the genus. There are still specimens, of hybrid origin or growing in unusual conditions or incompletely collected, that defy certain placement into a species (Schilling 2006). Since many of the species are wide-ranging geographically, they exhibit extensive phenotypic variation, which appears to include both heritable and non-heritable (environmental) components. Many species are also genetically quite variable, making rigorous identification and classification difficult.

1.4.7 Genome Size

Sunflower has an estimated haploid genome size of 3,000 Mb/1C (Arumuganathan and Earle 1991). There have been other estimates of the genome size of approximately 3.5 Gb by Baack et al. (2005) and Price et al. (2000) of approximately 3.6 Gb. The discovery of the influence of plant secondary compounds on the estimation of DNA content by both Fuelgen densitometry and flow cytometry (Greilhuber 1988; Price et al. 2000) has led to changes in practice and created uncertainty regarding many earlier reported C-values.

1.5 Genetic Resources

1.5.1 Primary Gene Pool

Aside from the local landraces of sunflower grown by the North American Indians, the first open-pollinated cultivars grown for significant commercial production were developed in the USSR. According to Pustovoit (1967), large numbers of local peasant cultivars were available by the 1880s, some of which had higher yield, greater uniformity in plant and seed type, and improved resistance to broomrape (*Orobanche cumana*) and the European sunflower moth *Homoeosoma nebulella*, compared to unimproved cultivars (Fick 1989). The largest group of local cultivars was known as "Zelenka", with "Cherenyankia", "Fuksinka", and "Puzanok" also widely grown. The first cultivars developed in the USSR were introduced early in the 20th century. "Kruglik A-41" and "Zhdanovsky 8281" were among the more popular cultivars grown. Kruglik A-41, introduced in 1927, was among the first cultivars with significantly higher oil percentage, confirming that sunflower could in fact be improved for this trait.

The creation of the oilseed type of sunflower is usually credited to the research program of V.S. Pustovoit. Achene oil percentages among Soviet cultivars increased from less than 300 g/kg in the early 1900s to over 400 g/kg during the 1930s and to over 500 g/kg by the early 1960s. "Peredovik" and "Armavirsky 3497" were two of the most widely grown cultivars in the USSR, each accounting for about 1.2 million hectares of production in 1966. Peredovik, which was introduced in 1960, also was grown extensively in the USA, Canada, Europe and other parts of the world prior to the introduction of hybrids. The open-pollinated cultivar Peredovik was licensed in Canada in 1964. It had a yield similar to the widely grown cultivar "Advent", but averaged 436 g/kg oil compared with 328 g/kg for Advent (Putt 1965). The high oil content of Peredovik greatly increased the efficiency of the processing operation, and interest in sunflower as a crop increased significantly. Peredovik is a very widely adapted cultivar, high in oil content, of medium height and maturity, moderately resistant to rust (*Puccinia helianthi* Schwein) and *Verticillium* wilt (*Verticillium dahliae* Kleb.), and resistant to broomrape (*Orobanche cernua* Loefl.), and the European sunflower moth. Cultivars that have been developed relatively recently include "Sputnik", "Voshod", and "Mayak" with very high oil percentages and "Progress" and "Novinka" with disease resistance transferred from *H. tuberosus*; and "Pervenets" with high oleic acid concentration (Fick 1989).

Open-pollinated cultivars were not without problems including uniformity of height, flowering, maturity, and dry-down rate after maturity. Effective use of insecticides required head uniformity and flowering, as well as non-shattering of achenes at harvest, all of which occur when plants mature unevenly. Differential plant height also causes harvest problems. Open- pollinated cultivars were often susceptible to many of the prevalent diseases, and also lacked autogamy, and if grown in areas that lacked bees for pollination, yields were seriously reduced. These problems, as well as the possibility of hybrid heterosis providing increased yields led to a search for a hybrid sunflower.

The first hybrid sunflower cultivars were introduced for commercial production in Canada in 1946 (Putt 1962). "Advance", and later "Advent" and "Admiral", were grown for most of the oilseed production in that country during the 1950s. These hybrids were produced by natural crossing in seed production fields, with two parents planted in alternating rows. Because self- and sib-pollination occurred on the female parent in addition to the desired cross pollination, hybridization percentages were often less than 50% and the full yield potential of the hybrids was not realized.

Through hybridization between *H. petiolaris* and *H. annuus* and backcrossing with *H. annuus*, Leclercq (1969) transferred the *H. annuus* genome into *H. petiolaris* cytoplasm and obtained the first cytoplasmic male sterile (CMS) plants that became the PET1 (French) CMS cytoplasm currently used in most all hybrids. In the USA, the first hybrids produced using CMS and a nuclear fertility restorer pollinator were introduced in 1972 and within five years were grown on about 90% of the production region. The best hybrids yielded over 20% more than open-pollinated cultivars available at that time. Much of the currently available germplasm had its origin in this limited early germplasm, resulting in a crop with an extremely narrow genetic base. Due to the use of a single male sterile cytoplasm for worldwide hybrid sunflower production and its consequence of genetic vulnerability, as shown by the catastrophic southern corn leaf blight disease caused by *Bipolaris maydis* race T in the early 1970s (Tatum 1971), a large portion of the interspecific hybridization research in sunflower has focused on the identification of additional unique CMS sources and their fertility restoration genes.

More than 300 maintainer, CMS, and restorer lines have been developed by the USDA-ARS programs or other public researchers (Miller et al. 1992). Several of these lines have plant introductions (PI) in their pedigrees and trace their origins back to cultivars such as Peredovik, "VNIIMK 8931", Armavirsky 9345, Voshod, and "Smena" for high yield and high oil content. The use of a combination of inbred lines has led to the development of uniform, high yielding, high oil hybrids with resistance to major pathogens and a high degree of autogamy.

Improved open-pollinated cultivars for non-oilseed or confectionery production also trace their origins to the USSR. Non-oilseed types known as "Giant" and "Mammoth Russian" were introduced into North America during the 19th century, and were grown in gardens for food by the early Russian immigrants to the USA and Canada. These types served as the basis for the development of improved cultivars for commercial production. "Mennonite", "Mingren", "Commander", and "Sundak", all of which had relatively good achene size, shape and color suitable for the non-oilseed markets, were the principal cultivars grown during the 1950s through the 1970s. Non-oilseed hybrids were introduced in the USA in 1974 and currently account for all confectionery production in the USA. Non-oilseed production in the USA is currently estimated to be about 20% of the total sunflower production.

1.5.2 Secondary and Tertiary Gene Pools

Genetic resources of a crop consist of the total pool of genetic variability that exists in the species or within species with which the crop plant is sexually compatible (Holden et al. 1993). Wild relatives of crop plants typically are genetically much more diverse than related cultivated lineages. Genetic diversity is thought to contribute to long-term preservation of species by allowing them to adapt quickly to changes in their environment.

Although many secondary and tertiary gene pools may appear to have no immediate use in breeding and genetic programs (Burton 1979), they may contain unidentified genes that will protect crops against new pests in the future. Hopefully, the present germplasm collection will contain the necessary germplasm. Although we cannot predict with acceptable levels of confidence the occurrence, severity, or even the nature of future stresses, germplasms with as wide a range of genetic diversity as possible should be developed for breeding programs (Jones 1983). Diversity in germplasm is also critical to successful crop breeding programs, but to date it has been rather limited (Harlan 1976). For the domesticated sunflower, this includes most species of *Helianthus*. Sunflower germplasm resources can be categorized as in situ resources (i.e., wild populations and landraces) or ex situ resources (accessions preserved in seed banks).

1.5.3 Ex situ Helianthus Collection

The USDA-ARS National Plant Germplasm System (NPGS) sunflower collection is maintained at the North Central Regional Plant Introduction Station (NCRPIS) in Ames, Iowa, USA. The mission of the NCRPIS is to conserve genetically diverse crop germplasm and associated information, to conduct germplasm-related research, and to encourage the use of germplasm and associated information for research, crop improvement and product development. The collection contains 37 perennial species, 14 annual species, and the cultivated species, *Helianthus annuus* (Schilling 2006). This NPGS sunflower collection is a diverse assemblage of 3,850 accessions: 1,708 cultivated *Helianthus annuus* accessions (44%) from 59 countries, 932 wild *Helianthus annuus* accessions (25%), 437 accessions representing 11 other wild annual *Helianthus* species (11%), and 773 accessions representing 37 perennial *Helianthus* species (20%). This collection is one of the largest and most genetically diverse ex situ sunflower collections in the world, and it is vital to the conservation of *Helianthus* germplasm. Over 14,000 samples of wild sunflower accessions from this collection have been distributed to more than 365 researchers from 34 different countries over the last 28 years.

The distributed accessions have become the basis of wild species genebanks and research programs in Argentina, France, Italy, Spain,

Germany, Bulgaria, Romania, Czech Republic, Hungary, Russia, Serbia, India, China, and Mexico. Notable is the collection at the Institute of Field and Vegetable Crops, Novi Sad, Serbia, which contains 39 of the 51 wild species (IBPGR 1984; Cuk and Seiler 1985; Atlagić et al. 2006). The wild species collection of the Dobroudja Agricultural Institute (DAI) at General Toshevo, Bulgaria, is also notable, containing 428 accessions representing 37 of the 51 species of *Helianthus* (Christov et al. 2001). The wild species collection maintained at INRA, Montpellier, France, has more than 600 accessions of 45 of the 51 wild sunflower species (Serieys 1992). The Institudo de Agricultura Sostenible (CSIC), Cordoba, Spain, maintains 44 annual and perennial accessions of *Helianthus* (Ruso et al. 1996). The genetic diversity of the wild species can make a significant contribution to sunflower in many countries by providing genes for resistance (tolerance) to pests and environmental stresses, allowing the crop to become and remain economically viable.

Diversity in germplasm also is critical to crop breeding programs, but to date it has not been fully exploited (Harlan 1976). Several species have contributed specific characteristics for sunflower improvement. The wild species are adapted to a wide range of habitats and possess considerable variability for most agronomic, and achene quality characters, and reaction to insects and disease pathogens (Thompson et al. 1981; Jan and Seiler 2007). More detailed discussions about the use and potential value of wild species for sunflower breeding programs can be found in Seiler (1988, 1992, 1996, 2002), Seiler and Rieseberg (1997), Skoric (1988, 1992), and Jan and Seiler (2007).

The wild species have provided many agronomically important traits for the cultivated sunflower. The estimated economic contribution of the wild species to the cultivated sunflower is US$384 million per year (Prescott-Allen and Prescott-Allen 1986). Another estimate is US$269.5 million per year (Phillips and Meilleur 1998). The greatest value is derived from the PET 1 (French) CMS cytoplasm from *H. petiolaris*. Wild *Helianthus* species have been a good source of genes for resistance to pathogens of economically important diseases. Much of the value is derived from disease resistance genes for rust, downy mildew, *Verticillium* wilt, *Alternaria* leaf spot, powdery mildew, *Phomopsis* stem canker, *Sclerotinia* wilt/rot, broomrape and more recently, salt tolerance genes that have been identified and transferred into cultivated sunflower. One trait not accounted for in the estimates mentioned above is herbicide tolerance. A wild population of *H. annuus* from Kansas has been identified as a source for resistance to imidazolinone and sulfonylurea herbicides, and these traits have been transferred into cultivated sunflower (Al-Khatib et al. 1998; Al-Khatib and Miller 2000). In addition, these two herbicides control broomrape in areas of the world where this parasitic weed attacks sunflower (Alonso et al. 1998). Thus, herbicide resistant sunflower hybrids could be used with these herbicides to combat broomrape infection.

Collection of germplasm not only serves a valuable purpose in preserving germplasm, but it also provides valuable information about the diverse habitats occupied by wild sunflowers and associated species. This information is particularly important for the genus *Helianthus* because of the co-evolution of its species and associated native insects and pathogens. Knowledge of a particular habitat and adaptations of a species occurring therein can often help to identify potential sources for a desired trait. Based on the habitat of a species and its immediate environment, selection of potential species for a particular characteristic may become easier, more accurate, and more efficient.

In situ preservation of genetic diversity within wild sunflower populations in their native habitat is critical because we lack the necessary resources to preserve locally adapted sunflower populations of all wild species in gene banks. The primary obstacle for long-term preservation of wild sunflower populations is human activity. Unfortunately, the long-term outlook for survival of many sunflower species is not promising; some species are already rare and endangered, or in the case of *H. nuttallii* ssp. *parishii*, probably extinct. Additional potential threats to the preservation of rare sunflower populations include their small population sizes and subsequent loss of genetic diversity.

The genus *Helianthus*, besides constituting the basic genetic stock from which cultivated sunflower originated, continues to contribute specific characteristics for cultivated sunflower improvement. However, there is a continued need to collect, maintain, evaluate, and enhance wild *Helianthus* germplasm for future utilization in cultivated sunflower.

1.5.4 Core Collection

Frequently, researchers are uncertain about the criteria or information required to select germplasm materials needed for their specific research objectives. The assembly of a core subset of the cultivated sunflower collection may provide an efficient means of identifying useful traits. This will enable researchers to sample the diversity within the collection without testing excessively large numbers of accessions. A core subset of the cultivated sunflower collection was established by Brothers and Miller (1999). Twenty descriptors were used in the construction of the core subset. The sunflower core subset of cultivated sunflower consists of 112 accessions (approximately 7% of the 1,708 available accessions) grouped into 10 clusters. Accessions within the same cluster should be more genetically similar than accessions between clusters. Accessions in the core subset represent 38 of the 57 countries of origin for the total cultivated sunflower collection. The core subset contains two ornamental accessions, seven breeding lines, 12 landraces, and 91 cultivars. Researchers may initially use the core collection to determine traits of interest and, pending the results of their research, request additional accessions from one or more clusters to explore indepth at a later date.

1.5.5 Wild Species Utilization

1.5.5.1 Pathogen Resistance

Wild sunflower species have been a valuable source of resistance genes for many of the common pathogens of cultivated sunflower. *Helianthus annuus, H. petiolaris,* and *H. praecox* are the major sources of genes for *Verticillium* wilt resistance (Hoes et al. 1973). These species plus *H. argophyllus* are also the major sources of resistance genes for downy mildew and rust in cultivated sunflower. Resistance genes for these pathogens occur frequently in the wild annual species (Tan et al. 1992; Quresh et al. 1993). Resistance to broomrape has been observed in most of the wild perennial species (Fernández-Martínez et al. 2000). Early reports of broomrape resistance were from the FSU where they developed cultivars Progress and Novinka using the "Group Immunity" breeding approach (Pustovoit and Gubin 1974). *Phoma* black stem (*Phoma macdonaldii*) resistance has been reported in several perennial species, *H. decapetalus, H. eggertii, H. hirsutus, H. resinosus* and *H. tuberosus* (Skoric 1985). *Phomopsis* stem canker (*Phomopsis helianthi* Munt-Cvet. et al.) resistance has been found in perennials *H. maximiliani, H. pauciflorus, H. hirsutus, H. resinosus, H. mollis,* and *H. tuberosus* (Skoric 1985; Dozet 1990). *Alternaria* leaf spot [*Alternaria helianthi* (Hansf.) Tubaki and Nishihara] resistance was observed in perennials *H. hirsutus, H. pauciflorus,* and *H. tuberosus* (Morris et al. 1983). *Rhizopus* head rot (*Rhizopus arrhizus* Fischer) resistance was observed in perennials *H. divaricatus, H. hirsutus, H. resinosus,* and *H.* x *laetiflorus* (Yang et al. 1980). Powdery mildew (*Erysiphe cichoracearum*) resistance was observed in annuals *H. debilis* ssp. *debilis, H. bolanderi,* and *H. praecox* (Saliman et al. 1982; Jan and Chandler 1985). *Sclerotinia* head rot (*Sclerotinia sclerotiorum*) tolerance was observed in perennials *H. resinosus, H. tuberosus, H. decapetalus, H. grosseserratus, H. nuttallii,* and *H. pauciflorus* (Pustovoit and Gubin 1974; Mondolot-Cosson and Andary 1994; Ronicke et al. 2004). *Sclerotinia* root rot tolerance was observed in perennials *H. mollis, H. nuttallii, H. resinosus,* and *H. tuberosus* (Skoric 1987). *Sclerotinia* stalk rot tolerance was observed in annual *H. praecox,* and perennials *H. pauciflorus, H. giganteus, H. maximiliani, H. resinosus,* and *H. tuberosus* (Skoric 1987). Stalk rot resistance has been identified in hexaploid perennial *H. californicus* and is being transferred into cultivated sunflower (Feng et al. 2006). Five interspecific amphiploids derived from perennial

H. strumosus, H. grosseserratus, H. maximiliani, H. nuttallii and *H. divaricatus* appear to have stalk rot resistance (Jan et al. 2006a).

1.5.5.2 Viral and Bacterial Resistance

Bacterial foliar diseases, including apical chlorosis (*Pseudomonas syringae* pv. *Tagetis*) and bacterial blight (*P. syringae* pv. *helianthi*), generally have little economic impact on sunflower (Gulya 1982). Sunflower can be infected by over 30 viruses, but viral diseases are generally of concern only in tropical or subtropical climates, such as India, where tobacco streak virus is a problem. In North America, sunflower mosaic virus and sunflower chlorotic mottle virus are rarely seen on sunflower, with only sunflower mosaic virus noted on wild sunflower in Texas. Sources of resistance are available from wild *H. annuus* to produce resistant hybrids if necessary (Gulya et al. 2002). Jan and Gulya (2006) have released three germplasms with sunflower mosaic virus resistance, SuMV-1, SuMV-2, and SuMV-3.

1.5.5.3 Insects

Wild sunflowers are native to North America where their associated insect herbivores and entomophages co-evolved in natural communities. This provides the opportunity to search for insect resistance genes in the wild species. North America has the largest losses due to insect pests. In the major production area there are about 15 principal insect pests of cultivated sunflower, and of this total, about six are considered important economic pests from year to year (Charlet and Brewer 1997). Sunflower moth (*Homoeosoma electellum*) tolerance was observed in annual *H. petiolaris*, and perennials *H. maximiliani, H. ciliaris, H. strumosus,* and *H. tuberosus* (Rogers et al. 1984). Stem weevil (*Cylindrocopturus adspersus*) tolerance was found in perennials *H. grosseserratus, H. hirsutus, H. maximiliani, H. pauciflorus, H. salicifolius,* and *H. tuberosus* (Rogers and Seiler 1985). Sunflower beetle (*Zygogramma exclamationis*) tolerance was observed in annual, *H. agrestis* and *H. praecox,* and perennials *H. grosseserratus, H. pauciflorus, H. salicifolius,* and *H. tuberosus* (Rogers and Thompson 1978, 1980). Charlet and Seiler (1994) found indications of resistance to the red sunflower seed weevil (*Smicronyx fuluus*) in several native *Helianthus* species.

1.5.5.4 Oil and Oil Quality

Variability for oil concentration exists in the wild species. While oil concentration is lower in the wild species than in cultivated sunflower, backcrossing to cultivated lines quickly raises the oil concentration to an acceptable level. Annual *H. anomalus* has the highest oil concentration of

460 g/kg, the highest ever observed in a wild sunflower species (Seiler 2007), followed by *H. niveus* ssp. *canescens* with 402 g/kg, *H. petiolaris* with 377 g/kg, and *H. deserticola* with 343 g/kg. Perennial *H. salicifolius* has a concentration of 370 g/kg (Seiler 1985; Seiler and Brothers 2003). Cultivated sunflower generally contains 450 to 470 g/kg. The linoleic fatty acid concentration in the oil of *H. anomalus* populations was uncharacteristically high for a desert environment, approaching 700 g/kg (Seiler 2007). A linoleic acid concentration of 540 g/kg in *H. deserticola* is more typical for a desert environment. Cultivated sunflower grown at northern latitudes generally have linoleic acid contents of over 680 g/kg, while southern latitudes have approximately 550 g/kg. Reduced concentrations of saturated palmitic and stearic fatty acids have been observed in a population of wild *H. annuus* that had a combined palmitic and stearic acid concentration of 58 g/kg (Seiler 1998). This is 50% lower than in oil of cultivated sunflower. A combined palmitic and stearic acid concentration of 65 g/kg was observed in a wild perennial species, *H. giganteus* (Seiler 1998).

1.5.5.5 Cytoplasmic Male Sterility

Sunflower is the only Asteraceae in which the cytoplasmic male sterile (CMS) system is known. Cytoplasmic male sterility is sporophytic in sunflower (Pearson 1981). Sunflower plants that are CMS usually have anthers one-half the normal length, which does not project from the corolla (Leclercq 1969). Anthers in male sterile lines are fused only at their bases, and not at their tips. In some lines, anthers are of normal length and project from the corolla, although devoid of pollen. Female flowers of male sterile plants appear normal and are fertile when pollinated.

A single male sterile cytoplasm, PET1, derived from *H. petiolaris* ssp. *petiolaris* (Leclercq 1969) and the identification of dominant fertility restoration genes (Enns et al. 1970; Kinman 1970; Vranceanu and Stoenescu 1971) advanced sunflower production from the use of open-pollinated cultivars to hybrid production 35 years ago. This source of cytoplasmic male sterility and a few fertility restoration genes, including the widely used Rf_1 and Rf_2 genes, have been used exclusively for sunflower hybrid production worldwide (Fick and Miller 1997).

Seventy CMS sources have been identified from progenies of crosses between wild *Helianthus* accessions and cultivated lines, from wild accessions grown in observation nurseries, or from induced mutation. Fertility restoration genes have been reported for 34 CMS sources, and detailed inheritance studies have been conducted for 19 of the 34 sources (Serieys 2002). A universal coding system was proposed by Serieys (1991) to accommodate the ever increasing number of CMS sources. This 3-letter coding abbreviation of the cytoplasm donor species and/or subspecies followed by a numerical number starting from the number 1, depending on the time of its discovery and its reaction to restoration testers, is widely accepted among sunflower researchers.

1.5.5.6 Salt Tolerance

Several species of *Helianthus* are native to salt-impacted habitats. Interspecific germplasm with high salt tolerance withstanding salt concentrations up to EC 24.7 d/Sm have been identified from annual *H. paradoxus*. It appears that one major gene controls salt tolerance, although a modifier gene may also be present, possibly recessive in control (Miller 1995). Two salt-tolerant parental oilseed maintainer lines, HA 429 and HA 430, have been released (Miller and Seiler 2003).

1.5.5.7 Herbicide Tolerance

A wild population of annual *H. annuus* from a soybean field in Kansas, USA, that had been repeatedly treated with imazethapyr chemical for seven consecutive years developed resistance to the imidazolinone and sulfonylurea herbicides (Al-Khatib et al. 1998). Resistance to imazethapyr and imazamox herbicides has great potential for producers in all regions of the world for controlling several broadleaf weeds. Several populations of wild sunflower (*H. annuus* and *H. petiolaris*) from the USA and Canada have been screened for resistance to these two herbicides. Eight percent of 50 wild sunflower populations had some resistance to imazamox and 57% had some resistance to tribenuron in the central USA (Olson et al. 2004). In Canada, 52% of 23 wild *H. annuus* populations had some resistance to tribenuron (Miller and Seiler 2005). Genetic stocks IMISUN-1 (oil maintainer), IMISUN-2 (oil restorer), and IMISUN-3 (confection maintainer) have been developed and released (Al-Khatib and Miller 2000). Miller and Al-Khatib (2002) also released one oilseed maintainer and two fertility restorer breeding lines with imidazolinone herbicide resistance. Genetic stocks SURES-1 and SURES-2 with resistance to the sulfonylurea herbicide tribenuron have been developed and released by Miller and Al-Khatib (2004). In addition, the two herbicides may control broomrape in areas of the world where this parasitic weed attacks sunflower (Alonso et al. 1998).

1.5.6 Cytogenetic Stocks

Induction of polyploidy in sunflower has been accomplished using colchicine. Heiser and Smith (1964) grew seedlings for 8 h on filter paper saturated with a 2.0 g/kg solution of colchicine and obtained some chromosome doubling. Using this technique, they obtained tetraploids from the perennial hybrid cross *H. giganteus* x *H. microcephalus*. Colchicine-induced tetraploids of *H. annuus* have also been obtained (Dhesi and Saini 1973). Jan et al. (1988) subjected growing points of young plants of inbred lines P21 and HA 89 to a 5-h colchicine treatment at 1.5 g/kg, pH 5.4, with 20 g/kg dimethyl sulfoxide, resulting in a high frequency of chromosome doubling and the production of autotetraploid P21 and HA 89. Tetraploidy in P21 was not stable, with plants having $2n = 4x = 65$ to 70 chromosomes. Tetraploid plants of HA 89 had reduced vigor and did not produce achenes. Reciprocal crosses of diploid and tetraploid P21 produced four triploid plants. Backcrossing triploids to P21 produced 137 plants with $2n = 34$ to $47 + t$. Thirty-one of these plants were trisomics having $2n = 35$.

Trisomic progenies of the 31 originally identified trisomic P21 plants (Jan et al. 1988) only displayed limited variation in morphological characteristics when grown in the greenhouse. Few trisomic groups appear to have distinctive features such as unusual plant height, flowering dates, leaf texture, and stem size. In order to quickly identify the first set of sunflower trisomics, future research will need to be focused on the mitotic metaphase and meiotic pachytene karyotyping, supplemented by the fluorescence in situ hybridization (FISH) technique using linkage-group-specific restriction fragment length polymorphism (RFLP) and BAC clones.

The same colchicine treatment on interspecific F_1 hybrids also resulted in high frequencies of chromosome doubling and the production of amphiploids (Jan and Fernández-Martínez 2002). The tetraploid amphiploids produced included crosses of P21 x *H. bolanderi* (Jan and Chandler 1989), *H. gracilentus* x P21, *H. grosseserratus* x P21, *H. cusickii* x P21, *H. mollis* x P21, *H. maximiliani* x P21, and *H. nuttallii* x P21. These amphiploids have restored fertility, and provide additional genetic diversity for the improvement of cultivated sunflower. They can be backcrossed with cultivated lines without the use of embryo culture to produce progenies. An amphiploid of *H. hirsutus* x P21 represents the first hexaploid amphiploid in sunflower. Like the autotetraploid P21, the amphiploids were not completely stable. After sib-pollination of $2n = 68$ amphiploids, progenies could have $2n$ chromosome numbers from 64 to 71. The sib-pollination of $2n = 102$ *H. hirsutus* x P21 amphiploid produced progenies with $2n$ chromosome numbers from 94 to 103. Additional hexaploid amphiploids of *H. hirsutus* x P21 and *H. strumosus* x P21 were also produced.

It is expected that interspecific amphiploids will enable the establishment of a number of chromosome addition lines for genetic studies of specific chromosomes of both cultivated and wild *Helianthus* species. With the available amphiploids and some specific interspecific crosses, the potential exists to establish addition lines with HA 89 chromosomes in *H. californicus* cytoplasm and the chromosomes of *H. hirsutus, H. angustifolius, H. cusickii, H. gracilentus, H. grosseserratus, H. nuttallii, H. strumosus,* and *H. giganteus* in HA 89 cytoplasm (Jan and Seiler 2007).

1.5.7 *Mutagenesis*

Mutagens have the potential to generate new genetic variants, but often cause chromosome aberrations and abnormalities in meiotic cell division. Gundaev (1971) reports that air-dried sunflower achenes treated with 40 Gy of X-rays, and moist, swollen achenes treated with 10 Gy produced nearly equal frequencies of chromosomal aberrations. The frequency of abnormal anaphase and telophase divisions was about 20%. Georgieva-Todorova (1969) also observed cytological aberrations in anaphase following treatment of achenes with X-rays. Chromosome bridges and fragments were the most common defect. Gamma radiation from a cobalt-60 source has been reported to affect length and stainability of sunflower chromosomes (Kurnik et al. 1971). However, mutagenesis has been successfully applied to generate useful traits for the improvement of sunflower, which is especially important if the trait is not found in wild *Helianthus* species or other germplasm sources.

Twenty-two cytoplasmic male sterile mutants and seven nuclear male sterile mutants were produced by Jan and Rutger (1988) using the chemical mutagens streptomycin and mitomycin C. These lines are *H. annuus* cytoplasm which represents a new CMS source and are all stable with degenerated anthers and fertility restored using the same restorers as CMS-PET1. These CMS sources have the potential of being quickly utilized for future hybrid sunflower production using the currently available restoration lines for the CMS-PET1 cytoplasm (Jan et al. 2006c).

New variations of the tocopherol profile have been created by the use of the chemical mutagen ethylmethane sulfonate (EMS) in the search for a partial substitution of α-tocopherol by another tocopherol derivative that would improve the oxidative stability of sunflower oil (Velasco et al. 2004c). From a total of 2,000 treated achenes of four Peredovik accessions, IAST-540 was selected in M_3 and IAST-1 in M_5 with γ-tocopherol content of 940 g/kg.

The development of sunflower oil with high oleic acid content was report by Soldatov (1976). It was produced by treating cultivar VNIIMK 8931 with a 0.5% solution of dimethyl sulfate, a chemical mutagen. By bulking the plants with high oleic content, the Pervenets variety was created and released to producers in Russia by the VNIIMK Research Center (Pukhalsky and Dvoryadkin 1978). The Pervenets variety is the parental material for the modern-day high-oleic hybrids, as well as the NuSun® oilseed hybrids.

High concentrations of palmitic and stearic acid were achieved using the mutagens sodium azide, ethylmethane sulfate, and X-rays on mature achenes (Osorio et al. 1995). The resulting mutant lines were selected from single M_2 and M_3 achenes. The mutant line CAS-5 had 250 g/kg palmitic acid, a fivefold increase, and mutant line CAS-3 contained 260 g/kg, a five-fold increase compared to the non-treated line.

High palmitic acid concentration, up to 400 g/kg, was obtained using γ-ray irradiation on dry achenes (Ivanov et al. 1988). Fernández-Martínez et al. (1997) X-rayed dry achenes of high-oleic line BSD-42-3, producing line CAS-12, having 300 g/kg palmitic acid, but at the expense of the oleic acid concentration which was reduced from 880 to 560 g/kg.

Miller and Vick (1999) used N-nitroso-N-methylurea and ethyl methanesulfonate to produce three sunflower lines with reduced palmitic and stearic acids, RHA 274 LP-1, a low palmitic line with 47 g/kg, and two low stearic lines, HA 821 LS-1 and RHA 274 LS-2, with 41 and 20 g/kg, respectively. Using the same two chemical mutagens, Miller and Vick (2004) produced high-oleic line HA 435 with 856 g/kg of oleic acid.

1.5.8 Genetic Stocks

There are 50 sunflower genetic stocks registered by the Crop Science Society of America, Madison, WI, USA. The first of these is a tetraploid (4x) of "Peredovik 21" (Jan 1992a). There are also four nuclear male sterile lines of HA 89 created by chemical mutation (Jan 1992b). Two other nuclear male sterile lines with markers for sterile and fertile plants were developed by Miller (1997). A series of altered saturated fatty acid genetic stocks developed by Miller and Vick (1999) contain low palmitic and low stearic acids in RHA 274, HA 821 and HA 382 backgrounds. Mid-range oleic acid (650 g/kg) lines have also been developed (Miller and Vick 2002). Three other genetic stocks lines with reduced palmitic and stearic fatty acids have been developed (Vick et al. 2003, 2007). A series of herbicide-resistant genetic stocks have been developed. They include four genetic stocks having resistance to imazethapyr and imazamox (Miller and Al-Khatib 2002), and two sulfonylurea (tribenuron) tolerant stocks, SURES-1 and SURES-2 (Miller and Al-Khatib 2004). A dwarf parental line (Dw 271) near-isogenic to RHA 271 (about half the height of RHA 271) has also been developed and a dwarf parental line (Dw HA 89) near-isogenic to HA 89 (about one-third the height of HA 89) has been developed by Velasco et al. (2003). Miller et al. (2006) released three low cadmium confectionery lines, HA 448, HA 449, and RHA 450. Jan and Gulya (2006) released three genetic stocks with sunflower mosaic virus resistance, SuMV-1, SuMV-2, and SuMV-3. Two new cytoplasmic male sterile lines derived from wild *H. annuus* and eight complementary fertility restoration genetic stocks have been developed by Jan (2006). Jan et al. (2006b) also released one new cytoplasmic male sterile genetic stock, RIGX-HA 89, selected from wild perennial species, *H. pauciflorus* (*rigidus*), and a corresponding fertility restoration line, *Rf* RIGX-Luch. A mutant seedling from an interspecific cross bearing three cotyledons and three true leaves at each internode for the first few internodes was released as a genetic stock called 'Tricot' by Hu et al. (2006).

1.6 Importance as a Model Plant

1.6.1 Speciation

While the size and adaptive success of the Compositae have stimulated considerable research into its systematics and evolution, molecular characterization has lagged behind other plant families (Kesseli and Michelmore 1997). The sunflower genus *Helianthus* is recognized widely for the cultivated sunflower *H. annuus* and scientifically as a model organism for studying diploid and polyploid hybrid speciation, introgression, and genetic architecture (Timme et al. 2007). *Helianthus* has become an important genus for the study of speciation, especially hybrid speciation, thanks to the work of Loren Rieseberg and his colleagues. Over the last 20 years, annual hybrids have dominated the literature and made *Helianthus* the model system for studying speciation. Rieseberg and his coworkers pioneered much of this research, revealing that *H. annuus* and *H. petiolaris* have hybridized repeatedly to produce three homoploid species: *H. anomalus*, *H. deserticola*, and *H. paradoxus* (Rieseberg et al. 1990, 1991; Rieseberg 1991). Evidence provided by Rieseberg (1991) indicated that *H. bolanderi* may be a more ancient homoploid hybrid formed from the ancestors of *H. annuus* and *H. petiolaris*.

1.6.2 Interspecific Hybridization

Wild sunflower species and cultivated sunflower can generally be crossed, but the divergence and heterogeneity of the genus causes considerable difficulties, such as cross-incompatibility, embryo abortion, sterility, and reduced fertility in interspecific hybrids. Cytogenetic studies are used for determining chromosome number and structure, analysis of meiosis (microsporogenesis), and pollen viability, making it possible to establish phylogenetic relationships between wild species and the cultivated sunflower. Cytogenetic studies of the sunflower have evolved from cytology, through cytotaxonomy, and classical cytogenetics to cytogenetic-molecular studies (Atlagić 2004).

Cultivated sunflower is grown primarily as a single cross hybrid. It is the second largest hybrid crop in the world. As a hybrid crop, much effort has gone into creating genetically diverse inbred lines. A considerable amount of this diversity has come from the wild ancestors with agronomic traits introgressed into the crop species. Many early interspecific hybridization studies focused on species relationships, with discussion mostly about hybrid F_1 achene (seed) set, pollen fertility, meiosis abnormalities, and further crosses (Jan 1997). Using classical breeding methods, this research provided more taxonomic information for evolutionary studies than information about agronomic potential. Successful interspecific crosses among the wild species of *Helianthus* have been reviewed by Whelan (1978) and Miller et al. (1992). The development of a two-step embryo procedure by Chandler and Beard (1983) greatly facilitated interspecific hybridization in sunflower. They successfully produced 53 interspecific cross combinations without the exhaustive effort of endless pollination, with 21 of these combinations not previously produced. Using this method, Kräuter et al. (1991) obtained 33 interspecific hybrids with an overall success rate of 41%. This procedure allowed for the production of interspecific combinations not previously available and has facilitated additional studies of species relationships not previously possible.

In recent years, there has been greater interest in interspecific hybridization for transferring desired genes from wild species into cultivated lines to develop pre-breeding germplasm for sunflower improvement. Characteristics such as disease and insect resistance, salt tolerance, drought tolerance, fatty acid variation, CMS, and fertility-restoration diversity have been emphasized. Whelan (1980, 1981) and Whelan and Dorrell (1980) used interspecific hybridization to obtain cytoplasmic male sterility conditioned by the cytoplasm of three species, *H. petiolaris*, *H. giganteus*, and *H. maximiliani*. Sunflower has served as a model crop for the transfer of genes from the wild species into the cultivated crop.

1.6.3   Gene Flow

The sunflower crop is unique in being one of the few crops native to North America. This has facilitated the collection and preservation of the germplasm. Since wild sunflower species are native to the major sunflower production areas of North America, there is a concern about the flow of genes from the cultivated crop to the wild species. A widely acknowledged risk associated with transgenic crops is the possibility that hybridization with wild relatives will transfer fitness-related transgenes to persist in wild populations (Armstrong et al. 2005). If wild populations acquire transgenes for resistance to diseases, herbivory, environmental stress, and/or commonly used herbicides, they could become more abundant in their natural habits, or invade previously unsuitable habitats. Hybrids between cultivated and wild annual sunflower (*H. annuus*) are frequent. As high as 42% of progeny from wild plants near cultivated fields are hybrids with cultivar genes persist in the wild populations for at least five generations, and in certain areas up to 40 years (Linder et al. 1998). Moreover, there was morphological evidence of hybridization in 10–33% of the populations surveyed within a given year. Burke et al. (2002a) indicate that the opportunity for crop-wild hybridization exists throughout the range of sunflower cultivation where approximately two-thirds of all cultivated fields surveyed occurred in close proximity to, and flowered coincidentally with, common sunflower populations. In these populations, the phenological overlap was extensive, with 52-96% of all wilds flowering coincidentally with the adjacent cultivar field. These findings indicate that crop-wild hybridization is likely in all areas where sunflowers are cultivated in the USA.

Crop-to-wild gene flow with species other than *H. annuus* is far less likely due to infertility barriers and non-overlapping ranges. Although transgenes will often escape from cultivation, their rate of spread will be mainly governed by their fitness effects, not the migration rate. Thus, only highly advantageous transgenes will spread rapidly enough to have substantial ecological impact. Therefore, research on the risks associated with transgene escape should focus on the fitness effects of the genes in question and consequences of "escaped" traits. Snow et al. (2003) concluded that wild sunflower plants containing a Bt-toxin gene (*cry1Ac*) specific to lepidopterans exhibited decreased lepidopteran herbivory and produced on average 55% more achenes than non-transgenic controls. Burke and Rieseberg (2003) examined the fitness effects of a transgene, an oxalate oxidase (*OXO*) gene, conferring resistance to *Sclerotinia* white mold. They concluded that the *OXO* transgene will do little more than diffuse neutrally after its escape.

Wild-crop hybridization has the potential to influence the evolutionary ecology of related wild/weedy taxa such as sunflower, but little is known about the persistence or ecological effects of crop genes that enter wild populations via pollen movement. Snow et al. (1998) studied F_1 wild-crop hybrids of sunflower and observed that F_1 wild-crop hybrids had lower fitness than wild genotypes, especially when grown under favorable crop conditions, but the F_1 barrier to the introgression of crop genes is quite permeable.

High rates of hybridization and introgression have been reported between the cultivated sunflower and its wild progenitor, *H. annuus*. However, little consideration has been given to the possibility that other wild sunflower species may hybridize with cultivated sunflower. A closely related wild progenitor, *H. petiolaris*, was studied by Rieseberg et al. (1999) using selectable amplified fragment length polymorphism (AFLP) markers. They examined four sympatric populations of *H. petiolaris* and found a low rate of introgression ranging from 0.006 to 0.026, indicating that the *H. petiolaris* genome is differentially permeable to introgression and that escape is likely to be sporadic, occurring in some populations and not others and at different times. Thus, the risk assessment of wild *H. annuus* is of more immediate concern than that of *H. petiolaris*.

Kane and Rieseberg (2007) reported on the evolution of weedy populations of the common sunflower, *Helianthus annuus,* using microsatellites which showed that between 1 and 6% of genes were significant outliers with reduced variation in weedy populations, implying that a small but insignificant fraction of the genome may be under selection and involved in adaptation of weedy sunflowers. They concluded that weedy populations are more closely related to nearby wild populations than to each other, implying that weediness likely evolved multiple times within the species, although a single origin followed by gene flow with local populations cannot be ruled out. Together, the results point to the relative ease which weedy forms of this species can evolve and persist despite the potentially high levels of gene flow with nearby wild populations.

Gene flow between crops and wild relatives has occurred for many years and contributed to the evolution and extinction of weed species. Resistance to imidazolinone (IMI) herbicides was discovered in wild *H. annuus* and introduced into domesticated sunflower (Al-Khatib et al. 1998; Al-Khatib and Miller 2000; Miller and Al-Khatib 2002). Massinga et al. (2003) studied the gene flow of the IMI gene from domesticated sunflower to wild sunflower, concluding that domesticated sunflower outcrosses with common (*H. annuus*) and prairie (*H. petiolaris*) sunflower over distances typically encountered in the major sunflower production areas, and that backcross resistant hybrids with wild parents are successful, further increasing the potential spread of IMI-resistant feral sunflowers.

1.6.4 Genetic Diversity

Molecular markers provide an effective means for characterizing genetic variability and establishing phylogenetic relationships among cultivated and wild *Helianthus* species. Markers linked with both qualitative and quantitative traits and genes will facilitate marker-assisted selection (MAS), and eventually lead to the cloning and manipulation of desirable genes. Using RFLP markers on 17 sunflower inbred lines, Gentzbittel et al. (1994) reported a lower available genetic variability in cultivated sunflower than in other crops, suggesting that efforts to introgress new genes from wild sunflower species should be increased. More recently, simple sequence repeat (SSR) markers have been developed for sunflower, and used in one study to characterize the genetic diversity among 16 elite inbred lines, and in another study among 19 elite inbred lines and 28 domestic and wild germplasm accessions, including Native American landraces (Paniego et al. 2002; Yu et al. 2002; Tang and Knapp 2003).

Genetic variability to mid-stem infection of *Sclerotinia sclerotiorum* (white mold) among interspecific sunflower hybrids using arbitrarily primed-PCR (AP-PCR) was demonstrated by Köhler and Friedt (1999). Interspecific hybrid progenies exhibited substantial genetic distance from their parental sunflower inbred lines, and hybrids having reduced *Sclerotinia* infection were identified.

Hu and Vick (2003) developed a new marker technique, "target region amplification polymorphism" (TRAP), which uses bioinformatics tools and expressed sequence tag (EST) database information to generate polymorphic markers around targeted gene sequences. This technique has been successful in constructing a phylogenetic tree of 16 perennial *Helianthus* species (Hu et al. 2003) and assessing the genetic diversity and relationship among 177 USDA-ARS released public sunflower inbred lines (Yue et al. 2009).

1.7 Conclusions

Sunflower production continues to face challenges from both abiotic and biotic factors as well as from today's ever-changing market needs as production is shifting from areas of high productivity to marginal areas with lower yield potential. The challenge for the sunflower community is to breed sunflower adaptable to these marginal environments and at the same time increase seed yield. The crop has been faring quite well; however, the marked reduction in genetic diversity during domestication of the cultivated sunflower crop has placed the crop in a vulnerable position should any major shifts of disease races or pests occur. The uniform use of a single CMS PET1 (French) cytoplasm and a few fertility restoration genes for worldwide hybrid sunflower production makes the crop extremely vulnerable. Wild species of sunflower have been a source of many genes for pest resistance, especially for diseases. They also serve as the female parent for all hybrid sunflowers. Since wild sunflower and the sunflower crop are native to North America, associated pests have co-evolved in natural communities, thus providing the opportunity to search for pest resistance genes in the diverse wild species.

Significant advances have been made in understanding the origin, domestication, and organization of the genetic diversity, characterization, and screening methods for abiotic and biotic stresses. Useful germplasms have been identified for many agronomic traits and MAS is beginning to be used for selection of favorable alleles and QTLs. Molecular biology has added to the scope of plant breeding in sunflower, providing an option to manipulate plant expressions. The process has barely begun, but there is great opportunity to address all aspects of crop production, utilization, and food value. Similarly, more integrated linkage maps are becoming available. Researchers will have to strive to combine the best conventional and modern molecular approaches to improve sunflower germplasm to keep sunflower an economically viable global crop. This will require a multidisciplinary team approach and a commitment to a long-term integrated genetic improvement program.

References

Al-Allaf S, Godward MBE (1977) Karyotype analysis of four varieties of *Helianthus annuus* L. Cytologia 44: 319–323.

Al-Khatib K, Miller JF (2000) Registration of four genetic stocks of sunflower resistant to imidazolinone herbicides. Crop Sci 40: 869–870.

Al-Khatib K, Baumgartner JR, Peterson DE, Currie RS (1998) Imazethapyr resistance in common sunflower (*Helianthus annuus*). Weed Sci 46: 403–407.

Alonso LC, Rodriguez-Ojeda MI, Fernandez-Escobar J, Lopez-Calero G (1998) Chemical control of broomrape (*Orobanche cernua* Loefl.) in sunflower (*Helianthus annuus* L.) resistant to imazethapyr herbicide. Helia 21: 45–54.

Anashchenko AV (1974) On the taxonomy of the genus *Helianthus* L. Bot Zhurn 59: 1472–1481.

Anashchenko AV (1979) Phylogenetic relations in the genus *Helianthus* L. Bull Appl Bot Genet Plant Breed 64: 146–156.

Angadi SV, Entz MH (2002a) Root system and water use patterns of different height sunflower cultivars. Agron J 94: 136–145.

Angadi SV, Entz MH (2002b) Water relations of standard height and dwarf sunflower cultivars. Crop Sci 42: 152–159.

Arias DM, Rieseberg LH (1995) Genetic relationships among domesticated and wild sunflowers (*Helianthus annuus*, Asteraceae). Econ Bot 49: 239–248.

Armstrong TT, Fitajohn RG, Newstrom LE, Wilton AD, Lee WG (2005) Transgene escape: what potential for crop-wild hybridization? Mol Ecol 14: 2111–2132.

Arumuganathan K, Earle ED (1991) Nuclear DNA content of some important plant species. Plant Mol Biol Rep 9: 208–218.

Asch DL (1993) Common sunflower (*Helianthus annuus* L.): the pathway toward domestication. In: Proc 58th Annu Meet of Soc Am Archaeol, 17 May 1993, St. Louis, MO, USA, pp 1–15.

Atlagić J (2004) Roles of interspecific hybridization and cytogenetic studies in sunflower breeding. Helia 27: 1–24.

Atlagić J, Terzić S, Skorić D, Marinković R, Vasiljević LJ, Panković-Staftić D (2006) The wild sunflower collection in Novi Sad. Helia 13: 55–64.

Baack E, Whitney KD, Rieseberg LH (2005) Hybridization and genomic size evolution: timing and magnitude of nuclear DNA content increases in *Helianthus* homoploid species. New Phytol 167: 623–630.

Beard BH (1981) The sunflower crop. Sci Am 244: 150–161.

Boot RGA (1990) The significance of the size and morphology of root systems for nutrient acquisition and competition. In: H Labers, HL Cambridge, H Konings, TL Pons (eds) Causes and Consequences of Variation in Growth Rate and Productivity of Higher Plants. SPB Academic, The Hague, The Netherlands, pp 299–311.

Brothers ME, Miller JF (1999) Core subset for the cultivated sunflower collection. In: Proc 21st Sunflower Res Workshop, 14–15 Jan 1999, Fargo, ND, USA, pp 124–127.

Burke JM, Rieseberg LH (2003) Fitness effects of transgenic disease resistance in sunflowers. Science 300: 1250.

Burke JM, Gardner KA, Rieseberg LH (2002a) The potential for gene flow between cultivated and wild sunflower (*Helianthus annuus*) in the United States. Am J Bot 89: 1550–1552.

Burke JM, Tang S, Knapp S, Rieseberg LH (2002b) Genetic analysis of sunflower domestication. Genetics 162: 1257–1267.

Burton GW (1979) Handling cross-pollinated germplasm efficiently. Crop Sci 19: 685–690.

Brewer A (1973) Analysis of floral remains from the Higgs Site (4OLD45) In: MCR McCollough, CH Faulkner (eds) Excavation of the Higgs and Doughty Sites: I–75 Salvage Archaeology. Tennessee Archaeol Soc, Dep of Anthropol, Univ of Tennessee, Knoxville, USA, pp 141–147.

Carrera A, Poverene M, Rodríguez R (2004) Isozyme and cytogenetic analysis in *Helianthus resinosus* Small. In: Proc 16th Int Sunflower Conf, 29 Aug–2 Sept 2004, Fargo, ND, USA, vol 2, pp 685–691.

Chandler JM (1991) Chromosome evolution in sunflower. In: T Tsuchiya PK, Gupta (eds) Chromosome Engineering in Plants: Genetics, Breeding, Evolution, Part B. Elsevier Sci Publ, Amsterdam, The Netherlands, pp 229–249.

Chandler JM, Beard BH (1983) Embryo culture of *Helianthus* hybrids. Crop Sci 23: 1004–1007.

Chandler JM, Jan CC, Beard BH (1986) Chromosomal differentiation among the annual *Helianthus* species. Syst Bot 11: 354–371.

Charlet LD, Seiler GJ (1994) Sunflower seed weevils (Coleoptera: Curculionidae) and their parasitoids from native sunflowers (*Helianthus* spp.) in the Northern Great Plains. Ann Entomol Soc Am 87: 831–835.

Charlet LD, Brewer G (1997) Management strategies for insect pests of sunflower in North America. Recent Res Dev Entomol 1: 215–229.

Christov M, Nikolova L, Djambasova T (2001) Evaluation and use of wild *Helianthus* species grown in the collection of Dobroudja Agricultural Institute, General Toshevo, Bulgaria for the period 1999–2000. In: GJ Seiler (ed) FAO Sunflower Subnetwork Progr Rep 1999–2000, FAO, Rome, Italy, pp 30–31.

Cockerell TDA (1912) The red sunflower. Popul Sci Monthly 373–382.

Cockerell TDA (1915) Specific and varietal characters in annual sunflowers. Am Nat 49: 609–622.

Cockerell TDA (1918) The story of the red sunflowers. Am Mus J 18: 38–47.

Cockerell TDA (1919) Hybrid perennial sunflowers. J Roy Hort Soc 15: 26–29.

Crites GD (1991) Investigations into early plant domesticates and food production in middle Tennessee: A status report. Tennessee Anthropol 16: 69–87.

Crites GD (1993) Domesticated sunflower in fifth millennium b.p. temporal context: New evidence from middle Tennessee. Am Antiq 58: 46–148.

Cronn R, Brothers M, Klier K, Bretting PK, Wendel JF (1997) Allozyme variation in domesticated annual sunflower and its wild relatives. Theor Appl Genet 95: 532–545.

Cronquist A (1977) The Compositae revisited. Brittonia 29: 137–153.

Cuk L, Seiler GJ (1985) Collection of wild sunflower species: A collection trip in the USA. Zbornik-Radova 15: 283–289.

De Candolle AP (1836) *Helianthus*. In: AP de Candolle (ed) Prodromus Systematics Naturalis Regni Vegetabilis. Treuttel and Wurtz, Paris, France, vol II, pt V, pp 585–591.

Demurin Y (1993) Genetic variability of tocopherol composition in sunflower seeds. Helia 16: 59–62.

Dewer D (1893) Perennial sunflowers. J Roy Hort Soc 15: 26–39.

Dhesi JS, Saini RC (1973) Cytology of induced polyploids in sunflower. Nucleus 16: 49–52.

Dolde D, Vlahakis C, Hazebroek J (1999) Tocopherols in breeding lines and effects of planting location, fatty acid composition, and temperature during development. J Am Oil Chem Soc 76: 349–355.

Dorrell DG, Vick BA (1997) Properties and processing of oilseed sunflower. In: AA Schneiter (ed) Sunflower Technology and Production. CSSA, Madison, WI, USA, pp 709–745.

Dozet BM (1990) Resistance to *Diaporthe/Phomopsis helianthi* Munt.-Cvet. et al. in wild sunflower species. In: Proc 12th Sunflower Res Workshop, 8–9 Jan 1990, Fargo, ND, USA, pp 86–88.

Enns H, Dorrell DG, Hoes JA, Chubb WO (1970) Sunflower research, a progress report. In: Proc 4th Int Sunflower Conf, 23–25 Jun, 1970, Memphis, TN, USA, pp 162–167.

FAO (2008) FAOSTAT: *http://faostat.fao.org/site/567/DesktopDefault.aspx/*. Cited 20 April 2008.

Feng J, Seiler GJ, Gulya TJ, Jan CC (2006) Development of *Sclerotinia* stem rot resistant germplasm utilizing hexaploid *Helianthus* species. In: Proc 28th Sunflower Res Workshop, Jan 11–12, 2006, Fargo, ND, USA: *http://www.sunflowernsa.com* /research /research-workshop/documents/Feng_Sclerotinia_06.pdf

Fernández-Martínez J, Mancha M, Osorio J, Garces R (1997) Sunflower mutant containing high levels of palmitic acid in a high oleic background. Euphytica 97: 113–116.

Fernández-Martínez J, Melero-Vara JJ, Muñoz-Ruz J, Ruso J, Domínguez J (2000) Selection of wild and cultivated sunflower for resistance to a new broomrape race that overcomes resistance to *Or5* gene. Crop Sci 40: 550–555.

Fick GN (1976) Genetics of floral color and morphology in sunflowers. J Hered 67: 227–230.

Fick GN (1989) Sunflower. In: G Robbelen, RK Downey, A Ashri (eds) Oil Crops of the World. McGraw-Hill, New York, USA, pp 301–318.

Fick GN, Miller JF (1997) Sunflower breeding. In: AA Schneiter (ed) Sunflower Technology and Production. CSSA, Madison, WI, USA, pp 395–439.

Ford RI (1985) Patterns of prehistoric food production in North America. In: RI Ford (ed) Prehistoric Food Production in North America. Mus of Anthropol, Univ of Michigan, Ann Arbor USA, pp 341–364.

Funk VA, Bayer RJ, Keeley S, Chan R, Watson L, Gemeinholzer B, Schilling E, Panero JL, Baldwin BG, Garcia-Jagas N, Susanna A, Jansen RK (2005) Everywhere but Antarctica: Using a supertree to understand the diversity and distribution of the Compositae. Biol Skr 55: 343–374.

Georgieva-Todorova J, Lakova M (1978) Karyological study of *Helianthus salicifolius*. Nucleus 21: 60–64.

Georgieva-Todorova J, Bohorova NE (1979) Studies on the chromosomes of two *Helianthus* L. tetraploid species. Caryologia 32: 335–347.

Georgieva-Todorova J, Bohorova NE (1980) Karyological investigation of the hybrid *Helianthus annuus* L. (2n=34) × *Helianthus hirsutus* Ralf. (2n=68). CR Acad Agri G Dimitrov 7: 961–964.

Georgieva-Todorova Y (1969) The effect of X-rays on chromosome aberrations in *Helianthus annuus* L. (In Bulgarian). Genet Sel 2: 469–476.

Georgieva-Todorova Y, Lakova M, Spirkov D (1974) Karyologic study of *Helianthus mollis* L. CR Acad Agri G Dimitrov 7: 59–62.

Gentzbittel L, Zhang YX, Vear F, Griveau B, Nicolas P (1994) RFLP studies of genetic relationships among inbred lines of the cultivated sunflower, *Helianthus annuus* L.: evidence for distinct restorer and maintainer germplasm pools. Theor Appl Genet 89: 419–425.

Gimenez C, Fereres E (1986) Genetic variability in sunflower cultivars under drought: II. Growth and water relations. Aust J Agri Res 37: 583–597.

Gray A (1884) Synoptical flora of North America. Smithsonian Institution, Washington, DC, USA, vol I, part II, pp 271–280.

Gulya TJ (1982) Apical chlorosis of sunflower caused by *Pseudomonas syringae* pv. *tagetis*. Plant Dis 66: 598–600.

Gulya TJ, Shiel PJ, Freeman T, Jordan RL, Isakeit T, Berger PH (2002) Host range and characterization of sunflower mosaic potyvirus. Phytopathology 92: 694–702.

Gundaev AI (1971) Basic principles of sunflower selection. In: Genetic Principles of Plant Selection, Nauka, Moscow [Trans. Dep of the Secretary of State, Ottawa, Canada, 1972], pp 417–465.

Greilhuber J (1988) Self tanning—a new and important source of stoichiometric error in cytophotometric determination of nuclear-DNS content in plants. Plant Sys Evol 158: 87–96.

Harlan JR (1976) Genetic resources in wild relatives of crops. Crop Sci 16: 329–332.

Harter AV, Gardner E, Faulush D, Lentz D, Bye R, Rieseberg LH (2004) Origin of extant domesticated sunflowers in eastern North America. Nature 430: 201–205.

Heiser CB (1951) The sunflower among the North American Indians. Proc Am Phil Soc 95: 432–448.

Heiser CB (1954) Variation and sub-speciation in the common sunflower *Helianthus annuus*. Am Midl Nat 51: 287–305.

Heiser CB (1965) Sunflowers, weeds, and cultivated plants In: HG Baker, GL Stebbins (eds) The Genetics of Colonizing Species. Acad Press, NY, USA, pp 391–401.

Heiser CB (1976) The Sunflower. Univ of Oklahoma Press, Norman, OK, USA.

Heiser CB (1978) Taxonomy of *Helianthus* and origin of domesticated sunflower. In: JF Carter (ed) Sunflower Science and Technology, Agron Monogr 19. ASA, CSSA, SSSA, Madison, WI, USA, pp 31–53.

Heiser CB (1985) Some botanical considerations of the early domesticated plants north of Mexico. In: RI Ford (ed) Prehistoric Food Production in North America. Mus of Anthropol, Univ of Michigan, Ann Arbor, pp 57–82.

Heiser CB (1998) The domesticated sunflower in Mexico. Genet Resour Crop Evol 45: 447–449.

Heiser CB (2008) The sunflower (*Helianthus annuus*) in Mexico: Further evidence for a North American domestication. Genet Resour Crop Evol 55: 9–13.

Heiser CB, Smith DM (1964) Species crosses in *Helianthus*: II. Polyploid species. Rhodora 66: 344–358.

Heiser CB, Smith DM, Clevenger SB, Martin WC (1969) The North American sunflowers (*Helianthus*). Mem Torr Bot Club 22: 1–219.

Heywood VH (1978) Flowering Plants of the World. Prentice Hall, Englewood Cliffs, NJ, USA.

Hoes JA, Putt ED, Enns H (1973) Resistance to Verticillium wilt in collections of wild *Helianthus* in North America. Phytopathology 63: 1517–1520.

Holden J, Peacock J, Williams T (1993) Genes, Crops, and the Environment. Cambridge Press, NY, USA.

Hu J, Vick BA (2003) Target region amplified polymorphism: a novel marker technique for plant genotyping. Plant Mol Biol Rep 21: 1–6.

Hu J, Miller JF, Vick BA (2006) Registration of a tricotyledon sunflower genetic stock. Crop Sci 46(6): 2734–2735.

Hu, J, Seiler GJ, Jan CC, Vick BA (2003) Assessing genetic variability among sixteen perennial *Helianthus* species using PCR-based TRAP markers. Proc 25th Sunflower Res Forum, Fargo, ND, USA: *http://www.sunflowernsa.com/research/research-workshop/documents/88.PDF*

International Board for Plant Genetic Resources (IBPGR) (1984) Report of a Working Group on Sunflowers (First Meeting). Int Board for Plant Genet Resour, Rome, Italy.

Ivanov P, Petakov D, Nikolova V, Pentchev E (1988) Sunflower breeding for high palmitic acid content in the oil. In: Proc 12th Int Sunflower Conf, 25–29 Jul 1988, Novi Sad, Yugoslavia, pp 463–465.

Ivanov P, Stoyanoova Y (1978) Results from sunflower breeding directed to obtaining gene materials of high protein content in the kernel. In: Proc 8th Int Sunflower Conf, 23–27 Jul 1978, Minneapolis, MN, USA, pp 441–448.

Jain SK, Kesseli R, Olivieri A (1992) Biosystematic status of the serpentine sunflower, *Helianthus exilis* Gray In: AJ Baker, J Proctor, RD Reeves (eds) The Vegetation of Ultramafic (Serpentine) Soils. Intercept Ltd, Andover, England, UK, pp 391–408.

Jan CC (1992a) Registration of an induced tetraploid sunflower genetic stock line, Tetra P21. Crop Sci 32: 1520.

Jan CC (1992b) Registration of four nuclear male-sterile sunflower genetic stock lines. Crop Sci 32: 1519.

Jan CC (1997) Cytology and interspecific hybridization. In: AA Schneiter (ed) Sunflower Technology and Production. CSSA, Madison, WI, USA, pp 497–558.

Jan CC (2006) Two cytoplasmic male-sterile and eight fertility restoration sunflowers. Crop Sci 46(4): 1835.

Jan CC, Chandler JM (1985) Transfer of powdery mildew resistance from *Helianthus debilis* Nutt. into cultivated sunflower (*H. annuus* L.). Crop Sci 25: 664–666.

Jan CC, Chandler JM (1989) Interspecific hybrids and amphiploids of *Helianthus annuus* x *H. bolanderi*. Crop Sci 29: 643–646.

Jan CC, Fernández-Martinez JM (2002) Interspecific hybridization, gene transfer, and the development of resistance to broomrape race F in Spain. Helia 25: 123–136.

Jan CC, Gulya TJ (2006) Three virus resistant sunflower genetic stocks. Crop Sci 46(4): 1834.

Jan CC, Rutger JN (1988) Mitomycin C- and streptomycin-induced male sterility in cultivated sunflower. Crop Sci 28: 792–795.

Jan CC, Seiler GJ (2007) Sunflower. In: RJ Singh (ed) Genetics Resources, Chromosome Engineering, and Crop Improvement, vol 4, Oilseed Crops. CRC Press, NY, USA, pp 103–165.

Jan CC, Chandler JM, Wagner SA (1988) Induced tetraploidy and trisomics production of *Helianthus annuus* L. Genome 30: 647–651.

Jan CC, Feng J, Seiler GJ, Gulya TJ (2006a) Amphiploids of perennial *Helianthus* species x cultivated sunflower possess valuable genes for resistance to Sclerotinia stem and head rot. In: Proc 28th Sunflower Res Workshop, 11–12 Jan 2006, Fargo, ND, USA:*http://www.sunflowernsa.com/research/research-workshop/documents/Jan_Amphiploids_06.pdf*

Jan CC, Miller JF, Seiler GJ, Fick GN (2006b) Registration of one cytoplasmic male-sterile and two fertility restoration sunflowers. Crop Sci 46(4): 1835.

Jan CC, Miller JF, Vick BA, Seiler GJ (2006c) Performance of seven new cytoplasmic male-sterile sunflower lines from induced mutation and a Native American variety. Helia 29(44): 47–54.

Jansen RK, Michaels HJ, Palmer JD (1991) Phylogeny and character evolution in the Asteraceae based on chloroplast DNA restriction site mapping. Syst Bot 16: 98–115.

Jones Q (1983) Germplasm needs for oilseed crops. Econ Bot 37: 418–422.

Jones OR (1984) Yield, water use efficiency, and oil concentration and quality of dryland sunflower grown in the southern high plains. Agron J 76: 229–235.

Kane N, Rieseberg LH (2007) Genetics and evolution of weedy *Helianthus annuus* populations: adaptation of an agricultural weed. Mol Ecol 17: 384–394.

Kesseli RV, Michelmore RW (1997) The Compositae: Systematically fascinating but specifically neglected. In: AH Paterson (ed) Genome Mapping in Plants. RG Landes, Georgetown, TX, USA, pp 179–191.

Kinman ML (1970) New developments in the USDA and state experiment station sunflower breeding programs. In: Proc 4th Int Sunflower Conf, Memphis, 23–25 Jun 1970, Memphis, TN, USA, pp 181–183.

Kleingartner LK (2002) NuSun® sunflower oil: redirection of an industry. In: J Janick , A Whipkey (eds) Trends in New Crops and New Uses. ASHS Press, Alexandria, VA, USA, pp 135–138.

Kleingartner LW (2004) World outlook and future development of sunflower markets around the world. In: GJ Seiler (ed) Proc 16th Int Sunflower Conf, Fargo, ND, USA, 29 Aug–2 Sept, pp 69–77.

Knowles PF (1978) Morphology and anatomy. In: JF Carter (ed) Sunflower Science and Technology, Agron Monogr 19. ASA, CSSA, and SSSA, Madison, WI, USA, pp 55–88.

Köhler RH, Friedt W (1999) Genetic variability as identified by AP-PCR and reaction to mid-stem infection of *Sclerotinia sclerotiorum* among interspecific sunflower (*Helianthus annuus* L.) hybrid progenies. Crop Sci 39: 1456–1463.

Kräuter R, Steinmetz A, Friedt W (1991) Efficient interspecific hybridization in the genus *Helianthus* via "embryo rescue" and characterization of the hybrids. Theor Appl Genet 82: 521–525.

Kulshreshtha VB, Gupta PK (1981) Cytogenetic studies in the genus *Helianthus* L. II. Karyological studies in twelve species. Cytologia 46: 279–289.

Kurnik E, Parragh JM, Pozsar B (1971) Effect of g-ray dose (^{66}Co) on the size and stability of root chromosomes of sunflower (In Hungarian). Bot Kozl 58: 235–237.

Lentz DL, Pohl MED, Pope KO (2001) Prehistoric sunflower (*Helianthus annuus* L.) domestication in Mexico. Econ Bot 55: 370–376.

Lentz DL, Pohl MD, Alvarado, JL, Tarighat S, Bye R (2008) Sunflower (*Helianthus annuus* L.) as a pre-Columbian domesticate in Mexico. Proc Nat Acad Sci USA 105: 6232–6237.

Leclercq P (1969) Cytoplasmic male sterility in sunflower. Ann Amel Plant 19: 99–106.

Leclercq P, Cauderon Y, Dauge M (1970) Sélection pour la résistance au mildiou du tournesol a partir d'hybrides topinambour x tournesol. Ann Amel Plant 20: 363–373.

Levan A, Fregda K, Sandberg L (1964) Nomenclature for centromeric position on chromosomes. Hereditas 52: 201–220.

Linder CR, Taha I, Seiler GJ, Snow AA, Rieseberg LH (1998) Long-term introgression of crop genes into wild sunflower populations. Theor Appl Gen 96: 339–347.

Linnaeus CL (1753) Species *Plantarum*, Holmiae.

Luhs W, Friedt W (1994) Non-food uses of vegetable oils and fatty acids. In: DJ Murphy (ed) Designer Oil Crops: Breeding, Processing, and Biotechnology. VCH Press, NY, USA, pp 73–130.

Lusas EW (1982) Sunflower meals and food proteins in Sunflower. In: J Adams (ed) The Sunflower. Natl Sunflower Assoc, Bismarck, ND, USA, pp 25–36.

Massinga R, Al-Khatib K, St. Amand P, Miller JF (2003) Gene flow from imidazolinone-resistant domesticated sunflower to wild relatives. Weed Sci 51: 854–862.

Matthews JF, Allison JR, Ware RT, Nordman C (2002) *Helianthus verticillatus* Small (Asteraceae) rediscovered and redescribed. Castanea 67: 13–24.

Mensink RP, Temple EHM, Hornstra G (1994) Dietary saturated and trans fatty acids and lipoprotein metabolism. Ann Med 26: 461–464.

Miller JF (1995) Inheritance of salt tolerance in sunflower. Helia 18: 9–16.

Miller JF (1997) Registration of CMS HA 89 (PEF1) cytoplasmic male-sterile, RPEF1 restorer, and two nuclear male-sterile (NMS 373 and 377) sunflower genetic stocks. Crop Sci 37: 1984.

Miller JF, Fick GN (1997) The genetics of sunflower. In: AA Schneiter (ed) Sunflower Technology and Production. CSSA, Madison, WI, USA, pp 441–495.

Miller JF, Vick BA (1999) Inheritance of reduced stearic and palmitic acid content in sunflower seed oil. Crop Sci 39: 364–367.

Miller JF, Al-Khatib K (2002) Registration of imidazolinone herbicide-resistant sunflower maintainer (HA 425) and fertility restorer (RHA 426 and RHA 427) germplasms. Crop Sci 42: 988–989.

Miller JF, Vick BA (2002) Registration of four mid-range oleic acid sunflower genetic stocks. Crop Sci 42: 994.

Miller JF, Seiler GJ (2003) Registration of five oilseed maintainer (HA 429–HA 433) sunflower germplasm lines. Crop Sci 43: 2313–2314.

Miller JF, Al-Khatib K (2004) Registration of two oilseed sunflower genetic stocks, SURES-1 and SURES-2 resistant to tribenuron herbicide. Crop Sci 44: 1037–1038.

Miller JF, Vick BA (2004) Registration of two maintainer (HA 434 and HA 435) and three restorers (RHA 436 to RHA 438) high oleic oilseed sunflower germplasm. Crop Sci 44: 1034–1035.

Miller JF, Seiler GJ (2005) Tribenuron resistance in accessions of wild sunflower collected in Canada. In: Proc 27th Sunflower Res Workshop, 12–13 Jan 2005, Fargo, ND,

USA: *http://www.sunflowernsa.com/research/research- workshop/documents /miller_tribenuron_05.pdf, 2005*

Miller JF, Seiler GJ, Jan CC (1992) Use of plant introductions in cultivar development, part 2. In: H Shands, LE Weisner (eds) CSSA Spl Publ 20. ASA, CSSA, and SSSA, Madison, WI, pp 151–166

Miller J, Green C, Li Y-M, Chaney R (2006) Three low cadmium confection sunflower. Crop Sci 46(1): 489.

Miller JF, Seiler GJ, Jan CC (1992) Use of plant introductions in cultivar development, part 2. In: H Shands ,LE Weisner (eds) CSSA Spl Publ 20. ASA, CSSA, and SSSA, Madison, WI, pp 151–166

Mondolot-Cosson L, Andary C (1994) Resistance factors of wild species of sunflower, *Helianthus resinosus* to *Sclerotinia sclerotiorum*. Acta Hort 381: 642–645.

Morris JB, Yang SM, Wilson L (1983) Reaction of *Helianthus* species to *Alternaria helianthi*. Plant Dis 67: 539–540.

Morrison WH, Hamilton RJ, Kalu C (1995) Sunflower seed oil. In: RJ Hamilton (ed) Developments in Oils and Fats. Chapman and Hall, London, UK, pp 132–152.

Oliveri AM, Jain SK (1977) Variation in the *Helianthus exilis-bolanderi* complex. Madroño 24: 177–189.

Olson B, Al-Khatib K, Aiken RM (2004) Distribution of resistance to imazamox and tribenuron-methyl in native sunflowers. In: Proc 26th Sunflower Res Workshop, 14–15 Jan, Fargo, ND, USA: *http://www.sunflowernsa.com/research/research-workshop/documents/ 158.pdf*

Osorio J, Fernández-Martínez J, Mancha M, Garces R (1995) Mutant sunflowers with high concentration of saturated fatty acids in the oil. Crop Sci 35: 739–742.

Palmer JH, Phillips TDJ (1963) The effect of the terminal bud, indoleacetic acid, and nitrogen supply on the growth and orientation of the petiole of *Helianthus annuus*. Physiol Plant 16: 572–584.

Panero JL, Funk VA (2002) Toward a phylogenetic subfamilial classification for the Compositae (Asteraceae). P Biol Soc Wash 115: 909–922.

Paniego N, Echaide M, Muñoz M, Fernández L, Torales S, Faccio P, Fuxan I, Carrera M, Zandomeni R, Suárez EY, Hopp HE (2002) Microsatellite isolation and characterization in sunflower (*Helianthus annuus* L.). Genome 45: 34–43.

Paniego N, Heinz R, Hopp HE (2007) Sunflower. In: C Kole (ed) Genome Mapping and Molecular Breeding in Plants, vol 2: Oilseeds. Springer, Heidelberg, Berlin, New York, pp 153–178.

Pearson OH (1981) Nature and mechanisms of cytoplasmic male sterility in plants: A review. Hort Sci 16: 482–487.

Phillips OL, Meilleur BA (1998) Usefulness and economic potential of rare plants of the United States: A statistical survey. Econ Bot 52: 57–67.

Prescott-Allen CP, Prescott-Allen R (1986) The First Resource: Wild Species in the North American Economy. Yale Univ Press, New Haven, CN, USA.

Price HJ, Hodnett G, Johnston JS (2000) Sunflower (*Helianthus annuus*) leaves contain compounds that reduce nuclear propidium iodide fluorescence. Ann Bot 86: 929–934.

Pruski JF (1998) *Helianthus porteri* (A. Gray) Pruski, a new combination validated for the "Confederate Daisy." Castanea 63: 74–75.

Pryde EH, Rothfus JA (1989) Industrial and nonfood uses of vegetable oils. In: G Robbelen, RK Downey, A Ashri (eds) Oil Crops of the World: Their Breeding and Utilization. McGraw–Hill, NY, USA, pp 87–117.

Pukhalsky AV, Dvoryadkin NI (1978) Achievements of sunflower breeding in the USSR. In: Proc 8th Int Sunflower Conf, 23–27 Jul 1978, Minneapolis, MN, USA, pp 48–55.

Pustovoit VS (1967) Handbook of Selection and Seed Growing of Oil Plants (In Russian), English translation. US Dep of Commerce, Springfield, VA, USA.

Pustovoit VS (1975) The Sunflower (In Russian). Kolos Press, Moscow, USSR, pp 182–210.

Pustovoit GV, Gubin IA (1974) Results and prospects in sunflower breeding for group immunity by using the interspecific hybridization method. In: Proc 6th Int Sunflower Conf, 22–24 Jul 1974, Bucharest, Romania, pp 373–381.

Putt ED (1962) The value of hybrids and synthetics in sunflower seed production. Can J Plant Sci 42: 488–500.

Putt ED (1965) Sunflower variety Peredovik. Can J Plant Sci 45: 207.

Putt ED (1997) Early history of sunflower. In: AA Schneiter (ed) Sunflower Technology and Production. CSSA, Madison, WI, USA, pp 1–20.

Quresh Z, Jan CC, Gulya TJ (1993) Resistance of sunflower rust and its inheritance in wild sunflower species. Plant Breed 110: 297–306.

Raicu P, Vranceanu V, Mihailescu A, Popescu C, Kirillova M (1976) Research of the chromosome complement in *Helianthus* L. Caryologia 29: 307–316.

Rieseberg LH (1991) Homoploid reticulate evolution in *Helianthus* (Asteraceae): evidence from ancient ribosomal genes. Am J Bot 78: 1218–1237.

Rieseberg LH, Seiler GJ (1990) Molecular evidence and the origin and development of the domesticated sunflower (*Helianthus annuus*). Econ Bot 44: 79–91.

Rieseberg LH, Soltis DE, Palmer JD (1988) A molecular re-examination of introgression between *Helianthus annuus* and *H. bolanderi* (Compositae). Evolution 42: 227–238.

Rieseberg LH, Cartrer R, Zona S (1990) Molecular tests of the hypothesized hybrid origin of two diploid *Helianthus* species (Asteraceae). Evolution 44: 1498–1511.

Rieseberg LH, Beckstrom-Sternberg S, Liston A, Arias D (1991) Phylogenetic and systematic inferences from chloroplast DNA and isozyme variation in *Helianthus* sect. *Helianthus*. Syst Bot 16: 50–76.

Rieseberg LH, Kim MJ, Seiler GJ (1999) Introgression between the cultivated sunflower and a sympatric wild relative, *Helianthus petiolaris* (Asteraceae). Int J Plant Sci 160: 102–108.

Robinson A (1979) Studies in the *Heliantheae* (Asteraceae). XVIII. A new genus *Helianthopsis*. Phytologia 44: 257–259.

Rogers CE, Thompson TE (1978) Resistance in wild *Helianthus* to the sunflower beetle. J Econ Entomol 71: 622–623.

Rogers CE, Thompson TE (1980) *Helianthus* resistance to the sunflower beetle. J Kansas Entomol Soc 53: 727–730.

Rogers CE, Seiler GJ (1985) Sunflower (*Helianthus*) resistance to stem weevil (*Cylindrocopturus adspersus*). Environ Entomol 14: 624–628.

Rogers CE, Thompson TE, Seiler GJ (1984) Registration of three *Helianthus* germplasms for resistance to the sunflower moth. Crop Sci 24: 212–213.

Ronicke S, Hahn V, Horn R, Gron I, Brahn L, Schnabl H, Freidt W (2004) Interspecific hybrids of sunflower as sources of Sclerotinia resistance. Plant Breed 123: 152–157.

Ruso J, Sukno S, Domínguez-Gimenez J, Melero-Vara JM, Fernández-Martínez JM (1996) Screening wild *Helianthus* species and derived lines for resistance to several populations of *Orobanche cernua*. Plant Dis 80: 1165–1169.

Sadras VO, Hall AJ, Trapani N, Vilella F (1989) Dynamics of rooting and root–length: leaf area relationships as affected by plant population in sunflower crops. Field Crops Res 22: 45–57.

Saliman M, Yang SM, Wilson L (1982) Reaction of *Helianthus* species to *Erysiphe cichoracearum*. Plant Dis 66: 572–573.

Schilling EE (2001) Phylogeny of *Helianthus* and related genera. Oleagineux Crops Gras Lipides 8: 22–25.

Schilling EE (2006) *Helianthus*. In: Flora of North America Editorial Committee (ed) Flora of North America North of Mexico. Oxford Univ Press, New York and Oxford, vol 21, pp 141–169.

Schilling EE, Heiser CB (1981) Infrageneric classification of *Helianthus* (Compositae) Taxon 30: 393–403.

Schilling EE, Panero JL (1996) Phylogenetic reticulation in subtribe Helianthinae. Am J Bot 83: 939–948.

Schilling EE, Panero JL (2002) A revised classification of subtribe Helianthinae (Asteraceae: Heliantheae). I. Basal lineages. Bot J Linn Soc 140: 65–76.

Schilling EE, Linder CR, Noyes RD, Rieseberg LH (1998) Phylogenetic relationships in *Helianthus* (Asteraceae) based on nuclear ribosomal DNA internal transcribed spacer regions sequence data. Syst Bot 23: 177–187.

Schneiter AA, Miller JF (1981) Description of sunflower growth stages. Crop Sci 21: 901–903.

Schneiter AA (1992) Production of semidwarf and dwarf sunflower in the Northern Great Plains of the United States. Field Crops Res 30: 391–401.

Schrader O, Ahne R, Fuchs J, Schubert I (1997) Karyotype analysis of *Helianthus annuus* using giemsa banding and fluorescence in situ hybridization. Chrom Res 5: 451–456.

Seiler GJ (1985) Evaluation of seeds of sunflower species for several chemical and morphological characteristics. Crop Sci 25: 183–187.

Seiler G J (1988) The genus *Helianthus* as a source of genetic variability for cultivated sunflower. In: Proc 12th Int Sunflower Conf, 25–29 Jul 1988, Novi Sad, Yugoslavia, vol 1, pp 17–58.

Seiler GJ (1992) Utilization of wild sunflower species for the improvement of cultivated sunflower. Field Crops Res 30: 195–230.

Seiler GJ (1996) The USDA-ARS sunflower germplasm collection. Helia 19: 44–45.

Seiler GJ (1998) The potential use of wild *Helianthus* species for selection of low saturated fatty acids in sunflower oil. In: AM de Ron (ed) Int Symp on Breeding of Protein and Oil Crops. EUCARPIA Congr, 1–4 April 1998, Pontevedra, Spain, pp 109–110.

Seiler GJ (2002) Wild sunflower germplasm: A perspective on characteristics of use to sunflower breeders in developing countries. In: Proc 2nd Int Symp on Sunflower in Developing Countries, 18–21 Feb 2002, Benoni, South Africa: *http://www.isa. cetiom.fr/symposium/seiler.htm*

Seiler GJ (2007) Wild annual *Helianthus anomalus* and *H. deserticola* for improving oil content and quality in sunflower. Ind Crop Prod 25: 95–100.

Seiler GJ, Brothers ME (2003) Exploration for wild *Helianthus anomalus* and *H. deserticola* in the desert southwest USA. In: Proc 25th Sunflower Res Workshop, 16–17 Jan 2003, Fargo, ND, USA: *http://www.sunflowernsa.com/research/research_workshop/documents /90.pdf*

Seiler GJ, Rieseberg LH (1997) Systematics, origin, and germplasm resources of wild and domesticated sunflower. In: AA Schneiter (ed) Sunflower Technology and Production. CSSA, Madison, WI, USA, pp 21–65.

Serieys H (1991) Note on the codification of sunflower CMS sources, FAO sunflower research subnetwork. In: Proc FAO Sunflower Subnetwork Progr Rep, 1990. FAO, Rome, Italy, pp 913.

Serieys HA (1992) Sunflower: A Catalogue of the Wild Species of the Genus *Helianthus*, ENSAM and INRA, Montpellier, France.

Serieys H (2002) Report on the Past Activities of the FAO Working Group Identification, Study and Utilization in Breeding Programs of New CMS Sources for 1999–2001. In: Proc FAO Sunflower Subnetwork Progr Rep, 1999–2001, FAO, Rome, Italy, pp 1–54.

Skoric D (1985) Sunflower breeding for resistance to *Diaporthe/Phomopsis helianthi* Munt.-Cvet. et al. Helia 8: 21–23.

Skoric D (1987) FAO sunflower sub-network report 1984–1986. In: D Skoric (ed) Genetic Evaluation and Use of *Helianthus* Wild Species and Their Use in Breeding Programs. FAO, Rome, Italy, pp 1–17.

Skoric D (1988) Sunflower breeding. J Edible Oil Ind 25: 1–90.

Skoric D (1992) Achievements and future directions of sunflower breeding. Field Crops Res 30: 195–230.

Smith BR (2006) Eastern North America as an independent center of plant domestication. Proc Natl Acad Sci USA 103: 12223–12228.

Snow AA, Moran-Palma P, Rieseberg LH, Wszlaki A, Seiler GJ (1998) Fecundity, phenology, and seed dormancy of F_1 wild-crop hybrids in sunflower (*Helianthus annuus*, Asteraceae). Am J Bot 85: 794–801.

Snow AA, Pilson D, Rieseberg LH, Paulsen MJ, Pleskac N, Reagon MR, Wol DE, Selbo SM, (2003) A Bt transgene reduces herbivory and enhances fecundity in wild sunflowers. Ecol Appl 13: 279–286.

Soldatov KI (1976) Chemical mutagenesis in sunflower breeding. In: Proc 7th Int Sunflower Conf, 27 Jun–3 Jul 1976, Krasnodar, USSR, pp 352–357.

Sossey-Alaoui K, Serieys H, Tersac M, Lambert P (1998) Evidence of several genomes in *Helianthus*. Theor Appl Genet 97: 422–430.

Suslov VM (1968) Economic significance of sunflowers in the USSR. In: Proc 3rd Int Sunflower Conf, 13–15 Aug, 1968, Crookston, MN, USA, pp 1–11.

Tan AS, Jan CC, Gulya TJ (1992) Inheritance of resistance to race 4 of sunflower downy mildew in wild sunflower accessions. Crop Sci 32: 949–952.

Tang SX, Knapp SJ (2003) Microsatellites uncover extraordinary diversity in native American land races and wild populations of cultivated sunflower. Theor Appl Genet 106: 990–1003.

Tatum LA (1971) The southern corn leaf blight epidemic. Science 171: 1113–1116.

Thompson TE, Zimmerman DC, Rogers CE (1981) Wild *Helianthus* as a genetic resource. Field Crops Res 4: 333–343.

Timme RE, Simpson BB, Linder CR (2007) High-resolution phylogeny for *Helianthus* (Asteraceae) using the 18S–26S ribosomal DNA external transcribed spacer. Am J Bot 94: 1837–1852.

Torrey J, Gray A (1842) A Flora of North America, vol II. Wiley and Putnam, New York, USA, pp 318–333.

U.S. Department of Agriculture (2007) Foreign Agricultural Service, Oilseeds: World Market and Trade, Circular Series FOP 02–05: *http://www/fas.usda.gov/oilseeds/circular/2005*

Velasco L, Domínguez J, Muñoz-Ruz J, Pérez-Vich B, Fernández-Martínez FM (2003) Registration of Dw 89 and Dw 271 dwarf parental lines of sunflower. Crop Sci 43: 1140–1141.

Velasco L, Pérez-Vich B, Fernández-Martínez JM (2004a) Evaluation of wild sunflower species for tocopherol content and composition. Helia 27(40): 107–112.

Velasco L, Domínguez J, Fernández-Martínez JM (2004b) Registration of T589 and T2100 sunflower germplasm with modified tocopherols. Crop Sci 44: 363.

Velasco L, Pérez-Vich B, Fernández-Martínez JM (2004c) Novel variation for the tocopherol profile in a sunflower created by mutagenesis and recombination. Plant Breed 123: 490–492.

Vick BA, Miller JF (2002) Strategies for the development of NuSun® sunflower hybrids. In: TM Kuo, HW Gardner (eds) Lipid Biotechnology. Marcel Dekker, New York, USA, pp 115–128.

Vick BA, Jan CC, Miller JF (2003) Registration of two sunflower genetic stocks with reduced palmitic and stearic acids. Crop Sci 43: 747–748.

Vick BA, Jan CC, Miller JF (2007) Registration of sunflower genetic stock RS3 with reduced levels of palmitic and stearic acids. J Plant Reg 1(1): 80.

Vranceanu AV, Stoenescu FM (1971) Pollen fertility restorer genes from cultivated sunflower. Euphytica 20: 536–541.

Warner K (2005) Effects of the flavor and oxidative stability of stripped soybean and sunflower oils with added pure tocopherols. J Agri Food Chem 53: 9906–9910.

Warner K, Miller JF, Demurin Y (2008) Oxidative stability of crude mid-oleic sunflower oils from seeds with high gamma and delta tocopherol levels. J Am Oil Chem Soc 85(6): 529–533.

Watson EE (1929) Contributions to a monograph of the genus *Helianthus*. Pap Mich Acad Sci Arts Lett 9: 305–475.

Weaver JE (1926) Root hairs of sunflower. In: JE Weaver (ed) Root Development of Field Crops. McGraw-Hill, New York, USA, pp 247–252.

Whelan EDP (1978) Hybridization between annual and perennial diploid species of *Helianthus*. Can J Genet Cytol 20: 523–530.

Whelan EDP (1979) Interspecific hybrids between *Helianthus petiolaris* Nutt. and *H. annuus* L.: effect of backcrossing on meiosis. Euphytica 28: 297–308.

Whelan EDP (1980) A new source of cytoplasmic male sterility in sunflower. Euphytica 29: 33–46.

Whelan EDP (1981) Cytoplasmic male sterility in *Helianthus giganteus* L. × *H. annuus* L. interspecific hybrids. Crop Sci 21: 855–858.

Whelan EDP (1982) Trisomic progeny from interspecific hybrids between *Helianthus maximiliani* and *H. annuus*. Can J Genet Cytol 24: 375–384.

Whelan EDP, Dorrell DG (1980) Interspecific hybrids between *Helianthus maximiliani* Schrad. and *H. annuus* L.: effect of backcrossing on meiosis, anther morphology, and seed characteristics. Crop Sci 20: 29–34.

Willett WC (1994) Diet and health: What should we eat? Science 264: 532–537.

Yang SM, Morris JB, Thompson TE (1980) Evaluation of *Helianthus* spp. for resistance to Rhizopus head rot. In: Proc 9th Int Sunflower Conf, 8–13 Jun, 1980, Torremolinos, Spain, vol 1, pp 147–151.

Yu JK, Mangor J, Thompson L, Edwards KJ, Slabaugh MB, Knapp SJ (2002) Allelic diversity of simple sequence repeats among elite inbred lines of cultivated sunflower. Genome 45: 652–660.

Yue, B, Cai X, Vick BA, Hu J (2009) Genetic diversity and relationship among 177 public sunflower inbred lines assessed by TRAP markers. Crop Sci 49: 1242–1249.

2

Classic Genetics and Breeding

Felicity Vear

ABSTRACT

Sunflower hybrids were first developed in sunflower 60 years ago and with the discovery of cytoplasmic male sterility have become the main type of variety since 1970. Breeding programs use mostly pedigree, recurrent and backcross selection. The main characters selected for are seed yield, oil content and oil type, earliness, height, disease, insect and drought resistance. Except for a few major genes, until 1995, breeding depended only on phenotypic observations, but in the last 10 years genomics have improved knowledge of the genetics of quantitative characters and marker-assisted selection has started to become a practical tool.

Keywords: cytoplasmic male sterility; hybrid; inbred line; major genes; quantitative inheritance; recurrent selection; resistance

2.1 Introduction

Since sunflowers are cross-pollinated and show heterosis, the most vigorous plants, producing the best yields, are heterozygotes. The first sunflower varieties maintained by the Russian farmers were heterogeneous open-pollinated populations of mixtures of natural hybrids. From 1920 to 1970, more characterized populations were bred, especially in Russia, at the VNIIMK station at Krasnodar. The best known were VNIIMK 6540, VNIIMK 8931 and Peredovik. There were also Argentinean, French, Canadian and Romanian varieties. All these populations constituted the main genetic resources of cultivated sunflower until the mid-1970s. They are now

INRA, UMR 1095, Domaine de Crouelle, 234 Ave du Brezet, 63000 Clermont-Ferrand, France; e-mail: *vear@clermont.inra.fr* maintained by breeders either in isolated plots (where pollination is carried out by bees) or by a series of sib-crosses under paper tubes or cloth bags, with mixture of the seed of all plants harvested in order to maintain variability.

The last widely grown open-pollinated varieties were highly self-sterile, since self-sterility was in fact a favorable character to obtain a maximum of hybrid plants in the population and therefore the highest yields. However, the fact that these open-pollinated varieties were made up of a population of different hybrids meant that they were not uniform for important characters, such as height or earliness. Breeders followed the example of maize and, from 1950 onwards, became interested in hybrid varieties. The development of modern breeding programs to create hybrid varieties was a radical change since it was necessary to have homozygous, inbred parental lines that could be maintained and thus were self-fertile. It was not really possible to maintain breeding for both open-pollinated varieties and hybrids although there would have been a continued market for the former in some developing countries where farmers did not wish to buy seed each year.

2.2 Hybrids

The first studies of hybrid vigor in sunflowers were made during the late 1940s by Canadians, who found an increase of 60% in yield compared with open-pollinated varieties (Unrau and White 1944). They produced some hybrid varieties, such as "Advance" between 1946 and 1952, using self-sterility to facilitate crosses between parental lines. However, the proportion of hybrid plants was often only 50% as it was impossible to use highly self-sterile parents, since they could not be multiplied.

Many recessive genes causing male sterility are known, but, to be useable in hybrid production, it is necessary to be able to distinguish the plants that will be male sterile from those that will be male fertile, before flowering. The gene linked with only 1% recombination to a gene controlling anthocyanin production, reported by Leclercq (1966), fulfilled this condition since the plants that carried the male fertile allele produced anthocyanins at the seedling stage onwards and thus could be eliminated, leaving only non-anthocyanin male sterile plants at flowering to be pollinated by the male parent inbred (Fig. 2-1). This male sterility becomes apparent during pollen maturation, a few non-viable pollen grains being present in the anthers at flowering. Normal sunflower lines without the male sterile gene were used as male parents. Hybrid varieties using this system were developed by INRA from 1969 ("INRA6501") to 1975 ("Airelle") and grown up to about 1980. However, two important problems appeared as discussed below.

With a recessive gene, the maximum proportion of male sterile plants in a progeny is 50%, obtained from crossing the homozygous recessive *ms/ms*

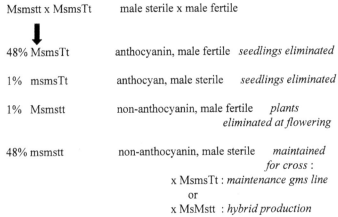

Figure 2-1 Use of marked genic male sterility in sunflower (genes *Ms/ms*: male fertile/male sterile *T/t*: anthocyanin/non-anthocyanin.

with a heterozygous male fertile *Ms/ms*. It was thus necessary to eliminate half of the plants in the rows of female parent, with a very dense sowing, followed by thinning, which was costly and irregular.

The 1% recombination rate gave rise to male fertile plants not showing anthocyanin, which produced pollen within the rows of females. Since bees tend to remain on one genotype, the pollen from these male fertile female plants was transferred by bees and pollinated up to 30% of the male sterile plants around them. If they were removed by hand at flowering, this was costly.

The discovery of a cytoplasmic male sterility (CMS) source by Leclercq (1969), now known as PET1, from a cross between *H. petiolaris* and *H. annuus*, simplified hybrid production since all female plants are male sterile (Fig. 2-2). Anthers of male sterile plants are very small and carry no visible pollen at flowering. Most cultivated sunflower genotypes are CMS sterility maintainers, but restorer genes were found in progenies from the same cross and also in wild *H. annuus* and some other annual *Helianthus* species (Kinman 1970; Leclercq 1971). The first hybrids (FRANSOL and RELAX) were registered in France in 1974 and cytoplasmic hybrids have been widely grown in the world since 1978, and are now almost the sole type of variety. This CMS source is used throughout the world for commercial hybrid production. From other inter- or intra-specific crosses, about 20 other forms of CMS have been obtained (Serieys 1999) with the idea that they would constitute an insurance should any specific problem arise with the first CMS. Studies on their possible agronomic interests indicated that specific breeding programs would be necessary to obtain lines better than those with the first CMS and these do not appear to have been undertaken.

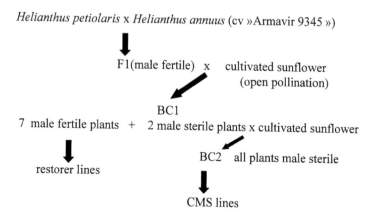

Figure 2-2 Origin of cytoplasmic male sterility in sunflower (from Leclercq 1969).

2.3 Breeding Methods

Mass or recurrent selection programs permit improvement of basic breeding populations, by increasing the frequencies of favorable genes. Then, to obtain inbred lines which are parents of hybrids, pedigree selection is used.

2.3.1 Mass Selection

In mass selection, each selection cycle represents only one generation. Plants, grown in an isolated plot, are allowed to intercross and selection of those that should be retained is made according to their phenotype, that is, their appearance. Such a method can be efficient for characters that can be observed before flowering, so that the undesirable plants can be eliminated before they produce pollen (for example, *Phomopsis* resistance, flowering date or height). It was used at Krasnodar to breed the early open-pollinated variety "Cernianka".

2.3.2 Recurrent Selection

One cycle of selection involves two or three generations, with two steps: intercrossing of material and testing of the progenies obtained. The first programs of recurrent selection in sunflowers were developed by Pustovoit at Krasnodar, following what he called the "method of reserves". The method was based on the studies of offspring and the creation of new populations from the remaining seed of the best individuals. This system is efficient when there is good genetic variability and a high level of heritability (good prediction from one generation to the next). It has been used for oil content and for capitulum resistance to *Sclerotinia*, for example (Vear et al. 2007).

Recurrent selection for yield is more complex, since it is necessary to make hybrids in order to determine the combining ability of inbred lines. Miller and Hammond (1985) used a system of reciprocal full-sib selection and concluded that the population could be improved at the same time as identifying interesting plants which could be selfed. However, the necessary effort was large, compared with the amount of improvement obtained. The use of tester genotypes simplifies operations and both general and specific combining abilities can be estimated. The system used at INRA, Clermont-Ferrand from 1980 to 2000 is given in Figure 2-3.

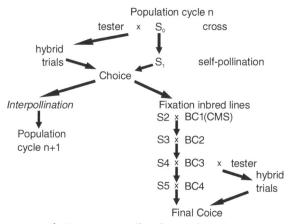

Figure 2-3 Recurrent selection program (female population).

2.3.3 Pedigree Selection

This method is used to obtain fixed, homozygous lines from recurrent selection programs, and to combine interesting characteristics from complementary lines. In the latter case, crosses between two male fertile lines are made either using emasculation by gibberellic acid (possible on a small-scale only) or by crossing two male fertile plants and distinguishing the F_1 hybrids from the selfed inbreds by their vigor. Plants are self-pollinated at each generation until they show complete fixation, by covering capitula with paper or cloth bags. Each progeny is followed separately and selection can be practised at each generation. An example of this method is given in Figure 2-4.

For the potential female lines, CMS must be introduced by backcrossing. This takes six to seven generations. Since it involves considerable crossing, it is usually started only after a test for combining ability has been completed. It was suggested that this could be accelerated by in vitro culture of immature embryos but generally additional generations are grown in the greenhouse or out-of-season nurseries.

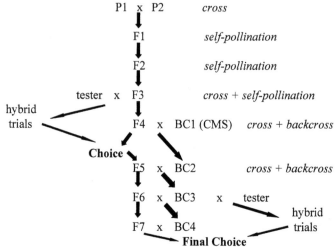

Figure 2-4 Pedigree selection program (female population).

2.4 Breeding Objectives

Sunflower seeds provide not only oil but also a high protein seed cake. In some countries, farmers are paid according to their oil yield per hectare, but in France they are paid according to their seed yield, with a variable premium according to oil content (the base being 44%) and recently according to oil quality. Present breeding criteria are thus the following.

2.4.1 Seed Yield

For sunflower, as for maize, the yield of a hybrid is not generally well correlated with the yield of its inbred parents. Seed yield has a low level of both broad and narrow sense heritability (Fick 1978), that is, it is highly influenced by the environment and specific reactions of hybrid combinations according to parental lines, are apparent. It has always been considered as being under multiple gene control, and is analyzed as a quantitative character. It is, therefore, the character which requires the most effort to obtain an accurate description for any genotype. The yield of sunflowers is determined by the number of seeds per head and the weight of 1,000-seed. These characters generally have a higher level of heritability than yield itself, but they may be negatively correlated, so, in the course of breeding seed yield of hybrids is generally measured in multilocational field trials.

Breeding for seed yield involves selecting plants, and then the inbred lines bred from them, for their "combining ability", that is their aptitude to give hybrids with good seed yields. In a recurrent selection program, hybrids are made with S_0 or S_1 plants obtained from intercrossing and then further hybrids are from the S_2 or S_3 generations. Selection of useful inbred lines is most commonly based on measurements of general combining ability with one or more tester lines. For new male (restorer parents) known inbred CMS lines (or unrestored single cross hybrids) can be used. For new female (sterility maintainers) lines, in the first generations, when no CMS version is available, it is more complicated although there are several possibilities.

Unrelated CMS lines: however, if the line belongs to the same group, heterosis may not represent that with restorer lines. In addition, the hybrid would be male sterile and Vear (1984) showed that such hybrids overestimated the value of the new line, especially for oil content.

A restorer line sterilized with gibberellin: very often the quantity of seed obtained is too small.

A restorer line with a cytoplasm different from PET1, so that it is male sterile. The validity of this depends on whether the hybrid produced is male fertile or male sterile.

A restorer line with genic male sterility: in this case the restorer acts as female parent and the hybrid is male fertile. Branches may be removed from the restorer line to increase the main capitulum and seed size.

To obtain an estimate of general combining abilities, several tester lines are used in the course of breeding. Then the best female (CMS) lines are crossed with the best male (restorer) lines and further yield trials are necessary to determine the best specific combining abilities. All these trials are an expensive part of breeding programs.

Recently, research has been carried out on possible methods of predicting hybrid yield from characteristics of inbred lines. Debaeke et al. (2004) and Triboi et al. (2004) observed that the maintenance after flowering, of a large area of actively photosynthesizing leaves is correlated with seed yield in hybrids. This character is measured by total leaf area at flowering, duration of leaf area after flowering and nitrogen contents of leaves at and after flowering (which represent enzyme activity). These characters were shown to be heritable between lines and their hybrids, so it would be possible to select between early generation plants or families for "residual leaf area", for example, and only make hybrids with the best material. However, the measurement of leaf area by hand (length × width × 0.71, for sunflower) requires a lot of time and some automation will be necessary to use it on a large scale.

In spite of the complications in judging seed yield, this character has been significantly improved over the last 40 years. In most European countries, hybrids have to show improved yields compared with the most widely grown varieties (103%) to be registered on the European and National Catalogs. In France, for a crop of 600–800 000 ha, 20–30 new varieties are registered each year. In 2000–2001, the French sunflower industry as a whole carried out trials with the most widely grown varieties from 1970 to 2000 to determine the genetic gain in breeding sunflowers over 30 years. Results from 27 locations over two years showed that the five most widely grown varieties in 2000 (all single cross CMS hybrids) had a mean of 140% of the yield of the five varieties grown from 1965–1975, which included one open-pollinated variety (Peredovik), three genic hybrids and one cytoplasmic hybrid (Vear et al. 2003; Fig. 2-5). This represents 1.3% gain in seed yield per year, similar to many results for maize, the other main hybrid crop. Yield improvement was at first due to an increase in seed number /m²but, since 1990, to an increase in seed size (Table 2-1).

Table 2-1 Mean characteristics of four groups of the five varieties most widely cultivated in France from 1970 to 2000, in 2000–2001 trials as % of the first group mean (from Vear et al. 2003).

	1970	1980	1990	2000
Grain yield	100	112.7	120.3	139.9
Oil yield	100	119.6	128.4	147.6
Oil content	100	106.0	106.6	105.8
1,000-seed wt	100	81.2	93.9	106.6
Seed number/m²	100	129.2	128.2	131.0
% moisture	100	94.7	98.8	99.6
Height	100	98.7	98.8	100.4
% lodging	100	57.9	35.0	24.7
% Phomopsis	100	95.4	76.0	63.7
% *Sclerotinia* head rot	100	94.7	59.3	36.3

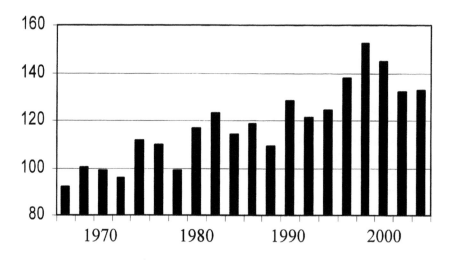

Figure 2-5 Yield of the most widely cultivated sunflower varieties in France from 1970 to 2000, measured in 2000–2001 trials, as a percentage of mean of the 1970 varieties (from Vear et al. 2003).

2.4.2 *Oil Content and Oil Quality*

The highest oil content known (60 to 65%) in cultivated sunflowers is probably close to biological limits. Oil content is determined by the plant on which the seeds are borne. The pollen has little effect. Oil content is a highly heritable character (Fick 1975), which can be selected on individual plants from the early generations of a breeding program. Oil content is now measured by nuclear magnetic resonance (NMR), a rapid, non-destructive method, which only requires 2 or 3 g of seed. It is therefore one of the easiest characters to select, but there is a tendency for the genotypes with the highest oil content not to have the highest yield. This could be due to the fact that more energy is required to produce 1 g oil than 1 g cellulose. Generally hybrids have higher oil content than that of their parents, but this is not true when inbred lines have 50% or more oil.

Importance of oil content depends on the economic question of whether the seed meal has any value. In France, the character "oil yield per hectare" is used in official trails, but farmers are not often paid for this. The genetic gain study in 2000–2001 showed that the first hybrids had lower oil content than the last open-pollinated variety, Peredovik, but then, in the 1980s, there was a return to the level of Peredovik. On average, there has been no further increase since then. There has been specialization with some varieties showing progress for oil content, but balanced by others with improved yield but lower oil content. The gain for oil yield per ha was 148% from 1970 to 2000 (Table 2-1).

The composition of "conventional" sunflower oil varies according to climate: in temperate conditions, it contains up to 75% linoleic acid and 20% oleic acid, whereas in hotter climates, up to 60% oleic acid and 30% linoleic acid is common. With no modification, this oil is valuable for direct use and in margarine production.

A mutant which blocked the activity of the enzyme desaturating oleic acid to linoleic acid in sunflower seed was discovered by Soldatov (1976), with the result that proportions of fatty acids are reversed giving up to 90% oleic acid. This character appears to be "semi-dominant" (Miller et al. 1987), a hybrid with one oleic parent having 45–65% oleic acid and both parents must be "high-oleic" to give a hybrid with at least 75%, the minimum level for international designation as a "high oleic variety". Varieties with up to 92% oleic acid have now been bred. Inheritance appears to follow the pattern of a major gene, but some genotypes with "suppresser genes" have been identified (Lacombe et al. 2002). Unlike oil content, which depends on mother plant genotype, oleic acid content is determined by the genotype of the developing seed. Therefore, high oleic crops must be grown isolated from conventional crops and, in first generations of breeding (F_2 or backcross populations), it is necessary to make analyses on single plants or seeds. In the case of backcross programs, following a cross with the recurrent parent, to distinguish the offspring which carry the high oleic allele, it may be necessary to analyze half seed and then grow on the other half, carrying the plumule and radicle. If lines carrying suppresser effects are eliminated, high oleic acid content can be bred rapidly, as a major gene character, but to obtain lines with more than 80% oleic acid (industry would prefer single fatty acid oils) breeding is quantitative, with selection of plants showing the highest levels.

Measurement of oleic acid content remains more complicated than measurement of total oil content. Generally, it is destructive, seed samples (of 1–2 g) need to be ground and then fatty acid content measured either by gas chromatography (GPC) or by near-infrared reflectance spectroscopy (NIRS). This is possible on plants with good seed production or on advanced generations. Special adaptations are necessary to measure single F_2 seed or half seeds of backcross programs.

At first, this type of variety appeared of marginal interest, being used to develop mixed oils for use in commercial food preparations since they can be heated to higher temperatures than conventional oil. However, in 1990, nutritional research suggested that an oil with about 60% oleic acid, and a reduced saturated acid content (as is the case for high oleic sunflower) was a good dietary element. Mid-oleic varieties were developed in the US, whereas in Europe, agri-food companies preferred to make their own mix from high oleic and conventional sunflower. Since 2002, oleic sunflower have been shown to be of interest not only for food uses, but also for biofuel production, with the best iodine number of all temperate crops (86, compared with conventional sunflower: 132, soybean: 130; rapeseed: 117). In many countries, there has been a considerable change in the type of sunflowers grown, and it may be suggested that high oleic sunflowers may become the normal type with what were conventional as "special crops", for margarine production or some industrial uses.

After oil extraction, a seed meal is obtained which may be used as animal food. Its use depends on the proportion of hull it contains, complete grains providing a meal rich in cellulose that can only be used for ruminants. It also depends on economic factors, especially the relative price of other seed meals, especially that of soybean. Studies have been made of seed protein content, but this character is generally negatively correlated with seed oil content. The only method of increasing the two is to breed for reduced hull. This is possible to a certain extent, but the hull must remain sufficient to protect the embryo in seed production.

The alternative method to obtain a seed meal rich in protein is to hull seed before oil extraction. In the 1990s, at the time when soybean meal was expensive, research showed that hullability was correlated with a large seed and quite high percent hull. It showed heritabilty of 0.8 (Denis et al.

1994). However, this character is generally negatively correlated with oil content, and often about 2% oil is lost in the hulling process, so it has not been included in breeding programs.

2.4.3 Cycle Length

The earliness of a genotype can be defined in several ways. The date of flowering is important in determining the period when the plant is most susceptible to drought and to *Sclerotinia* capitulum attack. However, for the farmer, the most important character is the date at which sunflowers can be harvested. Physiological maturity (maximum yield and oil content) are reached when the seed has about 35% water content (the capitulum is yellow with brown bracts, and more than half of the leaves are dry). There is a certain independence between drying of the seed and drying of the capitulum. The most usual measure of earliness is therefore the moisture content of seed at, or slightly before harvest, when the range between early and late varieties is at least 10 points. Differences in maturity date depend more on the length of the period flowering-maturity than on the pre-flowering duration. Broad sense heritability of both flowering and seed water content are moderate (0.4) but narrow sense heritability of flowering is higher (0.62) than for seed maturity (0.29–0.43) (Chervet and Vear 1989). Only the water content of female lines are correlated with those of their hybrids, restorer lines, which carry apical branching, have small capitula which dry quickly.

Conventional breeding is generally by development of inbred lines with acceptable flowering dates for the region and then choice among lines according to the seed moisture content at harvest of their hybrids, compared with their seed yield. In France since 1970, there has been little change in the mean cycle length of widely grown varieties (Table 2-1).

2.4.4 Disease Resistance

Resistance to diseases is one of the main factors determining the success of a sunflower crop in different parts of the world. The problems are different in various countries and in different environments. In addition, the situation is not the same in North America, where sunflower and its main pests are endemic and in countries where sunflower is a recent introduction. The relative newness of the crop means also that new diseases can appear, as parasites become adapted from other plant species.

Some diseases are of worldwide importance, others are more localized. From recent publications and papers at International Sunflower Conferences (I.S.C.), Vear (2004) defined the most important diseases according to the number of papers published on each. At the last four I.S.C., from 1992 to 2004, white rot, *Sclerotinia sclerotiorum*, was the most important disease of sunflower worldwide. In 1992, the second disease was Phomopsis (*Diaporthe helianthi*), followed by charcoal rot (*Macrophomina phaseolina*), downy mildew (*Plasmopara halstedii*) and broomrape (*Orobanche cumana*). However, in 1996, downy mildew overtook Phomopsis and in 2000 and 2004 there was a relative increase in papers on broomrape and black-stem (*Phoma macdonaldii*) (all of the last from France). The other main diseases reported were *Alternaria* leaf blight (*Alternaria helianthi*), white rust (*Albugo tragopogonis*) and rust (*Puccinia helianthi*).

Methods of breeding for resistance to disease are broadly of two types, according to whether resistance is controlled by major genes (downy mildew, broomrape, rust, *Verticilium*), or follows quantitative inheritance (white rot, Phomopsis, grey rot, *Alternaria* leaf blight). Below, two examples of each type are described, but it should be noted that, because of problems of pathogen changes resulting use of single or a few major genes, combinations of the two types of inheritance are now envisaged, to obtain high levels of durable resistance.

2.4.4.1 Diseases Controlled Mainly by Major Gene Resistance

2.4.4.1.1 Downy Mildew (*Plasmopara halstedii*)

Of worldwide importance, the disease is maintained in the soil up to 10 years, and transported occasionally on seed. The hypocotyls and roots of young plants are infected in humid conditions. The plants become dwarfed and produce no seed.

Major gene resistance to downy mildew has been obtained from both cultivated sunflowers and other *Helianthus* species. It is generally functionally complete but some genes give resistance reactions which stop progress of the pathogen more slowly than others, with the result that seedlings may show some symptoms but the plants grow normally. This type of resistance is determined by a test on germinated seeds which are soaked in suspensions of zoosporangia and then pricked out and grown for two weeks (Cohen and Sackston 1973). Then, after 48 h at 100% RH, susceptible plants show sporulation on cotyledons and true leaves and resistant plants no sporulation or only on cotyledons (Vear 1978). Major gene resistance is inherited as single dominant genes which have been shown to be situated in clusters (Vear et al. 1997). However, it is race-specific, many different *P. halstedii* races being known in North America (10 races), and Europe (14 races) for example (Gulya et al. 1991; Moinard et al. 2006).

For the pathogen, the center of origin appears to be North America, but from about 1960, downy mildew was reported in Russia, with a race (100) never reported in the USA, where the first race was 300. While new races appeared quite regularly in the USA (Gulya et al. 1991), in Europe, there was stability until 1988 and 1989 when races 710 and 703 were reported (Tourvieille de Labrouhe et al. 1991). It now appears that race 100 is extremely homogeneous and probably homothallic and that races 710 and 703 are more closely related to North American isolates (Roeckel-Drevet et al. 2003). It was presumed that they were introduced. However, since then, other new races have appeared specifically in France, in particular race 304, which is now observed each year (Tourvieille de Labrouhe et al. 2000). Occasional samples of several other races have been found (300, 307, 314, 700, 704 and 714; Penaud et al. 2003). From a geneticist's point of view, when their virulence pattern is studied, they look like an F_2 between races 710 and 304. Whether this is the case remains to be seen, but certainly there is now more variability among European populations of *Plasmopara halstedii*, raising questions as to the durability of major resistance genes.

Studies of resistance genes are made conventionally by crossing resistant lines and then making a test cross of the F_1 hybrid with a susceptible line. Mendelian segregation for resistance in this progeny then makes it possible to define whether the lines carry the same (or closely linked) or independently segregating genes. At least three clusters of genes have been found (a cluster contains genes from different origins giving resistance to different races, but which do not segregate independently) (Fig. 2-6). The cluster including *Pl1*, *Pl2*, *Pl6* and *Pl7*, has been shown to cover a large area, with about 0.5 cM between resistance to races 100 and 300 on one hand and 700, 703 and 710 on the other, so that occasional segregants are susceptible to race 100 and

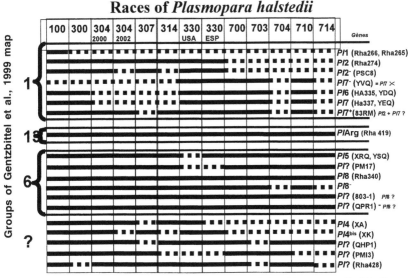

Figure 2-6 Efficiencies of *Pl* genes against *Plasmopara halstedii* races studied in France (from Vear 2004). ▪▪▪▪▪▪ susceptible ▬▬▬ resistant.

300 but resistant to race 710 (Vear et al. 1997). All the downy mildew resistance gene clusters appear to be able to confer resistance to all known races but different origins of resistance, perhaps with different blocks or structures of genes, may give the same or different race resistance patterns. For example, in the same cluster, *Pl6* (from *H. annuus*) and *Pl7* (from *H. praecox*) (Miller and Gulya 1991) both give resistance to races 710 but susceptibility to race 304, whereas, on group 6, *Pl5* (from *H. tuberosus*) is in the same cluster as *Pl8* (from *H. argophyllus*) but while the latter is resistant to all known races, *Pl5* does not give resistance to a US isolate of race 330 (Bert et al. 2001). At present the gene-for-gene hypothesis appears reasonable, but has not been proven since no segregants having lost resistance to some races have been reported for the Pl5/Pl8 cluster. It is still possible that some genes give resistance to several races. This organization in several independent clusters is quite comparable with resistance to downy mildew in lettuce, as studied by Michelmore and Meyers (1998).

These major genes are quite simple to use in breeding, it being possible to obtain homozygous families after two generations of selfing, or genes can be introduced into lines already developed by a backcross program. In France, for example, since 1995 many conversions for downy mildew resistance have been made. For example, for varieties originally carrying *Pl1* or *Pl2*, one of the parents was backcrossed to introduce *Pl6* or *Pl7*, to give resistance to races 703 and 710, then, in order that both parents should be resistant, and with the appearance of race 304, *Pl5* or *Pl8* have been added to the parent already carrying *Pl1* or *Pl2* (Fig. 2-7). However, the genes available must not be wasted and the question may be asked as to whether adding one new efficient gene each time a new race appears is the best solution.

1965-1988: Race 100: resistance: *Pl1*, *Pl2*
    hybrids: susceptible x *Pl1*
    susceptible x *Pl2*

1988-2000: Races 710, 703: resistance: *Pl5*, *Pl6*, *Pl7*, *Pl8*
    hybrids: 1990: *Pl6/Pl7 x Pl1/Pl2*   new or
        1996: *Pl5 x Pl2*   back crossed
                                            susceptible
2000-2002: Races 304, 314: resistance *Pl2*, *Pl5*, *Pl8*
    hybrids: *Pl6/Pl7* x *Pl2*
        Pl6/Pl7 x *Pl1* + *Pl5*   back crossed

2003: Races 307, 704, 714: resistance *Pl5*, *Pl8*   new or
    hybrids: *Pl6/Pl7* x *Pl2* + *Pl5/Pl8*   back crossed
                                                    susceptible

Figure 2-7 Breeding sunflower varieties for downy mildew resistance in France since 1970 (from Vear 2004).

Since 2003, studies have been made of partial, "horizontal" resistance, according to the definition of Robinson (1973). The reaction of cultivated sunflower genotypes, which do not have *Pl* genes effective against the predominant races present, to field attack by downy mildew was studied over four years in several environments. An experimental protocol with pre-emergence irrigation making it possible to observe downy protocol reaction whatever the weather conditions, was developed (Tourvieille et al. 2008). It was been shown (Fig. 2-8) that significant levels of partial resistance exist in cultivated sunflower lines (15% infection when the most susceptible lines show 80–90% infection). This resistance appears independent of race, at least concerning those known as 710 and 703 (Tourvieille de Labrouhe et al. 2004). Heredity is under additive control and the behavior of hybrids can be quite well predicted from inbred lines. This result suggests that it should be possible to select for non-race-specific downy mildew resistance and to incorporate it in modern varieties. However, since this non-specific resistance is partial, it may be necessary to combine it with major gene resistance.

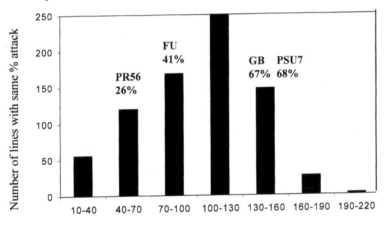

Figure 2-8 Reaction of nearly 800 inbred sunflower lines to downy mildew attack in the field (from Tourvieille de Labrouhe et al. 2008).

2.4.4.1.2 Broomrape (*Orobanche cumana*)

Resistance to broomrape appears to follow a similar pattern to that for downy mildew, but with the center of origin around the Black Sea, and more recently introduced into Spain. The first races (A and B) appeared in Russia and Ukraine about 1935, races C to E appeared in Eastern Europe and then in Spain in the 1980s and race F in Spain in 1996. It is not certain whether these six races are strictly the same in eastern Europe and in Spain, since the differential lines proposed by Vranceanu et al. (1980) do not give the expected reaction to all broomrape isolates (Melero-Vara et al. 2000). The resistance genes, denoted *Or1* to *Or5*, generally appear dominant (Dozet et al. 2000). In some cases, two genes may be involved (Dominguez 1996). Each source of resistance to a new race appears to give resistance to all the less virulent races (Melero-Vara et al. 2000; Jan and Fernandez-Martinez 2002), so it is not yet possible to know whether the gene-for-gene hypothesis applies. A paper by Venkov and Shindrova (2000) suggested that the efficiency of some of these genes was more durable than that of others.

Breeding is mainly the same as for downy mildew resistance, pedigree selection and backcross programs (using field or greenhouse tests) but search is also made for quantitative resistance (Perez-Vich et al. 2004). Some control can also be obtained with herbicides (Dominguez et al. 2004), but *Orobanche* produces an extremely large number of seeds and it appears likely that isolates with herbicide resistance will be found very rapidly if this control measure is widely used.

2.4.4.2 Quantitative Resistance

2.4.4.2.1 White rot (*Sclerotinia sclerotiorum*)

Of worldwide importance, except in very dry zones, *Sclerotinia* causes a soft humid rot on different parts of the plant: roots, stem base, terminal bud, leaves and capitulum. Although these different attacks are by exactly the same fungus, they may be considered almost as different diseases, of different importance in different parts of the world: root and stem base in North America and Europe; terminal bud in Europe and North Africa; capitulum in Argentina, China and Europe. Resistance tests require the use of adult plants, with infection of the plant part concerned, for example roots and stem base by sclerotia or mycelium, capitula by ascospores at flowering or by mycelium during maturation (Vear and Tourvieille 1988; Gulya 2004).

Sclerotinia sclerotiorum can attack everything except grasses and this is perhaps the reason why it is difficult to find high levels of resistance in cultivated sunflower. Vear et al. (2004) observed that when 16 sunflower lines were infected with ascospores of 10 isolates there were no interactions, so the resistances to *Sclerotinia* available at present can be considered as strictly horizontal. This means that they should be durable.

The inheritance of resistance to *Sclerotinia* has been shown to be partial, quantitative, generally additive, with a moderate level of heritability (Castaño et al. 2001). It is, therefore, necessary to carry out long-term breeding programs to assemble in one genotype many additive factors, the sum of which provides appreciable resistance to one or several forms of *Sclerotinia* attack. Pedigree selection gives the possibility of obtaining genotypes with increased levels of resistance (Vear et al. 2000) but is limited in the improvement obtained by combining two genotypes. Recurrent selection for resistance to *Sclerotinia* head rot was carried out for 15 cycles on a restorer sunflower population created in 1978. For the first three cycles a test measuring rate of extension of mycelium on the back of capitula was used; from the fourth cycle onwards, it was combined with a test based on ascospore infections, which repeat more closely to natural infections (Fig. 2-9). An 80% reduction in diseased area was obtained in four cycles, with the first test, thereafter the population remained stable and homogeneous for this character. In 12 cycles the latency index (measure of incubation period) of the ascospore test doubled, and the best relation with cycle was a simple regression, with a significant slope, indicating that further improvements should be possible. The hybrids made with the first, sixth and fifteenth generations of the population showed a halving of percentage attack in the field and hybrids with some of the best lines bred from several cycles presented even greater levels of resistance (Vear et al. 2007).

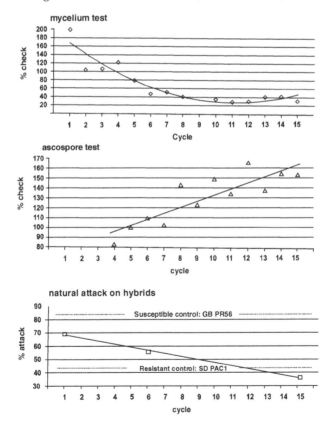

Figure 2-9 Reaction to *Sclerotinia* capitulum resistance tests over 15 generations of recurrent selection (from Vear et al. 2007).

Progress by such conventional breeding has provided a 60% reduction in attack in modern varieties grown in France, compared with those grown in 1960–70 (Vear et al. 2003; Table 2-1). Nevertheless, in extremely favorable conditions, most sunflower genotypes may show high levels of attack (Serre et al. 2004) and further work is necessary, particularly to introduce additional factors from perennial *Helianthus* species.

2.4.4.2.2 Phomopsis (*Diaporthe helianthi*)

This is a new fungal species, discovered in Yugoslavia in the early 1980s. It has since been found in neighboring countries and in France. The parasite attacks through leaves, spreading to the stem and causing wilting, premature drying and stem breakage. It overwinters in the remains of sunflower stems, and is thus particularly important in areas where sunflower crop residues are not ground and ploughed in the autumn.

Resistance sources found in naturally attacked breeding nurseries (Skoric 1985) gave good levels of resistance, and to start with, it appeared that they might contain oligogenic resistance (Vranceanu et al. 1994). However, with experience and observations of modern sunflower hybrids, it soon appeared that parental lines which were not particularly resistant could give hybrids good levels of resistance (Vear et al. 1996). Studies of factorial crosses showed that, when measured over several years or several locations, resistance was quantitative and strictly additive, with no interactions between different parental lines (Viguié et al. 2000). It could thus be concluded that combinations of inbred lines with the best levels of resistance give the best hybrids and that it should be possible to obtain increased resistance by selecting combinations from different sources.

Interactions between effects of parental lines have been reported but these all come from observations of Phomopsis reactions in single locations (Deglène et al. 1999). It has been suggested that this fungal species is still evolving quite fast and results in France are in agreement with this (Says-Lesage et al. 2002). In addition, when sunflower lines are infected with mycelium of different Phomopsis isolates, some small host/pathogen interactions appear, although these do not affect high levels of resistance (Viguié et al. 1999). It was presumed that observations of natural attack in several locations was the best method to assure that a given genotype would generally show a low level of susceptibility.

In the genetic gain trials in 2000–2001, there only appeared to have been a reduction of 30% in attack, but this may be because, in the specially infected field trials, the old varieties had about 60% attack and the recent, most widely grown hybrids, about 40% infection. Other varieties with higher levels of resistance exist, but, in France, the resistance level of the highest yielding varieties is sufficient to protect crops from yield loss and in breeding now, the requirement is only to maintain the present level of resistance.

2.4.5 Insects Affecting Seed Yield (from Vear and Miller 1992)

Sunflower is a host to a number of insect pests. In North America, approximately 15 species of sunflower insects cause plant injury and economic loss, depending upon the severity of infestation (Schultz 1978). The existence of this number of insect pests coincides with the evolution of wild sunflowers in North America. To date, these insect species have not been transferred to other production areas of the world. However, two species of insects attack sunflower other than in North America. These are the European Sunflower Moth in Europe and Russia, and the Rutherglen bug in Australia.

Head-infesting species of insects producing economic damage include: Sunflower moth, *Homoesoma electellum* (Lepidoptera: Pyralidae); European Sunflower Moth, *Homeosoma nebulella* (Lepidoptera: Pyralidae); Banded Sunflower Moth, *Cochylis hospes* (Lepodoptera: Cochylidae); Sunflower Budworm, *Suleima helianthana* (Lepidoptera: Tortrichidae); Seed Weevil, *Smicronyx fulvus* and *S. sordidus* (Coleoptera: Curculionidae); and Sunflower midge, *Contarinia schulzi* (Diptera: Cecidomyiidae). Foliage and stem feeding species include: Sunflower Beetle, *Zygogramma exclamationis* (Coleoptera: Chrysomelidae); Painted Lady, *Cynthia cardui* (Lepidoptera: Nymphalidae) and Stem Weevil, *Cylindropterus adspersus* (Coleoptera: Curculionidae).

The most important insects in North America are the Sunflower Moth, Banded Sunflower Moth, Seed Weevil and Sunflower Midge. Resistance to the Sunflower Moth and European Sunflower Moth has been associated with the armored layer, a pigmented substance between the outer layer and the adjoining sclerenchyma tissue in the sunflower hull. Few resistance mechanisms have been determined for the other species. Other mechanisms for decreasing damage have been investigated and include: biological control, chemical deterrents derived from wild *Helianthus* species, pheromones to trap insects, chemical feeding deterrents and cultural controls.

2.4.6 Drought Resistance

Sunflowers are most often grown in areas where irrigation is not possible or is restricted to use on maize crops. When water is not a limiting factor, sunflowers absorb and transpire relatively large amounts, with production of what may be an excessive leaf area if drought occurs later in the cycle (Merrien 1992). However, compared with many crops, sunflowers are relatively drought tolerant in the sense that they are able to extract water from a large depth of soil and at low soil water potentials.

Robelin (1967) showed that yield is affected if sunflowers lack sufficient water in the period of six weeks around flowering. Among cultivated hybrids, ratios of yield in dry and irrigated conditions may vary by 100%, but such trials are expensive to carry out, and it has been noted frequently that the genotypes with the highest ratios were not generally those with the highest yields under either drought or watered conditions. Breeders have searched for some simple characters that could be measured on inbred lines. An example was reported by Serieys (1989) who showed that the desert species *Helianthus argophyllus* had less permeable leaves than cultivated sunflower. He selected two populations which differed in their leaf transpiration rates and rapidity of wilting. Baldini et al. (1996) proposed the use of characters such as carbon exchange rate, transpiration rate and leaf relative water content in breeding for drought resistance. However, the cost of such measurements have meant that a few programs with intentional selection of drought resistance have been carried out, generally varieties which adapt well to dry conditions have been identified in yield trials in these conditions.

2.4.7 Resistance to High and Low Temperatures

The lowest soil temperature for germination and plant growth of sunflower is about 6°C, but seedlings at the cotyledon stage can resist slight frosts, whereas older plants may lose leaves or become branched because of partial destruction of the terminal bud. For sunflowers to be grown as a winter crop in Mediterranean type climates, resistance to slight frosts would be useful. Restarting of growth at low temperatures would also be useful, not only for autumn sown crops but also for very early spring sowings which could allow the crop to escape summer droughts. Some direct breeding work has been done (Gosset and Vear 1995), but large-scale studies have not been made on the variability available or the genetics of such characters.

Resistance to high temperatures often concerns sunflower crops in the same regions, in this case when they are sown in mid- or late-spring. Again most work has involved practical breeding rather than genetics, with elimination of genotypes which show symptoms of burnt dry capitular bracts.

2.4.8 Lodging Resistance and Reduced Height

Lodging may be at stem base or mid-stem, but it is always catastrophic. As a factor of resistance, reduced height may be selected, but often the two characters are independent. Breeding consists in eliminating the genotypes which show lodging. The behavior of inbred lines and their hybrids are generally significantly correlated.

Independently of lodging resistance, reduced height is of interest: to make possible increases in plant population, to reduce the proportion of dry matter in the stem and to facilitate surveillance and harvest of the crop. Inheritance of plant height is generally quantitative, with broad sense heritability of 6% and narrow sense of 20 to 40% (Fick 1978). In the 1980s, work was done on dwarf varieties, whose reduced height was determined by major genes. However, although many different sources were studied, in all cases, yield was reduced, with problems of grain filling in the center of capitula. In addition, reduced internode length may have led to problems in light absorption by leaves. In recent years, breeding has concerned selection of "short" varieties, but the tendency is still a positive correlation between height and seed yield, possibly related to greater leaf area.

2.5 Classical Genetic Mapping and Cytogenetics

Since research on sunflower genetics is relatively recent (1960 onwards), no chromosome map based on morphological characters was produced. With 17 pairs of chromosomes, the probability of finding linkages between the few major genes known in sunflower was quite low. The main exception was the recessive gene controlling male sterility that was found to be linked with only 1% recombination to a gene controlling anthocyanin production (Leclercq 1966). Most genetic studies, concerning male fertility restoration, branching, downy mildew, broomrape or rust resistance were limited to showing independence of the small number of loci controlling these characters. Studies of isoenzymes (Quillet et al. 1991) started only a few years before the first molecular markers were obtained, so that, although this system has been used to check hybrid parentage, isoenzyme markers were mapped only at the same time as restriction fragment length polymorphism (RFLP) markers. The existence of 17 different chromosomes has also meant that there have been a few cytogenetic studies and no real definition of karyotypes.

2.6 Interest of Molecular Maps and Knowledge of Genomics for Conventional Genetics and Breeding

For the few characters showing Mendelian inheritance, phenotypic studies allowed definition of dominance/recessivity and independence/linkage. However, as for most crops, other than these limited studies, sunflower breeding up to 1990 was based on field trials for characters which appeared quantitative, either because they were under polygenic control or because they were affected by the environment. It was necessary to make expensive factorial or diallel crossing plans to determine heritability and additivity (both parental lines contributing to the value of a hybrid) or dominance interactions (character not forecasted from parental values). With only phenotypic studies, it was difficult to say whether two parents had the same genotype for a complex character or produced the same effect from different genes.

The first interest of a molecular map, with a quite complete cover of linkage groups, was that the major genes could be located on it. Thus relations between quite different characters could easily be studied, and breeding programs were planned according to the position of genes on the same or different linkage groups. The first sunflower molecular maps became available in 1995 and in the 12 years since then, many characters have been mapped and breeders have got used to thinking at least in terms of "beads on strings", for example *Pl2* and *Pl6* downy mildew resistance genes are known to be in the same cluster (Mestries 1998) and so not easy to recombine, whereas broomrape resistance genes are clustered on a different linkage group (Gagne 2000) and so it is no problem to combine the two resistances.

Probably the greatest use to the breeder has been the definition of quantitative trait loci (QTL) for quantitative characters, since this makes it possible to transform statistics into "genes" that can be positioned along linkage groups almost as if they were major genes. It is then possible to infer that different lines have the same or different genes controlling a particular character and whether it will be possible to combine them and thus obtain a higher level of the character concerned, be it yield, earliness or quantitative disease resistance.

Even more important, positioning of QTLs on linkage maps makes it possible to compare the positions of genes (or QTLs) for different characters. It has become possible to distinguish between real pleiotropy and linkage between characters due to effects of origin. For example, the recessive branching gene *b1* (Putt 1964) has a pleotropic effect on seed size and oil content, whereas the linkage between *Sclerotinia* resistance and oil content observed in the first studies on this disease resulted from the genetic origin of resistance. Favorable linkages have also become evident: a *Sclerotinia* resistance QTL linked to a protene-kinase gene (Gentzbittel et al. 1998) has been shown to be involved also in resistance to other diseases such as *Phoma*, so the lines carrying the favorable allele at this QTL are particularly interesting to use in breeding (Bert et al. 2004). Of course, the efficiency of conclusions drawn from QTL positions depends on the closeness of molecular markers, but with fine mapping it will eventually be possible to determine the genes truly controlling facets of important characters. Comparison with genes whose effects are known in other species should help breeding for characters such as drought resistance, where so far little progress has been made because the overall phenotypic effect is difficult to measure. Perhaps, one day, breeding will be for a genotype that will have a predictable mean phenotype rather than depending on a large number of phenotypic observations to estimate this value.

2.7 The introduction of Genomics in Breeding Since 1995

In addition to changes in ideas about breeding programs, molecular studies and genomics have already been used to reduce the cost of some operations or to speed up conventional breeding programs. In the course of registration of a hybrid variety on official catalogs in Europe, checks are made to ensure that the hybrid tested comes from the parental inbred lines declared. In the past, the lines were crossed to reproduce the hybrid and its phenotype was compared with that of the seed provided by the breeder. For the last 10 years, this check has been made by genotypic analyses to show that the genotype of the hybrid is compatible with the genotypes of the parental lines. This was done first with isoenzyme systems, but, since 2007, microsatellite markers have been used. Breeders also make such analyses to check the homozygosity of their new lines before protection or commercialization.

The first real use of molecular markers in breeding has been in backcross programs to introduce downy mildew or broomrape resistance or the high oleic character, taking advantage of markers that are very closely linked or part of the genes concerned that have been identified in genetic and genomic studies of these characters (Lacombe and Berville 2001; Radwan et al. 2002). Molecular analyses on small pieces of young leaves of individual plants make it possible to determine those of interest (which may be only 1 in 10 in progeny from crosses between two male fertile plants), before flowering and thus to reduce the number of plants to be bagged and harvested and the need for phenotypic analyses before starting the following generation. Another use which is now starting is to follow segments of chromosomes from wild *Helianthus* in introgression programs, to make sure that selection for all the favorable characters of cultivated sunflower does not eliminate all the wild genome. Even if it may be some time before sunflower breeding for yield using QTL markers becomes a reality, it should be possible in the next few years to use molecular markers of simpler characters contributing to yield, such as leaf area, self-fertility or seed size, to select in early generations (F_2) and only make hybrids to be tested for yield in the field, with plants that have the favorable alleles for quite strong QTL.

References

Baldini M, Cecconi F, Cecchi A, Martorana F, Vanozzi GP, Benvenuti A (1996) Drought resistance in sunflower. Factors affecting yield and their variability under stress. Proc 13th Int Sunflower Conf, Pisa, Italy, vol 1, pp 513–521.

Bert PF, Tourvieille de Labrouhe D, Philippon J, Mouzeyar S, Jouan I, Nicolas P, Vear F (2001) Identification of a second linkage group carrying genes controlling resistance to downy mildew (*Plasmopara halstedii*) in sunflower (*Helianthus annuus* L.). Theor Appl Genet 103: 992–997.

Bert PF, Dechamp-Guillaume G, Serre F, Jouan I, Tourvieille de Labrouhe D, Nicolas P, Vear F (2004) Comparative genetic analysis of quantitative traits in sunflower

(*Helianthus annuus* L.). 3. Characterisation of QTL involved in resistance to *Sclerotinia sclerotiorum* and *Phoma macdonaldii*. Theor Appl Genet 109: 865–874.

Castaño F, Vear F, Tourvieille De Labrouhe D (2001) The genetics of resistance in sunflower capitula to *Sclerotinia sclerotiorum* measured by mycelium infections combined with ascospore tests. Euphytica 122: 373–380.

Chervet B, Vear F (1989) Evolution des caractéristiques de la graine et du capitule chez le tournesol en cours de maturation. Agronomie 9: 305–313.

Cohen Y, Sackston WE (1973) Factors affecting infection of sunflower by *Plasmopara halstedii*. Can J Bot 51:15–22.

Debake P, Triboi AM, Vear F, Lecoeur J (2004) Crop physiological determinants of yield in old and modern sunflower hybrids. Proc 16th Int Sunflower Conf, Fargo, ND, USA, vol 1, pp 267–274.

Denis L, Dominguez J, Baldini M, Vear F (1994) Genetical studies of hullability in comparison with other sunflower seed characteristics. Euphytica 79: 29–38.

Dominguez J (1996) R-41, a sunflower restorer inbred line carrying two genes for resistance against a highly virulent Spanish population of *Orobanche cernua*. Plant Breed 115: 203–204.

Dominguez J, Alvarado J, Espinosa JL, Falcon M, Mateos A, Navarro F (2004) Use of sunflower cultivars with resistance to imidazolinone herbicides to control broomrape (*Orobanche cumana*) infections. Proc 16th Int Sunflower Conf, Fargo, ND, USA, pp 181–186.

Deglène L, Alibert G, Lesigne P, Tourvieille de Labrouhe D, Sarrafi A (1999) Inheritance of resistance to stem canker (*Phomopsis helianthi*) in sunflower. Plant Pathol 48: 559–563.

Dozet B, Skoric D, Jovanovic D (2000) Sunflower breeding for resistance to broomrape (*Orobanche cernua* Loefl./*Orobanche cumana* Wallr.). Proc 15th Int Sunflower Conf, Jun 12–16, 2000, Toulouse, France, vol 2, pp J20–J25.

Fick G (1975) Heritability of oil content in sunflowers. Crop Sci 15: 77–78.

Fick G (1978) Breeding and genetics. In: J Carter (ed) Sunflower Science and Technology. Agronomy 19, ASA, Madison, WI, USA, pp 279–338.

Gagne G (2000) Variabilité de populations d'*Orobanche cumana*—Génétique de la résistance à l'orobanche chez le tournesol (*Helianthus annuus* L.). Doctoral Thesis, Univ of Clermont-Ferrand II, France.

Gentzbittel L, Mouzeyar S, Badoui S, Mestries E, Vear F, Tourvieille de Labrouhe D, Nicolas P (1998) Cloning of molecular markers for disease resistance in sunflower, *Helianthus annuus* L. Theor Appl Genet 96: 519–525.

Gentzbittel L, Mestries E, Mouzeyar S, Mazeyrat F, Badoui S, Vear F, Tourvieille de Labrouhe D, Nicolas P (1999) A composite map of expressed sequences and phenotypic traits of the sunflower (*Helianthus annuus* L.) genome. Theor Appl Genet 99: 218–234.

Gosset H, Vear F (1995) Comparaison de la productivité du tournesol au Maroc en semis d'automne et en semis de printemps. El Awamia 88: 5–20.

Gulya TJ (2004) Methods to evaluate sunflower germplasm for resistance to *Sclerotinia* root infection. Proc 3rd Aus Soilborne Dise Symp, Feb 8–11, 2004, Rowland Flat, Australia, pp 109–110.

Gulya T, Sackston W, Viranyi F, Rachid K (1991) New races of sunflower downy mildew pathogen (*Plasmopara halstedii*) in Europe and North America. J Phytopathol 132: 303–311.

Jan, CC, Fernandez-Martinez JF (2002) Interspecific hybridization, gene transfer and the development of resistance to the broomrape race F in Spain. Helia 25: 123–126.

Kinman ML (1970) Letter to Participants. Proc 4th Int Sunflower Conf, Memphis, TN, USA, June 23–25, 1970.

Lacombe S, Berville A (2001) A dominant mutation for high oleic acid content in sunflower (*Helianthus annuus* L.) seed oil is genetically linked to a single oleate-desaturase RFLP locus. Mol Breed 8(2): 129–137.

Lacombe S, Abbott AG, Berville A (2002) Repeats of an oleate desaturase region cause silencing of the normal gene explaining the high oleic Perevenets sunflower mutant. Helia 25: 95–104.

Leclercq P (1966) Une stérilité mâle utilisable pour la production d'hybrides simples de tournesol. Ann Amélior Plant 16: 135–144.

Leclercq P (1969) Une stérilité cytoplasmique chez le tournesol. Ann Amélior Plant 19: 99–106.

Leclercq P (1971) La stérilité mâle cytoplasmique du tournesol 1. Premières études sur la restauration de la fertilité. Ann Amélior Plant 21: 45–54.

Melero-Vara JM, Dominguez J, Fernandez-Martinez JM (2000) Update on sunflower broomrape situation in Spain: racial status and sunflower breeding for resistance. Helia 23: 45–56.

Merrian A (1992) Some aspects of sunflower crop physiology. Proc 13th Int Sunflower Conf, Pisa, Italy, pp 481–498.

Mestries E, Gentzbittel L, Tourvieille De Labrouhe D, Nicolas P, Vear F (1998) Analyses of quantitative trait loci associated with resistance to *Sclerotinia sclerotiorum* in sunflowers (*Helianthus annuus* L.) using molecular markers. Mol Breed 4: 215–226.

Michelmore RW, Meyers BC (1998) Clusters of resistance genes in plants evolve by divergent selection and a birth-and-death process. Genom Res 8: 1113–1130.

Miller J, Hammond J (1985) Improvement of yield in sunflower utilising reciprocal full-sib selection. Proc 11th Int Sunflower Conf, Mar-del-Plata, Argnetina, pp 715–720.

Miller J, Zimmerman D, Vick B (1987) Genetic control of high oleic acid content in sunflower. Crop Sci 27: 923–926.

Miller JF, Gulya TJ (1991) Inheritance of resistance to race 4 of downy mildew derived from interspecific crosses in sunflower. Crop Sci 31: 40–43.

Moinard, J, Mestries E, Penaud A, Pinochet X, Tourvieille de Labrouhe D, Vear F, Tardin MC, Pauchet I, Eychenne N (2006) An overview of sunflower downy mildew. Phytoma—La Défense des Végétaux 589: 34–38.

Penaud A, Moinard J, Molinero-Demilly V, Pauchet I, Bataillon C, Tourvieille de Labrouhe D (2003) Evolution du mildiou du tournesol en France: Le point sur les dernières données du réseau de surveillance. Proc 7th Int Conf on Plant Diseases 3–4 Dec, Tours, France, p 8.

Perez-Vich B, Aktouch B, Velasco L, Fernandez-Martinez J, Knapp SJ, Leon AJ, Berry ST (2004) Mapping QTLs controlling sunflower resistance to broomrape (*Orobanche cumana*). Proc 16th Int Sunflower Conf, Fargo, ND, USA, vol 2, pp 651–656.

Putt ED (1964) Recessive branching in sunflowers. Crop Sci 4: 444–445.

Quillet MC, Vear F, Branlard G (1992) The use of isoenzyme polymorphism for identification of sunflower (*Helianthus annuus*) inbred lines. J Genet Breed 46: 295–304.

Radwan O, Bouzidi MF, Vear F, Phillipon J, Tourvieille De Labrouhe D, Nicolas P, Mouzeyar S (2002) Identification of non-TIR-NBS-LRR markers linked to the *Pl5/Pl8* locus for resistance to downy mildew in sunflower. Theor Appl Genet 106: 1438–1446.

Robelin M (1967) Action et arrière-action de la secheresse sur la production du tournesol. Ann Agron 18: 579–599.

Robinson RA (1973) Horizontal resistance. Rev Plant Pathol 52: 483–501.

Roeckel-Drevet P, Tourvieille J, Gulya TJ, Charmet G, Nicolas P, Tourvieille de Labrouhe D (2003) Molecular variability of sunflower downy mildew, *Plasmopara halstedii*, from different continents. Can J Microbiol 49: 492–502.

Says-Leasage V, Roeckel-Drevet P, Viguié A, Tourvieille J, Nicolas P, Tourvieille de Labrouhe D (2002) Molecular variability within *Diaporthe helianthi/Phomopsis helianthi* from France. Phytopathology 92: 308–311.

Schultz J (1978) Insect pests. In: J Carter (ed) Sunflower Science and Technology. Agronomy 19, ASA, Madison, WI, USA, pp 169–223.

Serre F, Walser P, Tourvieille De Labrouhe D, Vear F (2004) *Sclerotinia sclerotiorum* capitulum resistance tests using ascospores: results over the period 1991–2003. Proc 16th Int Sunflower Conf, Fargo, ND, USA, vol 1, pp 129–134.

Serieys H (1989) Agrophysiological consequences of a divergent selection based on foliar dessication in sunflower. In: E Acevedo, AP Conesa, JP Srivastava (eds) Physiology-breeding of Winter Cereals for Stressed Mediterranean Environments. Montpellier, France, 3–6 July 1989, pp 25–28.

Serieys H (1999) Progress report of the working group "Identification, Study and Utilisation in Breeding Programs of New CMS Sources (1996–1999)". IX FAO Technical Consultation of the ECRN on Sunflower. Dobrich, Bulgaria, July 27–30, 1999. Helia 22: 71–116.

Skoric D (1985) Sunflower breeding for resistance to *Diaporthe helianthi/Phomopsis helianthi* Munt.-Cvet. Helia 8: 21–24.

Soldatov KI (1976) Chemical mutagenesis in sunflower breeding. Proc 7th Int Sunflower Conf, Krasnodar, Russia, pp 325–357.

Tourvieille de Labrouhe D, Mouzeyar S, Lafon S, Regnault Y (1991) Evolution des races de mildiou (*Plasmopara halstedii*) sur tournesol en France. Proc 3rd Int Conf on Plant Diseases, Bordeaux, France, pp 777–784.

Tourvieille de Labrouhe D, Lafon S, Walser P, Raulic I (2000) A new race of *Plasmopara halstedii*, sunflower downy mildew. Oléagineux, Corps Gras, Lipides 7: 404–405.

Tourvieille de Labrouhe D, Serre F, Walser P, Philippon J, Vear F, Tardin MC, Andre T, Castellanet P, Chatre S, Costes M, Cuk L, Jouve P, Madeuf JL, Mezzarobba A, Plegades J, Pauchet I, Mestries E, Penaud A, Pinochet X, Serieys H, Griveau Y, Moinard J (2004) Partial, non-race specific resistance to downy mildew in cultivated sunflower lines. Proc 16th Int Sunflower Conf, Fargo, ND, USA, vol 1, pp 105–110.

Tourvieille de Labrouhe D, Serre F, Walser P, Roche S, Vear F (2008) Non-race-specific resistance to downy mildew (*Plasmopara halstedii*) in sunflower (*Helianthus annuus*). Euphytica 164: 433–444.

Triboi AM, Messaoud J, Debaeke P, Lecoeur J, Vear F (2004) Heredity of sunflower leaf characters useable as yield predictors. Proc 16th Int Sunflower Conf, Fargo, ND, USA, vol 2, pp 517–524.

Unrau J, White WJ (1944) The yield and other characteristics of inbred lines and single crosses of sunflower. Sci Agri 24: 516–525.

Vear F (1978) Réaction de certains génotypes de tournesol résistants au mildiou (*Plasmopara helianthi*) au test de résistance sur plantule. Ann Amélior Plant 28: 327–332.

Vear F (1984) The effect of male sterility on seed yield and oil content in sunflowers. Agronomie 4: 901–904.

Vear F (2004) Breeding for durable resistance to the main diseases of sunflower. Proc 16th Int Sunflower Conf, Fargo, ND, USA, Aug 29 to Sept 3, 2004, vol 1, pp 15–28.

Vear F, Tourvieille de Labrouhe D (1988) Heredity of resistance to *Sclerotinia sclerotiorum* in sunflower. II. Study of capitulum resistance to natural and artificial ascospore infections. Agronomie 8: 503–508.

Vear F, Miller JM (1992) Sunflower. In: OECD (ed) Traditional Crop Breeding Practices: A Historical Review to Serve as a Baseline for Assessinbg the Role of Modern Biotechnology, pp 97–114.

Vear F, Garreyn M, Tourvieille de Labrouhe D (1996) Inheritance of resistance to phomopsis (*Diaporthe helianthi*) in sunflower. Plant Breed 11: 277–281.

Vear F, Gentzbittel L, Philippon J, Mouzeyar S, Mestries E, Roeckel-Drevet P, Tourvieille de Labrouhe D, Nicolas P (1997) The genetics of resistance to five races of downy mildew (*Plasmopara halstedii*) in sunflower (*Helianthus annuus* L.). Theor Appl Genet 95: 584–589.

Vear F, Serre F, Walser P, Bony H, Joubert G, Tourvieille de Labrouhe D (2000) Pedigree selection for sunflower capitulum resistance to *Sclerotinia sclerotiorum*. Proc 15th Int Sunflower Conf, Jun 12–16, 2000, Toulouse, France, pp K42–K47.

Vear F, Bony H, Joubert G, Tourvieille de Labrouhe D, Pauchet I, Pinochet X (2003) 30 years of sunflower breeding in France. Oléagineux, Corps Gras, Lipides 10: 66–73.

Vear F, Willefert D, Walser P, Serre F, Tourvieille de Labrouhe D (2004) Reaction of sunflower lines to a series of *Sclerotinia sclerotiorum* isolates. Proc 16th Int Sunflower Conf, Fargo, ND, USA, pp 135–140.

Vear F, Serre F, Roche S, Walser P, Tourvieille de Labrouhe D (2007) Improvement of *Sclerotinia sclerotiorum* head rot resistance by recurrent selection of a restorer population. Helia: 30: 1–12.

Venkov V, Shindrova P (2000) Durable resistance to broomrape (*Orobanche cumana* Wallr./*Orobanche cernua* Loefl.) in sunflower. Helia 23: 39–44.

Viguie A, Vear F, Tourvieille de Labrouhe D (1999) Interactions between French isolates of *Phomopsis/Diaporthe helianthi* Munt.-Cvet. et al. and sunflower (*Helianthus annuus* L.) genotypes. Eur J Plant Pathol 105: 693–702.

Viguie A, Tourvieille de Labrouhe D, Vear F (2000) Inheritance of several origins of resistance to Phomopsis stem canker (*Diaporthe helianthi* Munt.-Cvet. et al.) in sunflower (*Helianthus annuus* L.). Euphytica 116: 167–179.

Vranceanu AV, Tudor VA, Stoenescu FM, Pirvu N (1980) Virulence groups of *Orobanche cumana* Wallr., differential hosts and resistance sources and genes in sunflower. Proc 9th Int Sunflower Conf, Toremolinos, Spain, vol 1, pp 78–82.

Vranceanu AV, Craicui D, Soare G, Pacureanu M, Voinescu G, Sandu I (1994) Sunflower genetic resistance to *Phomopsis helianthi* (Munt.-Cvet.) attack. Rom Agri Res 1: 9–11.

3

Genetic Linkage Maps: Strategies, Resources and Achievements

Jinguo Hu

ABSTRACT

Sunflower genome mapping has come a long way. This chapter starts with a brief history of mapping efforts of using morphological and biochemical markers, and then describes all DNA-based marker techniques used in sunflower genome mapping. It also reviews the mapping populations developed and the mapping software packages available and summarizes the sunflower genome mapping achievement of approximately 20 published linkage maps. The chapter ends with listing the challenges to the sunflower research community in three areas: 1. handling the enormous amount of data of high-throughput sequence-based markers in sunflower genome mapping in the near future; 2. correlating DNA sequence alterations with the quantitative variation of economically important traits; and 3. applying the mapping achievement to sunflower improvement, the ultimate goal of genome mapping.

Keywords: Genome mapping; isozyme maker; protein marker; DNA marker; linkage groups

USDA Agricultural Research Service, Western Regional Plant Introduction Station, Pullman, WA 99164, USA; e-mail: *jinguo.hu@ars.usda.gov*

3.1 Introduction

The chromosomal theory of inheritance was proposed in the early years of the 20th century and supported by a large number of experiments in fruit fly (*Drosophila*) in the 1920s. The theory states that genes controlling different phenotypes are carried on the chromosomes and transmitted from parents to offspring. Now we know that in eukaryotic species, each chromosome contains a long piece of DNA that harbors numerous genes. The genes on the same chromosome tend to be transmitted together (linkage). However, crossing over between two chromatids in meiosis produces gametes of recombinant genotypes. The likelihood of crossing over depends on the distance between the two genes. Thus, the distance between two linked genes can be estimated by observing the frequency of recombinant phenotypes among the segregating progeny derived from a cross. The segregating data of three linked genes in a population permit estimation of the distance between genes and establishment of an order for the three genes. This distance between genes is called a genetic map unit (mu), or more commonly, a centiMorgan (cM). The success of constructing a linkage map for *Drosophila* with morphological markers stimulated gene mapping in various species. Among agricultural crops, corn (*Zea mays*) and tomato (*Solanum lycopersicum*, earlier known as *Lycopersicon esculentum*) were the first to have complete linkage maps with the number of linkage groups equal to the number of chromosome pairs.

A genetic linkage map provides knowledge of the organization of the genes of the species. This information eventually leads to the isolation (cloning) of the gene (by map-based or positional cloning) for structural and functional characterizations. It also helps breeders efficiently manipulate the genetic variation to create new varieties with improved quality and increased productivity via marker-assisted selection (MAS). This chapter reviews the history, techniques and strategies, achievement and challenges of the sunflower genome mapping efforts.

3.2 Brief History of Mapping Efforts: Morphological and Biochemical Markers

3.2.1 Morphological Markers

Although sunflower has been cultivated as a crop for a long time, the genetic study of sunflower lagged behind other major crop plants. Working towards improving this crop, individual researchers have observed and recorded many morphological variations in leaf, flower, and seed characteristics, as well as resistance to numerous diseases. Miller and Fick (1997) summarized the research and identified about 40 morphological traits in sunflower, prior to the advent of molecular markers that were reported to be controlled by one or two genes. The cultivated sunflower is a diploid species with 17 pairs of chromosomes, and early efforts to establish linkage relationships among the genes controlling the aforementioned traits met with little success. As a result, no classical linkage map of morphological markers was available for this crop. Instead, only a few pairs of linked genes governing morphological traits were reported. Leclercq (1966) discovered a tight linkage between a nuclear male sterility gene (Ms_{10}) and an anthocyanin pigment gene (T) and proposed a scheme to use male sterility to produce sunflower hybrids without the tedious emasculation process of the seed-bearing female plants. This enabled commercial hybrid seed production in several European countries including France and Romania prior to the discovery of the cytoplasmic male sterility system. Hockett and Knowles (1970) reported that the single gene Y for green versus yellow growing point color was found to be linked with the gene Br_3 for branching with a recombination frequency of $11.6 \pm 1.0\%$.

3.2.2 Biochemical Markers

3.2.2.1 Isozyme Markers

Isozymes are enzymes that have different amino acid sequences but catalyze the same chemical reaction in living organisms. The technique of electrophoresis of tissue extracts on starch gel followed by specific staining for enzyme activity in the gel was first described by Hunter and Merkert (1957). Markers generated by this technique were widely used in many biological research applications. In plants, isozyme markers had been extensively used by population geneticists to investigate the causes and effects of genetic variation within and between populations before the advent of DNA-based markers.

During the 1970s, plant geneticists started to map isozyme loci in the genetic maps of different crop plant species. Corn was the leading crop plant for which isozyme markers were mapped. Harris and Harris (1968) located an esterase locus, $E4$, on chromosome 3. Schwartz (1971) placed an alcohol dehydrogenase locus, Adh_1, on the long arm of chromosome 1, approximately 1.6 map units from lemon white gene, lw, conditioning the white seedling, pale yellow endosperm phenotype (Robertson 1961). By 1980, thirty isozyme loci had been mapped to nine of the ten chromosomes (Goodman et al. 1980). The chromosome locations of these mapped isozyme markers shed light on gene duplication in the maize genome evolution. Additionally, isozyme studies in tomato substantially influenced the application of molecular markers to the genetic improvement of crop plants. After discovering the association of an acid phosphatase 1 locus, $Aps-1$, with the root-knot nematode resistant gene, *Mi*, on chromosome 6 of tomato, Rick and Fobes (1974) reasoned that evaluation for nematode resistance can be done more efficiently by electrophoresis than by parasite inoculation. Furthermore, the co-dominant fashion of the zymogram of *Aps-1* will allow tomato breeders to distinguish between heterozygous and homozygous resistant seedlings. Based on the availability of a linkage map of 22 isozyme loci mapped to nine of the 12 tomato chromosomes, Tanksley and Rick (1980) discussed many potential uses of isozyme markers in basic plant genetics research and applied plant breeding including gene mapping and marker-assisted selection (MAS).

The first study on a sunflower isozyme was carried out by Torres (1974a, b, c) who investigated the genetics of alcohol dehydrogenase (ADH) in great detail. A total of 12 distinct bands of sunflower alcohol dehydrogenase were resolved electrophoretically with starch gels. The slowest- and the fastest-migrating sets of three bands were allozymic dimers, which are the products of two genes designated Adh_1 and Adh_2 respectively. There were two co-dominant alleles, F (for fast) and S (for slow) at each locus, and each heterozygote produced three bands as expected with a dimer molecule: Adh_1^{FF}, Adh_1^{FS}, Adh_1^{SS} and Adh_2^{FF}, Adh_2^{FS}, Adh_2^{SS}. The homozygotes produced just one band each, consisting of FF or SS homodimers (Torres 1974a, b). The remaining six bands were intergenic dimers or developmental artifacts (Torres 1974c). It seemed possible that a third allele, E (for early), exists at the Adh_1 locus and the expression of Adh_1 allele is under developmental control since it only appears in the developing seeds but not in the mature seeds. A survey on allele frequency indicated that Adh_1^S is a rare allele in the cultivated sunflower. Out of over 6,000 seeds belonging to 70 collections surveyed, no Adh_1^{SS} homozygote was found and only three heterozygotes (Adh_1^{FS}) were identified among 422 seeds of one collection. Heterozygote Adh_1^{FS} was seen infrequently and homozygote Adh_1^{SS} was even rarer in the seven wild populations sampled. The dissociation-recombination experiments using bands excised from starch gels revealed that an intermediately migrating isozyme dimeric intergenic product consisting of Adh_1^F and Adh_2^S subunits could be formed. However, the hybrid isozyme was unstable in vitro because its monomers spontaneously dissociated and recombined to produce Adh_1^{FF} and Adh_2^{SS} isozymes (Torres 1974b). Genetic studies indicated that Adh_1 and Adh_2 were not linked.

In sunflower, isozymes have been used for investigating genetic variability in both domesticated and wild populations (Dry and Burdon 1986; Rieseberg and Seiler 1990; Cronn et al. 1997) for assessing phylogenetic relationships and speciation mechanisms within the genus *Helianthus* (Rieseberg 1991, 1998) and for identifying interspecific hybrids (Carrera et al. 1996). There have been several reports on genetic and linkage studies of various isozymes in sunflower. Kahler and Lay (1985) studied the inheritance of six isozyme marker loci in the selfed progenies of cultivated annual sunflower lines collected by the United States Department of Agriculture-Agricultural Research Service (USDA-ARS). They found that Chi-square goodness-of-fit tests verified the genetic segregation for the six enzyme marker loci including a peroxidase (PRX) locus, *Prx3*, a malate dehydrogenase (MDH) locus, *Mdh1*, a 6-phosphogluconate dehydrogenase (PGD) locus, *Pgd1*, a glucose-phosphate isomerase (PGI) locus, *Gpi2*, a phosphoglucomutase (PGM) locus, *Pgm4* and an isocitrate dehydrogenase (IDH) locus, *idh2*. They also determined that *Prx3* is linked to *Pgm4* with a recombination value of 0.14 ± 0.02. Therefore, their results suggested that the six enzyme loci could be mapped to five different linkage groups. Lay et al. (1988) reported the investigation of nine isozyme loci in six F_2 populations. They found: 1) a distorted segregation for two loci, *Mdh1* and an acid phosphatase (ACP) locus, *Acp2*, 2) a confirmation of the linkage relationship between *Pgm4* and *Prx3* reported by Kahler and Lay (1985), but the recombination value dropped to 0.055 ± 0.01, and 3) a newly identified linkage relationship between *Acp1* and *Pgd1* with a recombination value of 0.05 ± 0.01. Quillet et al. (1995) followed six isozyme loci in the BC_1 population derived from an interspecific cross (*H. argophyllus* x *H. annuus* cv. RHA 274). Two loci, *Mdh2* and *Acp1* were unlinked to other markers; another two, *Pgm1* and a malic enzyme (ME) locus, *Me1*, were mapped to two separate linkage groups; and the remaining two, *Mdh1* and a shikimate dehydrogenase (SDH) locus, *Sdh1*, were linked at a map distance of 47.2 cM. Fambrini et al. (1997) reported a co-dominant locus controlling the two distinct achromatic bands for Cu/Zn superoxide dismutase (Cu/ZnSOD$_{Chl}$). The variant was identified in an ABA-deficient mutant, *w-1*, on which the linkage study was conducted. Mestries et al. (1998) reported the mapping of six isozyme loci onto five linkage groups of a restriction fragment length polymorphism (RFLP) linkage map constructed from a single F_2 population generated by crossing two inbred lines GH and PAC2. The six mapped loci included a glutamate-oxaloacetate transaminase (GOT) locus, *Got1*, on linkage group (LG) D, *idh1* and *pgi3* on LG H, *Me* on LG H, *Pgd2* on LG J and *Sdh* on LG N. The mapping distance between *idh1* and *pgi3* was 45 centiMorgans (cM). Carrera et al. (2002) mapped five isozyme loci in an $F_{2:3}$ population from which a linkage map of restriction fragment length polymorphism (RFLP) markers had been constructed and released to the public (Barry et al. 1997). Three of the loci, *Est1*, *Gdh2*, and *Pgi2* were mapped to linkage groups 3, 14 and 9, respectively. The other two, *acp1* and *pgd3*, were linked in LG 2 of the map with a mapping distance of 11.3 cM.

Since there is no classical linkage map for sunflower, the linkage group numbers were independently assigned in different laboratories. The relatively small number of mapped isozyme loci limits these loci in genome mapping of the sunflower crop. However, these studies did reveal an important feature of the sunflower genome, the extensive duplication, since most of the isozymes, such as ACP, ADH, EST, IDH, MDH, ME, PGI, PGM and SDH, had two or more loci. These observations support the hypothesis of the allopolyploid origin of the sunflower genome (Heiser and Smith 1955; Jackson 1983).

3.2.2.2 Storage Protein Markers

Sunflower seeds contain two major protein classes, albumins and globulins. Several seed storage protein markers were also reported in sunflower. Kortt and Caldwell (1990) resolved eight distinct components of the low molecular weight 2S albumins. The complete amino acid sequence of sunflower albumin 8 (SFA8), the major methionine-rich 2S protein of sunflower seed, was determined alone with a cDNA clone which codes for this protein (Kortt et al. 1991). Anisimova et al. (1995) demonstrated that SFA8 is polymorphic among sunflower breeding lines. Serre et al. (2001) used SDS-PAGE to separate the albumins into eight to ten bands in the molecular weight range of 10 to 18 kDa. Two of these bands, 14.5 or 15.5 kDa segregated in a co-dominant fashion in the 150 $F_{2:3}$ families derived from ZENB8 × HA 89. This 2S albumin locus was mapped in the interval between the markers ZVG0051 and ZVG0052 of linkage group 11 of a public sunflower RFLP map. Polymorphisms of the high molecular weight 11S globulins have been detected in both molecular weight and charge among sunflower breeding lines (Anisimova et al. 1996) and between species (Anisimova et al. 1993). The genes coding for this protein family were isolated and characterized (Vonder Haas et al. 1988). Unfortunately, no genetic mapping was conducted for these genes.

Konarev et al. (2000) characterized six groups of trypsin inhibitors in sunflower seeds. These included three groups of trypsin/subtilisin inhibitors, one major and two minor groups of trypsin inhibitors. Each group showed polymorphism among breeding lines and trypsin/subtilisin inhibitors also varied among wild species. Genetic analysis showed that the major trypsin inhibitor and three groups of trypsin/subtilisin inhibitors are each controlled by a single locus, and three loci coding for trypsin/subtilisin inhibitors (groups C, D_a and D_b) were loosely linked with recombination values ranging from 0.23 to 0.40.

3.3 DNA-based Marker Techniques

Numerous techniques have been developed to detect polymorphism at the DNA level in animals, plants and microbes. A DNA marker visualized with a special technique reveals the DNA variation among the samples and is analogous to a traditional heterozygous allelic pair. This section briefly discusses the DNA marker techniques and strategies that the sunflower community adopted for genome mapping.

3.3.1 RFLP

RFLP (restriction fragment length polymorphism) is the first DNA marker technique developed for genome mapping in humans (Botstein et al. 1980). This technique detects the gain or loss of a restriction enzyme recognition site and the insertion or deletion of a DNA fragment between two restriction sites. The RFLP assay includes isolating high molecular weight DNA samples from experimental individuals, digesting the DNA with a restriction endonuclease, electrophoresis of an agarose gel to separate the DNA fragments according to their size, blotting the DNA fragments onto a nylon membrane, hybridizing the membrane with a labeled DNA probe that will only bind to a specific sequence that is complementary to the probe, and the detection of the hybridization signals. The probe is a cloned fragment from a genomic or cDNA library. In the early days of RFLP analysis, the hybridization signals were detected with ^{32}P labeled probes. It would often take 2 or 3 weeks to complete a single round of hybridization. In the mid-1990s, the advent of enzyme-linked probes and chemiluminescent detection expedited the process dramatically and enabled one round of hybridization to be completed within about a week. Even so, RFLP remained a slow, labor-intensive technique and required a relatively large amount of high quality DNA for the experiment. However, RFLP produced repeatable, excellent results and the markers are highly informative. It was RFLP that revolutionized the genome mapping effort for many economically important animal and plant species.

The implication of RFLP to sunflower genome mapping was first reported by the USDA-ARS Sunflower Research Unit (Jan et al. 1992). The involvement of a private party in the map construction not only delayed slightly the publication of the map, but also impeded distribution of the probes to other research laboratories. The first sunflower RFLP marker map was published in the mid-1990s (Barry et al. 1995; Gentzbittel et al. 1995).

3.3.2 RAPD

It is necessary to talk briefly about polymerase chain reaction (PCR) here since the next several marker techniques are PCR-based. PCR is one of the most powerful and revolutionary procedures used today to analyze DNA sequences. It is not an exaggeration to say that the impact of this simple procedure on modern biology is beyond description. PCR is based merely on the unique biochemistry of DNA replication and uses a thermostable DNA polymerase to amplify a DNA fragment that is flanked by known sequences. Two synthetic oligonucleotides called primers corresponding to the known sequences are used to initiate the amplification. PCR can be modified in numerous, complex ways to achieve special results. The majority of DNA marker techniques nowadays are based on PCR amplifications.

RAPD (random amplified polymorphic DNA) was a novel DNA polymorphism assay based on the amplification of random DNA segments with single primers of arbitrary nucleotide sequence (Welsh and McClelland 1990; Williams et al. 1990). The amplification depends on the correct orientation and proper distances of the randomly distributed sequences complementary to the decanucleotide primers in the genome being tested. The polymorphic markers revealed by RAPD are due to the gain or loss of priming sites and insertions or deletions between the priming sites. Therefore, it does not require prior DNA sequence information to run RAPD, since the primers will bind somewhere in the template and enable the DNA polymerase to amplify fragments. This was an obvious advantage at that time when DNA sequence information was scarce for most of the species of economic importance. In addition, RAPD is easy to perform, requires a low investment for equipment (a thermal cycler and a horizontal gel electrophoresis system), and the universal random 10-mer primers are inexpensive. These made the RAPD assay the most popular in the plant research community. Sunflower researchers used RAPD for genome mapping and for tagging resistance genes to rust (Lawson et al. 1998) and broom rape (Lu et al. 2000). In recent years, RAPD has given way to other marker techniques because it has lower resolution due to being dominant in nature, because it is anonymous and not specific to the target sites, and it has a less satisfactory reproducibility among laboratories where different equipment and reagents are used (Jones et al. 1997). In addition, Rieseberg (1996) observed that only 91% of 220 pairs of RAPD fragments amplified from related species had homology as tested by cross-hybridization and restriction digestion.

3.3.3 AFLP

AFLP (amplified fragment length polymorphism) is another novel DNA fingerprinting technique based on PCR (Vos et al. 1995). It converts the detection of RFLP from the hybridization-based system to an amplification-based system. In other words, AFLP methodology employs PCR to detect RFLP indirectly and collectively. The procedures for AFLP assay are lengthy and complicated. The first step is the digestion of the DNA samples with two different restriction enzymes, usually, one six-base cutter and one four-base cutter. The second step is the ligation of the carefully designed, synthetic oligonucleotides (adapters) to the ends of the restriction fragments. The third step is the pre-amplification of the fragment with primers with sequences complementary to that of the adapters. The fourth step is the selective amplification of sets of restriction fragments, which is achieved by adding one to three selective nucleotides to the 3' ends of the primers in the pre-amplification step. Only those fragments in which the sequence variation matches the nucleotides flanking the restriction sites are amplified in this step. The final step is analysis of the amplified fragments by electrophoresis with a denaturing polyacrylamide gel and visualizing the fragment with silver-staining or autoradiography techniques. Typically, each primer combination can amplify 50–100 fragments that produce a multi-banded profile on the sequencing gel. AFLP has proven to be a powerful genotyping technique for almost all organisms with different levels of genome complexity. Sunflower researchers adapted AFLP for fingerprinting sunflower germplasm (Hongtrakul et al. 1997) and genome mapping (Peerbolte and Peleman 1996; Gedil et al. 2001) and for quantitative trait loci (QTL) analysis of various important traits such as in vitro organogenesis (Flores Berrios et al. 2000a, b), partial resistance to downy mildew (Al-Chaarani et al. 2002) and leaf chlorophyll concentration and stomatal conductance (Hervé et al. 2001). Like the RAPD markers, most of the AFLP markers are dominant in nature and not specific to the target sites. However, AFLP can reveal a much greater number of polymorphic loci and has a better reproducibility than RAPD. As a high-throughput genotyping technique, it is inevitable that artifacts will be generated. As a result, there are always 10 to 15% of the AFLP markers that cannot be assigned to linkage groups in map construction. This is well compensated for by a large number of markers amplified in each assay.

3.3.4 DALP

DALP (direct amplification of length polymorphism) was proposed by Desmarais et al. (1998). In this technique, several selective primers are made by extending the 3' end of the M13–40 universal sequencing primer with two or four nucleotides. These selective primers can be used with another universal sequencing primer M13 to generate multi-banded DNA fingerprints resolved with a sequencing gel. A different fingerprint can be obtained when a different selective primer is used. This technique is very easy to perform and offers the possibility to sequence each of the amplified fragments with the universal sequencing primers. Since the procedure requires the use of radioisotope to label one of the primers, this technique was not widely used. However, Langer et al. (2003) used it in combination with AFLP and specific PCR and produced a sunflower linkage map.

3.3.5 SSR

SSR (simple sequence repeat), also known as microsatellite or STR (short tandem repeats), is a class of DNA sequence stretches characterized by highly repeated di-, tri-, tetra-, or penta-nucleotide motifs (e.g., CACACACA..., CATCATCAT..., ACGTACGTACGT..., or TAAAATAAAATAAAA) and spread throughout the genomes. SSR polymorphisms are assayed by specifically amplifying a small DNA fragment that contains the repeats via PCR and then by sizing the amplified fragments on agarose, polyacrylamide gel, or capillary electrophoresis. The number of repeats varies among genotypes under investigation and results in a length variation of the amplified fragments. Being highly polymorphic, highly abundant, co-dominant inheritance, analytically simple and readily transferable, makes SSR the marker of choice over many other markers for genomic studies.

Abundant interspersed repetitive DNA elements in different eukaryotes were observed from the limited amount of DNA sequence information in the early 1980s (Miesfeld et al. 1981; Hamada and Kakunaga 1982). Weber and May (1989) first realized that this class of sequence represents a vast new pool of potential genetic markers to be used to fill the gaps in the existing human genetic map and to improve map resolution. The use of STR in human genome mapping turned out to be a great success. Gyapay et al. (1994) mapped 2,066 $(AC)_n$ STR markers to the human genetic linkage map and Dib et al. (1996) added 5,264 $(AC/TG)_n$ STR markers to the map.

SSR polymorphism was first reported in crop plants in soybean (Akkaya et al. 1992). Since then SSR markers have been developed in a number of crop species such as rice (Morgante and Olivieri 1993; Wu and Tanksley 1993), barley (Becker and Heun 1995), corn (Senior and Heun 1993) and tomato (Broun and Tanksley 1996). SSR markers rapidly displaced RFLP markers not only in plant genome mapping but also in other applied areas such as genotype identification and variety protection (Smith and Helentjaris 1996), seed purity evaluation and germplasm conservation (Brown and Kresovich 1996), germplasm diversity studies (Xiao et al. 1996), quantitative trait loci (QTL) analysis (Blair and McCouch 1997), pedigree analysis and marker-assisted breeding (Ayres et al. 1997), and as anchor points in screening of large insert libraries for positional gene cloning.

Since sequence information is needed for designing specific primer pairs for individual SSR markers, the initial development of SSR started with a laborious and costly approach of screening genomic libraries by hybridizing with SSR probes and sequencing the hybridized positive clones. The SSR-enrichment procedure of using biotin-labeled SSR-containing probes in combination with magnetic beads coated with streptavidin or a nitrate filter (Edwards et al. 1996) in small-insert libraries considerably improved the efficiency of isolating SSR sequences from the genome and hence reduced the time and cost for SSR development. The growing availability of expressed sequence tag (EST) sequences from a range of crop plants allows large-scale search for SSR motif and design of SSR primers using pipelined computer programs. Numerous SSR markers have been developed from the EST sequences in the public databases of several crop species (Holton et al. 2002; Thiel et al. 2003).

The laboratory of Steven Knapp at the Oregon State University (currently at the University of Georgia, Athens) was among the pioneers in sunflower SSR marker development and applying them to genome mapping (Huestis et al. 1996; Tang et al. 2002; Yu at al. 2002, 2003), although the first report appeared in 1994 (Brunel 1994), not much later than that in other crop plants. However, the progress was slow in the following several years. The total number of public sunflower SSR markers did not reach 100 before 2002 (Whitton et al. 1997; Hongtrakul et al. 1998a, b; Gedil 1999). Yu et al. (2002) developed 131 unique SSRs from 970 clones isolated from the $(CA)_{n-}$, $(CT)_{n-}$, $(CAA)_{n-}$, $(CATA)_{n-}$, or $(GATA)_{n-}$ enriched genomic DNA libraries. The test of allelic diversity among elite breeding lines revealed 3.7, 3.6, and 9.5 alleles per locus for dinucleotide, trinucleotide, and tetranucleotide repeats, respectively, and that the mean polymorphic information content (PIC) scores were 0.53 for dinucleotide, 0.53 for trinucleotide, and 0.83 for tetranucleotide repeats. These values are much higher than those for RFLP in sunflower. Tang et al. (2002) mapped 462 of the 1,089 SSR markers (with a prefix OSU) to a recombinant inbred line (RIL) population derived from a cross of two USDA-ARS released sunflower lines, RHA 280 and RHA 801. This map has been used as a public domain for linkage group assignment of mapped QTL and genes.

CARTISOL, a collaborative research consortium of several public research organizations and private seed companies in France also developed SSR markers for sunflower. Mokrani et al. (2002) used 61 out of 465 SSR primers from the CARTISOL source that were polymorphic between the two inbred lines of the mapping population to locate QTLs underlying grain oil content and agronomic traits in combination with 215 AFLP markers. Yu et al. (2003) mapped 91 out of 156 screened for polymorphism to the public RHA 280 × RHA 801 RIL map. Zhang et al. (2005) reported that they identified 78 SSR markers from the 1,111 primer pairs provided by CARTISOL as an effective set of SSR markers for sunflower variety identification and diversity assessment. These 78 primers detected a total of 276 alleles across the 124 elite inbred lines, with an average of 3.5 alleles per SSR locus.

3.3.6 TRAP

TRAP (target region amplification polymorphism) is a fairly new technique (Hu and Vick 2003). TRAP is designed to harness the existing sequence information and explores the bioinformatics tools to rapidly develop markers associated with the traits of interest. In TRAP, the fixed primers are designed against known sequences of annotated putative genes and used in combination with other primers of arbitrary sequence to trap polymorphism in the targeted genomic region. Since the fixed primers are targeting the candidate gene sequences, TRAP offers a potentially higher probability of amplifying markers linked to the phenotype under investigation than RAPD and AFLP, which generate random polymorphic markers across the genome. TRAP has been successfully used in developing markers for five important agronomic traits of sunflower: a dominant gene conferring resistance to several races of downy mildew (Hu et al. 2004b), a recessive gene controlling apical branching (Rojas-Barros et al. 2008), a recessive gene governing complete male sterility (Chen et al. 2006), a recessive gene governing light green leaf color caused by reduced chlorophyll content, one of the two genes controlling lemon ray flower color (Yue et al. 2008b) and QTL for resistance to *Sclerotinia* head rot (Yue et al. 2008a). TRAP was also used to generate markers in defining the sunflower linkage group ends with fixed primers designed against the *Arabidopsis*-type telomere sequence repeat (Hu 2006).

3.3.7 SNP

SNP (single nucleotide polymorphism) refers to a DNA sequence variation at a single nucleotide position in the genome among homozygous individuals (or between paired chromosomes of a heterozygous individual) and is the most common type of genetic variation. Theoretically, a SNP could have four alleles because there are four possible nucleotides at every nucleotide position. In practice, nearly all reported SNPs have only two alleles and occur in genes as well as in intergenic regions. Compared to the typical multi-allelic RFLP, SSR and SSLP (simple sequence length polymorphisms), SNPs are less informative. However, this disadvantage can be offset by a greater marker density and by the stability of SNPs. In plant genomes, the usual frequency of SNPs is one in every 100–300 bp (Gupta et al. 2001). In sunflower, a relatively high frequency of SNPs has been reported. Liu and Burke (2006) investigated nine genomic loci of a total length of 8,207 bp in 32 individuals sampled from wild populations, primitive and improved germplasm accessions. They found 444 polymorphic sites among the samples with an average of 1 SNP for every 16.8 bp of sequence. The wild sunflowers harbored 392 polymorphic sites (1 SNP/19.1 bp), whereas the cultivated sunflower harbored 194 polymorphic sites (1 SNP/38.8 bp). Kolkman et al. (2007) surveyed the nucleotide diversity of 82 previously mapped RFLP marker loci in two wild and 10 elite inbred lines. They identified 1,078 SNPs in 49.4 kbp of DNA/genotype with an average of 1 SNP/45.7 bp. Non-coding sequences harbored two-fold more SNPs (1/32.1 bp) than coding sequences (1/62.8 bp). Fusari et al. (2008) reported that one SNP was found in every 69 bp the 14,348-bp aligned sequences of 28 candidate genes related to biotic and abiotic stresses in 19 sunflower inbred lines. The abundance and interspersed nature of SNPs in the sunflower genome together with the upcoming automatic high-throughput analytic technology will make the construction of high-density sunflower genetic maps possible in the near future. SNPs are also ideal for fine mapping of QTL, marker-assisted selection and association mapping.

3.3.8 Markers from DNA Sequence Databases

The advent of high-throughput sequencing technology produced a large volume of DNA sequence data, which provide a rich source for DNA marker development. The Compositae Genome Program (CGP; *http:// cgpdb.ucdavis.edu/)*, supported jointly by grants from the National Science Foundation Plant Genome Program and the United States Department of Agriculture Plant Genome Program, has generated over two million ESTs for many biologically and economically important plant species in the Compositae (Asteraceae) family. The majority of the 284,745 sunflower ESTs in GenBank were deposited by the CGP. Heesacker et al. (2008) described a catalog of 16,643 EST-SSRs, a collection of 484 EST-SSR and 43 EST-INDEL markers developed from common sunflower ESTs by data mining. They found that approximately 11% (1,956/17,904) of the unigenes in the transcript assembly harbored one or more SSRs with repeat counts of n ≥ 5. Among the primers tested for 43 INDELs and 484 SSRs, 39 and 427 produced high quality genotype data (Heesacker et al. 2008).

3.4 Mapping Populations and Tools

Linkage maps of molecular markers are constructed from segregating populations with the aid of computer software packages. This section briefly discusses the mapping populations, tools, and strategies that have been used in sunflower genome mapping.

3.4.1 Mapping Populations

All the published sunflower linkage maps were constructed from two types of mapping populations: F_2 and RIL (recombinant inbred line) populations each having advantages that are useful in different situations for different purposes. The two parental plants were usually selected by maximum polymorphism based on molecular marker screening or by the difference in the phenotypic trait of interest. After creating the F_1 hybrid, an F_2 or a RIL population can be generated and used for molecular mapping. An F_2

population is developed by selfing the F_1 hybrids. The major disadvantage of using F_2 populations is that the populations are not immortal. Therefore, the sample tissue to isolate DNA for genotyping is finite. Another problem is that each individual plant in the F_2 population possesses one unique genotype and cannot be replicated in measuring quantitative traits. Since the F_2 populations are easy to develop, they were used in the construction of the early sunflower linkage maps (Barry et al. 1995: Gentzbittel et al. 1995, 1999; Jan et al. 1998).

Recombinant inbred lines (RILs) were first constructed for the model plant *Arabidopsis thaliana* as permanent mapping populations (Reiter et al. 1992; Lister and Dean 1993) to add more markers to the existing genome map. RIL populations are developed by self-pollinating individual F_2 plant followed by three additional cycles of single-seed descent (SSD) to advance the generation to F_5. Since each of the resulting RIL can be traced back to the individual F_2 plants, these lines are also called F_2-derived lines. Sunflower plants need large space to grow from seed to seed, generation advance in greenhouse is not practical. Only two generations (one in growing season and one in off-season winter nursery) can be achieved. Therefore, an RIL population needs at least four years to develop. After several generations of self-pollination, a RIL population becomes an eternal mapping population since each individual line possesses a practically homozygous genotype and can be propagated generation after generation as a clone. RILs increase the mapping accuracy of the high-throughput, dominant markers such as AFLP and TRAP, which cannot easily differentiate heterozygous from homozygous genotypes in the F_2 populations. RILs are also powerful tools for analyzing quantitative traits because lines of identical genotypes can be used in replicated trials to generate quantitative trait data, which is used to determine the chromosomal location of the molecular markers closely associated with the traits of interest. There are a number of sunflower linkage maps constructed with RIL populations (Flores Berrios et al. 2000a; Tang at al. 2002; Yu at al. 2003; Langer et al. 2003; Hu et al. 2004a; Poormohammad Kiani et al. 2007).

3.4.2 *Mapping Software Packages*

Classical linkage analysis consists of three steps. The first step is to establish a linkage relationship between two markers by a simple statistical test such as a Chi-square test. The second step is to calculate the genetic distance between the two markers from the recombinant frequency of the phenotypes in the segregating population under investigation. The third step is to determine the order of the markers, if three or more markers are in the same linkage group, by three-point test. These were manually done when working with a few morphological markers. This process has been computerized since the advent of molecular markers. Many computer packages have been developed for constructing linkage maps with molecular marker data collected from various mapping populations. The sunflower genome mapping community has used the following three mapping computer programs.

3.4.2.1 *MapMaker*

MapMaker is the most popular program for linkage map construction for crop plant species. An earlier version was released in 1987 (Lander et al. 1987) and was upgraded MAPMAKER/EXP v3.0 in 1992 (Lincoln et al. 1992). MAPMAKER/EXP uses an efficient algorithm to perform full multipoint linkage analysis for dominant, recessive, and co-dominant markers segregating in experimental mapping populations such as F_2, BC_1F_1 and RILs. MAPMAKER includes an interactive command language that makes it easy to use. MapMaker has been applied to the construction of most published sunflower linkage maps. Almost all researchers used the default LOD value of 3.0 or higher for the grouping and three-point commands and the Kosambi mapping function (default is Haldane) in the mapping. The weak point of MapMaker is that it lacks an MS-Windows operating system interface. But who can complain about such a remarkable, highly utilized program when it has been free?

3.4.2.2 *JoinMap*

JoinMap was developed to combine ("join") data derived from several sources into an integrated map (Stam 1993, 1995; Van Ooijen et al. 2001). The newest release of this commercial program is JoinMap 4.0, which has an MS-Windows user interface and much added practical functionality. JoinMap can generate publication-ready map charts from experimental data collected from a variety of population types: BC_1, F_2, RIL, F_1-derived and F_2-derived doubled haploid (DH), out-crossing full-sib families. JoinMap is powerful in linkage group determination, and has the ability to automatically determine linkage phases for markers segregating in the out-breeding full-sib family. JoinMap also embodies several diagnostic tools, such as testing segregation distortion, checking similarity of loci, checking similarity of individuals, and calculating genotype probabilities conditional on map and flanking genotypes to discover double recombination before and after the actual map construction. Detailed information on JoinMap can be found at *http://www.kyazma.nl/index.php/mc.JoinMap*. JoinMap has been used by several researchers in constructing their linkage maps (Gedil et al. 2001; Mimic et al. 2004, 2005; Lai et al. 2005).

3.4.2.3 GMendel

GMendel is a useful tool for constructing linkage maps from both molecular marker and Mendelian phenotypic data. The unique feature of GMendel is its multiple pairwise locus ordering methods. In GMendel, a simulated annealing algorithm is programmed for fast linkage analysis runs, and Monte Carlo and bootstrap methods are employed to validate the order of mapped markers (Holloway and Knapp 1993). GMendel has been used by Gedil et al. (2001), Yu et al. (2003), Slabaugh et al. (2003) and Lai et al. (2005) in sunflower genome mapping.

3.5 Achievement

During the past 15 years, genome mapping has been an active and highly successful area in the sunflower research community, including both public research institutions and proprietary seed companies. Linkage maps of various types of molecular markers have been constructed from over 20 mapping populations. These maps revealed the genome organization and expanded our knowledge of this important oil crop. Some representative maps are discussed below.

3.5.1 RFLP Maps

The first published linkage map of cultivated sunflower was a proprietary one (Berry et al. 1995). It spanned 1,380 cM on 17 linkage groups and contained 234 RFLP loci detected by 213 cDNA probes. The mapping population was an F_2 population of 289 individuals derived from a cross between two inbred lines, HA 89 and ZENB8. HA 89 is a public sunflower inbred line released by USDA and ZENB8 is proprietary inbred maintainer line. Twenty-three of the 234 loci showed significant segregating distortion. This map was later "joined" with maps constructed from eight other F_2 populations to form a high density, composite map with 635 marker loci and was 1,472 cM in length (Berry et al. 1996). The second sunflower RFLP linkage map was published by a French group (Gentzbittel et al. 1995). This map of 237 loci spanning 1,150 cM and consisting of 16 linkage groups of more than 20 cM and seven groups of less than 20 cM was constructed from three F_2 and two BC_1 populations. The length of individual population map ranged from 71 ((HA 89 x CX) x CX BC_1) to 763 (PAC2 x RHA 266 F_2) cM. The maps from three F_2 populations were further integrated into a composite map with the maps from four other F_2 populations (Gentzbittel et al. 1999). Although the length of the individual maps varied from 774 cM to 1,060 cM with an average of 14 major linkage groups, the composite map (CARTISOL map) comprised 273 marker loci and had a length of 1,573 cM. A third sunflower RFLP map was published by the USDA sunflower research group (Jan et al. 1998). This map had 271 marker loci detected by 232 cDNA probes and covered 1,164 cM in 20 linkage groups. Two public inbred lines, HA 234 and RHA 271, were used for developing the F_2 mapping population.

There are obvious common features of these maps: 1) marker distribution was not uniform throughout the genomes, 2) large gaps of more than 20 cM existed in a few linkage groups, 3) distorted segregation was frequently observed, and 4) a considerable proportion (30 to 35%) of the mapped loci showed dominant segregation. From this, we could infer that the sunflower genome has extensive duplication. Since there are few obvious conserved linkage blocks among chromosomes, enormous reshuffling must have occurred after the duplication. The common distorted segregation could have resulted from selection against certain genotypes during gametogenesis, fertilization (the existence of self-incompatibility), seed development germination and plant growth. Some of these maps were used later for mapping a downy mildew resistance gene (Mouzeyar et al. 1995; Roeckel-Drevet et al. 1996; Vear et al. 1997) and *Sclerotinia sclerotiorum* resistance QTL (Mestries et al. 1998), and for identifying candidate-genes for downy mildew and *S. sclerotiorum* resistance (Gentzbittel et al. 1998). Some of the mapped single or low-copy probes were recently used to identify chromosome-specific bacterial artificial chromosome (BAC) clones (Feng et al. 2006) and for assessing SNP frequencies, nucleotide diversity, and linkage disequilibrium (LD) in modern cultivated sunflower lines (Kolkman et al. 2007).

3.5.2 AFLP Maps

Peerbolte and Peleman (1996) applied AFLP to sunflower genome mapping. They used 11 primer combinations on the DNA of 92 individuals from two CARTISOL F_2 populations, CX x RHA 266 and PAC2 x RHA 266. Two hundred forty-one (188 co-dominant) and 282 AFLP (243 co-dominant) markers were mapped to the two populations, respectively. Using JoinMap, the two maps were integrated in a combined map consisting of 437 markers (146 RFLPs and 291 AFLPs) spread over 19 linkage groups (3 markers or more) with a total length of 1,144 cM. Gedil et al. (2001) reported an integrated RFLP-AFLP linkage map constructed from the 180 F_2 progeny derived from two public sunflower lines, HA 370 and HA 372. This map consisted of 296 AFLP and 104 RFLP markers in 17 linkage groups and was 1,326 cM long. They observed densely clustered AFLP markers on several linkage groups, and presumably reside in centromeric regions. Since the RFLP probes were mapped by Berry et al. (1995), 14 of 17 linkage groups of the two independent RFLP maps were able to be aligned by common RFLP markers. Kusterer et al. (2004) constructed a linkage map from 183 F_2 individuals from the cross RHA 325 (cms) × HA 342. The map covered 1,751.5 cM with 202 AFLP and

19 SSR markers in 18 linkage groups. Langar et al. (2003) used both AFLP and DALP for constructing a linkage map for an RIL population derived from a cross of HA 89 x LR4. The map contained seven allele-specific PCR markers, 64 DALP, and 301 AFLP markers, and covered 2,168.6 cM in 18 linkage groups. This RIL population displayed the lowest percentage (about 4%) of markers showing distorted segregation.

3.5.3 SSR Maps

The publication of an SSR map (Tang et al. 2002) marked the most significant milestone in sunflower genome mapping. The first sunflower SSR map was constructed from a mapping population of 94 RILs derived from a cross between two public inbred lines: a confectionery, RHA 280 (Fick et al. 1974), and a single-headed oilseed restorer line, RHA 801 (Roath et al. 1981). In the map, the 459 loci were amplified with 408 SSR primer pairs, each of which produced one to three loci, and coalesced into 17 linkage groups with an accumulated length of 1,368 cM. Although the mean density was 3.1 cM per locus, there were four large gaps longer than 30 cM in four linkage groups and two gaps longer than 20 cM in two linkage groups. Since there were three polymorphic loci unlinked, the completeness of this map remained questionable. Nevertheless, these mapped single or low-copy, co-dominant DNA markers became the markers of choice for establishing a genome-wide framework to anchor and cross reference genetic linkage maps constructed from different mapping populations with a universal linkage group nomenclature.

   This public SSR map was soon expanded to 1,432 cM adding 118 new SSR and INDEL markers and was aligned with other proprietary maps constructed from the PHA × PHB RIL population and from the HA 370 × HA 372 F_2 map with 80 RFLP marker loci (Yu et al. 2003). Lai et al. (2005) further advanced the map by adding SNP markers identified from sunflower ESTs to this map. Of the 273 amplified polymorphic loci, 243 were mapped to the 17 established linkage groups. One observation was that several markers derived from ESTs with putatively related functions were co-located with previously mapped QTLs for traits such as salt tolerance, stem diameter, shattering, flowering time, and achene size. The mapped SSR markers have been used by numerous researchers in their mapping projects and made it possible for all published maps to have the same linkage group numbers. For example, Kusterer et al. (2004) used 19 mapped SSRs in their AFLP map and Yue et al. (2008a) used 44 mapped SSRs in their TRAP marker map constructed for locating *Sclerotinia* head rot resistance QTL. Recently, the original RHA 280 × RHA 801 RIL map has been expanded to 1,161 SSR loci (Steven Knapp, personal communication).

3.5.4 TRAP Maps

Hu et al. (2004a) described a TRAP marker map constructed with 129 RILs derived from the cross 83HR4 × RHA 345. The map consisted of 160 markers on 17 linkage groups plus four pairs of linked markers with a total length of 1,140 cM. These markers were amplified with 38 primer combinations in 23 sets of PCR. To demonstrate the applicability of TRAP in rapid genome mapping, Hu (2006) added 183 TRAP markers to the public SSR map (Tang et al. 2002) and defined 21 of the 34 linkage group ends by using 94 individuals of the RIL mapping population developed by Steven Knapp's laboratory and nine fixed primers containing the conserved *Arabidopsis*-type telomeric repeats (Hu 2006). An additional 220 TRAP markers were integrated to the same map and expanded the length from 1,523 to 1,920 cM (Hu et al. 2007). These markers were amplified by fixed primers designed against selected sunflower ESTs showing homology with components of plant disease resistance genes, homeobox genes, and other functional genes. It is worth mentioning that seven polymorphic markers amplified by the fixed primer designed against a sunflower EST that has homology with RPS2 were mapped. The mapped positions of two markers, TRAP018749 and TRAP017445, were close to HaRGC1 on LG8 and Ha-1W41 on LG13, respectively. Both HaRGC1 and Ha-1W41 were mapped to the regions conferring resistance to different races of downy mildew (Slabaugh et al. 2003). These results suggested that TRAP has the potential to target genomic regions harboring functional genes controlling the phenotype of interest. Figure 3-1 shows the public sunflower linkage map. TRAP has been used to map important traits of sunflower such as nuclear male sterility (Chen et al. 2006), ray flower color gene (Yue et al. 2008b), apical branching gene (Rojas-Barros et al. 2008), and *Sclerotinia* head rot resistance QTL (Yue et al. 2008a).

3.6 Published Sunflower Linkage Maps

Table 3-1 summarizes the published sunflower maps with various types of markers. The earlier mapping populations were developed solely for mapping purposes and the parents were chosen based on the highest polymorphism level in preliminary screening with molecular markers (Gentzbittel et al. 1995; Jan et al. 1998). Later the mapping populations were developed from two parental lines showing differences in disease resistances or important agronomic traits for the purpose of mapping QTLs underlying these traits (Bert et al. 2004; Micic et al. 2004; Rönicke et al. 2005; Poormohammad et al. 2007; Yue et al. 2008b).

It is worth mentioning that the earlier RFLP marker maps were independently constructed and comparisons between maps were not possible because there were no common markers among the maps. A total of

68 linkage group codes were generated for the 17 sunflower chromosome pairs in four different linkage group nomenclatures. By using probes from three independent sources, Gedil et al. (2001) integrated 14 of 17 linkage groups from two independent RFLP maps with an F_2 population derived from two public sunflower lines, HA 370 × HA 372. Thanks to the availability of the mapped public SSR markers, almost all recently published maps are now anchored by linkage group specific SSR markers and can be cross-referenced.

3.7 Challenges

Satisfactory results have been achieved for sunflower genome mapping and characterization, yet several challenges remain for the sunflower research community. The first is data handling. Implementation of high-throughput sequence-based markers in sunflower genome mapping in the near future will generate enormous amount of data, which will be hard for individual researchers to comprehend without help from a database. Therefore, there is an urgent need to develop a public database of mapping information from all mapping populations with cross-references.

The second challenge is the correlation of DNA sequence alterations with the quantitative variations of economically important traits. Progress has been made in locating many QTLs for traits of agronomic importance in sunflower. However, previous QTL mapping was merely the identification of anonymous markers in or near the chromosome regions affecting the traits under study. The use of candidate gene-based markers in mapping has just begun. With the rapidly advancing sequencing technology and the fast reducing sequencing cost, most of the sunflower genes, or even the whole genome, will eventually be sequenced. The observed phenotypic variation of the mapping population is influenced by genotypes (G), environments (E), and the interactions of G and E. The collection of phenotype data will depend strongly on traditional quantitative genetic evaluation systems which are time-consuming and labor-intensive. A coordinated, cooperative effort on observing large mapping populations in various environments is crucial to the sunflower crop. The ideal map will comprise

---

Figure 3-1 contd....

Figure 3-1 The public sunflower linkage map consisting of 577 SSR markers and 403 TRAP markers. The names of the SSP markers are in black. The names of the TRAP markers amplified with fixed primers targeting the conserved telomeric repeat and mapped to interstitial positions sequences are in green and mapped to the termini positions are in red. The TRAP markers amplified with fixed primers designed against selected sunflower ESTs showing homology with components of plant disease resistance genes and homeobox genes are in blue (reprinted with permission from Dr. Dragan Škorić, Editor-in-Chief of Helia, an international scientific journal published in Serbia).

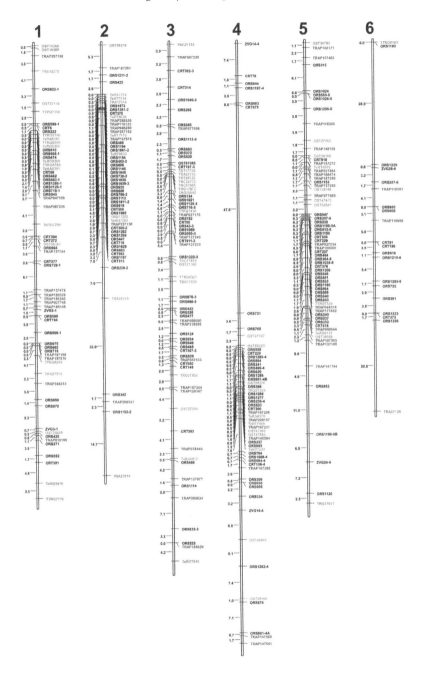

Figure 3-1 contd....

Figure 3-1 contd....

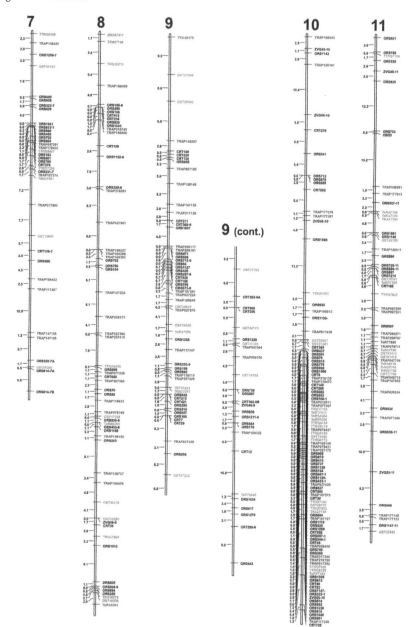

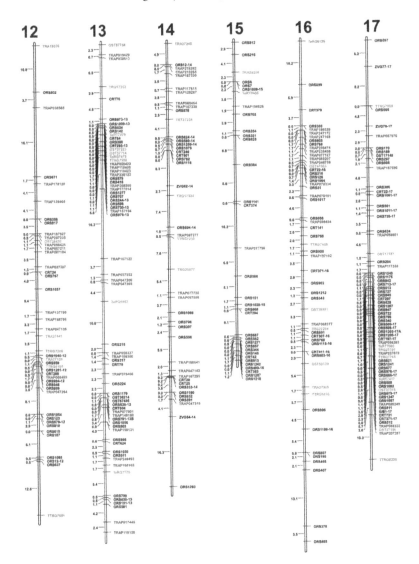

Table 3-1 Key details about published linkage maps for the cultivated sunflower genome.

Author	Year	Marker type	Marker #	Length (cM)	Linkage groups (parents)
Berry et al.	1995	RFLP	234	1,380	17 (ZENB8 x HA 89 F_2)
Gentzbittel et al.	1995	RFLP	237	1,150	16 + 7 (from 3 F_2 and 2 BC_1)
Berry et al.	1996	RFLP	633	1,472*	17 (composite of nine F_2)
Berry et al.	1997	RFLP	81	1,200	17 (selected mapped probes)
Jan et al.	1998	RFLP	271	1,164	20 (RHA 271 x HA 234 F_2)
Gentzbittel et al.	1999	RFLP	238	1,573*	17 (composite of seven F_2)
Flores Berrios et al.	2000	AFLP	264	2,558	18 (PAC2 x RHA 266) RIL
Gedil et al.	2001	RFLP/AFLP	400	1,326	17 (HA 370 x HA 372 F_2)
Leon et al.	2001	RFLP	205	1,380	17 (ZENB8 x HA 89)
Bert et al.	2002	RFLP/AFLP	290	2,318	19 (XRQ x PSC8)
Mokrani et al.	2002	AFLP	170	2,539	20 (L1 R- x L2 B-line)
Pérez-Vich et al.	2002	RFLP/AFLP	134	1,642	17 (HAOL-9 × CAS-3)
Pérez-Vich et al.	2002	RFLP/AFLP	154	1,808	17 (HA-89 × CAS-3)
Rachid Al-Chaarani	2002	AFLP	254	2,560	18 (PAC2 x RHA 266) RIL
Tang et al.	2002	SSR	459	1,368	17 (RHA 280 x RHA 801) RIL
Langar et al.	2003	DALP/AFLP	305	2,169	18 (LR4 x HA 89) RIL
Yu et al.	2003	SSR/RFLP	1044	1,523*	17 (composite of three maps)
	2003	SSR/RFLP	200	1,275	17 (HA 370 x HA 372 F_2)
	2003	SSR	264	1,199	17 (PHA x PHB)
	2003	SSR	577	1,423	17 (RHA 280 x RHA801)RIL
Bert et al.	2004	RFLP/AFLP	216	1,937	18 (FU × PAZ2)
Kusterer	2004	AFLP/SSR	221	1,752	18 (RHA 325(*cms*) × HA 342 F_2)
Hu et al.	2004	TRAP	160	1,140	17 + 4p (83HR4 x RHA 345) RIL
Micic et al.	2004	SSR	117	962	16 (NDBLOS x CM625)
Pérez-Vich et al.	2004a	SSR/RFLP	103	1,144	17 (P-21 x P-96)
Pérez-Vich et al.	2004b	SSR/RFLP	99	1,235	17 (HA 89 x CAS20)
Rachid Al-Chaarani	2004	AFLP	367	2,916	21 (PAC2 x RHA 266) RIL
Lai et al.	2005	SNP, SSR	243	1,349	17 (RHA 280 x RHA 801) RIL
Rönicke et al.	2005	AFLP/SSR	215	2,273	17 (SWS-B-04 × SWS-B-01)
Hu	2006	TRAP/SSR	760	1,747	17 (RHA 280 x RHA 801)RIL
Poormohammad et al.	2007	SSR/AFLP	495	1,825	21 (PAC2 x RHA 266) RIL
Hu et al.	2007	TRAP/SSR	980	1,920	17 (RHA 280 x RHA 801) RIL
Yue et al.	2008	TRAP/SSR	214	1,798	18 (HA 441 x RHA 439 F_2)

*: cM values not calculated from a single segregating population.

of both sequence-based markers and phenotypic variations and will be used to locate not only the genes coding for proteins to produce a specific phenotype, but also the controlling element such as transcription factors that turn the genes on and off in response to environment changes. The third challenge is the application of the mapping achievement to sunflower improvement, the ultimate goal of genome mapping. For sunflower, the breeding goal is the profitable and sustainable production of the crop with improved quality to satisfy consumers. Marker-assisted selection, which has been a proven and useful tool for increasing the breeding efficiency for other crop plants, will expedite the process of producing cultivars with resistance to multiple diseases and insects, with superior horticultural performance, and with widely optimized chemical components to meet the needs of end-users. Achieving this goal will strongly rely on the close interaction between public research and commercial breeders to integrate modern genomics tools with traditional breeding techniques.

References

Akkaya MS, Bhagwat AA, Cregan PB (1992) Length polymorphism of simple sequence repeat DNA in soybean. Genetics 132: 1131–1139.

Al-Chaarani GR, Roustaee A, Gentzbittel L, Mokrani L, Barrault G, Dechamp-Guillaume G, Sarrafi A (2002) A QTL analysis of sunflower partial resistance to downy mildew (*Plasmopara halstedii*) and black stem (*Phoma macdonaldii*) by the use of recombinant inbred lines (RILs). Theor Appl Genet 104: 490–496.

Anisimova IN, Georgieva-Todorova J, Vassileva R (1993) Variability of helianthinin, the major seed globulin in the genus *Helianthus* L. Helia 16: 49–58.

Anisimova IN, Fido RJ, Tatham AS, Shewry PR (1995) Genotypic variation and polymorphism of 2S albumins of sunflower. Euphytica 83: 15–23.

Anisimova IN, Gavrilova V, Loskutov A, Tolmachev V (1996) Identification of gene loci encoding for 11S globulin of sunflower seed. Proc 14th Int Sunflower Conf, Beijing, Shenyang, China, pp 1167–1170.

Ayres NM, McClung AM, Larkin PD, Bligh HFJ, Jones CA, Park WD (1997) Microsatellites and a single nucleotide polymorphism differentiate apparent amylose classes in an extended pedigree of US rice germplasm. Theor Appl Genet 94: 773–781.

Becker J, Heun M (1995) Barley microsatellites: allele variation and mapping. Plant Mol Biol 27: 835–845.

Berry ST, León AJ, Hanfrey CC, Challis P, Burkholz A, Barnes SR, Rufener GK, Lee M, Caligari PDS (1995) Molecular marker analysis of *Helianthus annuus* L. 2. Construction of an RFLP linkage map for cultivated sunflower. Theor Appl Genet 91: 195–199.

Berry ST, Leon AJ, Challis P, Livini C, Jones R, Hanfrey CC, Griffiths S, Roberts A (1996) Construction of a high density, composite RFLP linkage map for cultivated sunflower (*Helianthus annuus* L.). Proc 14th Int Sunflower Conf, Beijing/Shenyang, China, vol 2, pp 1155–1160.

Berry ST, León AJ, Peerbolte R, Challis P, Livini C, Jones R, Feingold S (1997) Presentation of the Advanta sunflower RFLP linkage map for public research. Proc 19th Sunflower Res Workshop, Fargo, ND, USA, pp 113–118.

Bert PF, D Tourvieille de Labrouhe, J Philippon, S Mouzeyar, I Jouan, P Nicolas, and F Vear. (2001) Identification of a second linkage group carrying genes controlling resistance to downy mildew (*Plasmopara halstedii*) in sunflower (*Helianthus annuus* L.). Theor Appl Genet 103: 992–997.

Bert PF, Dechamp-Guillaume G, Serre F, Jouan I, Tourvieille de Labrouhe D, Nicolas P, Vear F (2004) Comparative genetic analysis of quantitative traits in sunflower (*Helianthus annuus* L.). 3. Characterisation of QTL involved in resistance to *Sclerotinia sclerotiorum* and *Phoma macdonaldi*. Theor Appl Genet 109: 865–874.

Blair M, McCouch SR (1997) Microsatellite and sequence-tagged site markers diagnostic for the bacterial blight resistance gene, *xa-5*. Theor Appl Genet 95: 174–184.

Botstein D, White RL, Skolnick M, Davis RW (1980) Construction of a genetic linkage map in man using restriction fragment length polymorphisms. Am J Hum Genet 32: 314–331.

Broun P, Tanksley SD (1996) Characterization and genetic mapping of simple repeat sequences in the tomato genome. Mol Gen Genet 250: 39–49.

Brown, SM, Kresovich S (1996) Molecular characterization for plant genetic resources conservation. In: AH Paterson (ed) Genome Mapping in Plants. RG Landes, New York, USA, pp 85–93.

Brunel D (1994) A microsatellite marker in *Helianthus annuus* L. Plant Mol Biol 24: 397–400.

Carrera A, Poverene M, Rodriguez RH (1996) Isozyme variability *in Helianthus argophyllus*. Its application in crosses with cultivated sunflower. Helia 19: 19–28.

Chen J, Hu J, Vick BA, Jan CC (2006) Molecular mapping of a nuclear male-sterility gene in sunflower (*Helianthus annuus* L.) using TRAP and SSR markers. Theor Appl Genet 113: 122–127.

Cronn R, Brothers M, Klier K, Bretting PK, Wendel JF (1997) Allozyme variation in domesticated annual sunflower and its wild relatives. Theor Appl Genet 95: 532–545.

Desmarais E, Lanneluc I, Lagnel J (1998) Direct amplification of length polymorphisms (DALPs) or how to get and characterize new genetic markers in many species. Nucl Acids Res 26: 1458–1465.

Dib C, Fauré S, Fizames C, Samson D, Drouot N, Vignal A, Millasseau P, Marc S, Hazan J, Seboun E, Lathrop M, Gyapay G, Morisette J, Weissenbach J (1996) A comprehensive genetic map of the human genome based on 5,264 microsatellites. Nature 380: 152–154.

Dry PJ, Burdon JJ (1986) Genetic structure of natural populations of wild sunflowers (*Helianthus annuus* L.) in Australia. Aust J Biol Sci 39: 255–270.

Edwards KJ, Barker JHA, Daly A, Jones C, Karp A (1996) Microsatellite libraries enriched for several microsatellite sequences in plants. Biotechniques 20: 758–760.

Fambrini M, SebastianiL, Rossi VD, Cavallini A, Pugliesi C (1997) Genetic analysis of an electrophoretic variant for the chloroplast-associated form of Cu/Zn superoxide dismutase in sunflower (*Helianthus annuus* L.). J Exp Bot 48: 1143–1146.

Feng J, Vick BA, Lee MK, Zhang HB, Jan CC (2006) Construction of BAC and BIBAC libraries from sunflower and identification of linkage group-specific clones by overgo hybridization. Theor Appl Genet 113: 23–32.

Fick GN, Zimmer DE, Kinman ML (1974) Registration of six sunflower parental lines. Crop Sci 14: 912.

Flores Berrios E, Gentzbittel L, Kayyal H, Alibert G, Sarrafi A (2000a) AFLP mapping of QTLs for in vitro organogenesis traits using recombinant inbred lines in sunflower (*Helianthus annuus* L.). Theor Appl Genet 101: 1299–1306.

Flores Berrios E, Gentzbittel L, Alibert G, Sarrafi A (2000b) Genotypic variation and chromosomal location of QTLs for somatic embryogenesis revealed by epidermic layers culture of recombinant inbreds lines in sunflower (*Helianthus annuus* L.). Theor Appl Genet 101: 1307–1312.

Fusari, CM, Lia VV, Hopp HE, Heinz RA, Paniego NB (2008) Identification of single nucleotide polymorphisms and analysis of linkage disequilibrium in sunflower elite inbred lines using the candidate gene approach. BMC Plant Biol 8: 7.

Gedil, MA (1999) Marker development, genome mapping, and cloning of candidate disease resistance genes in sunflower, *Helianthus annuus* L. PhD Thesis, Oregon State Univ, Corvallis, OR, USA.

Gedil MA, Wye C, Berry ST, Seger B, Peleman J, Jones R, Leon A, Slabaugh MB, Knapp SJ (2001) An integrated RFLP-AFLP linkage map for cultivated sunflower. Genome 44: 213–221.

Gentzbittel L, Vear F, Zhang Y-X, Bervillé A, Nicolas P (1995) Development of a consensus linkage RFLP map of cultivated sunflower (*Helianthus annuus* L.). Theor Appl Genet 90: 1079–1086.

Gentzbittel L, Mouzeyar S, Badaoui S, Mestries E, Vear F, Tourvieille de Labrouhe D, Nicolas P (1998) Cloning of molecular markers for disease resistance in sunflower, *Helianthus annuus* L. Theor Appl Genet 96: 519–525.

Gentzbittel L, Mestries E, Mouzeyar S, Mazeyrat F, Badaoui S, Vear F, de Labrouhe Tourvieille D, Nicolas P (1999) A composite map of expressed sequences and phenotypic traits of the sunflower (*Helianthus annuus* L) genome. *Theor Appl Genet* 99: 218–234.

Goodman MM, Stuber CW, Newton K, Weissinger HH (1980) Linkage relationships of 19 enzyme loci in maize. Genetics 96: 697–710.

Gupta, PK, Roy JK, Prasad M (2001) Single nucleotide polymorphisms: A new paradigm for molecular marker technology and DNA polymorphism detection with emphasis on their use in plants. Curr Sci 80: 524–535.

Gyapay G, Morissette J, Vignal A, Dib C, Fizames C, Millasseau P, Marc S, Bernardi G, Lathrop M, Weissenbach J (1994) The 1993–94 Généthon human genetic linkage map. Nat Genet 7: 247–249.

Hamada H, Kakunaga T (1982) Potential Z-DNA forming sequences are highly dispersed in the human genome. Nature (Lond) 298: 396–398.

Harris JW, Harris JH (1968) Location of the E_4 esterase locus on chromosome 3. Maize Genet Co-op Newsl 42: 72–74.

Heiser CB, Smith DM (1955) New chromosome numbers in *Helianthus* and related genera. Proc Indiana Acad 64: 250–253.

Heesacker A, Kishore VK, Gao W, Tang S, Kolkman JM, Gingle A, Matvienko M, Kozik A, Michelmore RM, Lai Z, Rieseberg LH, Knapp SJ (2008) SSRs and INDELs mined from the sunflower EST database: abundance, polymorphisms, and cross-taxa utility Theor Appl Genet 117: 1021–1029.

Hervé D, Fabre F, Flores Berrios E, Leroux N, Al Chaarani G, Planchon C, Sarrafi A, Gentzbittel L (2001) QTL analysis of photosynthesis and water status traits in sunflower (*Helianthus annuus* L.) under greenhouse conditions. J Exp Bot 52: 1857–1864.

Hockett EA, Knowles PF (1970) Inheritance of branching in sunflowers, *Helianthus annuus* L. Crop Sci 10: 432–436.

Holloway JL Knapp SJ (1993) G-MENDEL 3.0 user's guide. Oregon State Univ, Corvallis, OR, USA, pp 1–130.

Holton TH, Christopher JT, McClure L, Harker N, Henry RH (2002) Identification and mapping of polymorphic SSR markers from expressed gene sequences of barley and wheat. Mol Breed 9: 63–71.

Hongtrakul V, Huestis GM, Knapp SJ (1997) Amplified fragment length polymorphisms as a tool for DNA fingerprinting sunflower germplasm: genetic diversity among oilseed inbred lines. Theor Appl Genet 95: 400–407.

Hongtrakul V, Slabaugh MB, Knapp SJ (1998a) A seed specific D12 oleate desaturase is duplicated, rearranged, and weakly expressed in high oleic sunflower lines. Crop Sci 38: 1245–1249.

Hongtrakul V, Slabaugh MB, Knapp SJ (1998b) DFLP, SSCP, and SSR markers for D9- stearoyl-acyl-carrier protein desaturase strongly expressed in developing seeds of sunflower: intron lengths are polymorphic among elite inbred lines. Mol Breed 4: 195–203.

Hu J (2006) Defining the sunflower (*Helianthus annuus* L.) linkage group ends with the *Arabidopsis*-type telomere sequence repeat-derived markers. Chrom Res 14: 535–548.

Hu J, Vick BA (2003) Target region amplification polymorphism, a novel marker technique for plant genotyping. Plant Mol Biol Rep 21: 289–294.

Hu J, Chen J, Berville A and Vick BA (2004a) High potential of TRAP markers in sunflower genome mapping. Proc 16th Int Sunflower Conf, Aug 29–Sept 2, Fargo, ND, USA, pp 665–671.

Hu J, Chen J, Gulya TJ, Miller JF (2004b) TRAP markers for a sunflower downy mildew resistance gene from a new *Helianthus annuus* source, PI468435. Proc 16th Int Sunflower Conf, Aug 29–Sept 2, Fargo, ND, USA, pp 623–629.

Hu J, Yue B, Vick BA (2007) Integration of TRAP markers into a sunflower SSR marker linkage map constructed from 92 recombinant inbred lines. Helia 30(46): 25–36.

Huestis GM, Kimura Y, Knapp SJ (1996) Simple sequence repeat (SSR) markers for sunflower (*Helianthus unnuus* L.). Proc 18th Sunflower Res Forum, January 21–22, Fargo, ND, USA, pp 120–123.

Hunter RL, Merkert CL (1957) Histochemical demonstration of enzymes separated by zone electrophoresis in starch gels. Science 125: 1294–1295.

Jackson RC (1983) Colchicine induced quadrivalent formation in *Helianthus*: Evidence of an ancient polyploidy. Theor Appl Genet 64: 219–222.

Jan CC, Vick BA, Miller JF, Kahler AL (1992) Progress in the development of a genomic RFLP map of cultivated sunflower (*Helianthus annuus*). In: Plant & Anim Genom Conf 1, San Diego, CA, USA, p 30.

Jan CC, Vick BA, Miller JF, Kahler AL, Butler ET (1998) Construction of an RFLP linkage map for cultivated sunflower. Theor Appl Genet 96: 15–22.

Jones CJ, Edwards KJ, Castaglione S, Winfiefd MO, Sala F, vande Wiel C, Bredemeijer G, Vosman B, Matthes M, Daly A,Brettschneider R, Bettini P, Buiatti M, Maestri E, MalcevschiN, Aert R, Volckaert G, Rueda J, Linacero R, Vazquez A, Karp A (1997) Reproducibility testing of RAPD, AFLP and SSR markers in plants by a network of European laboratories. Mol Breed 3: 381–390.

Kahler AL, Lay CL (1985) Genetics of electrophoretic variants in the annual sunflower. J Hered 76: 335–340.

Konarev AV, Anisimova IN, Gavrilova VA, Rozhkova VT, Fido R, Tatham AS, Shewry PR (2000) Novel proteinase inhibitors in seeds of sunflower (*Helianthus annuus* L.): polymorphism, inheritance and properties. Theor Appl Genet 100: 82–88.

Kortt AA, Caldwell JB (1990) Low molecular weight albumins from sunflower seed: identification of a methionine-rich albumin. Phytochemistry 29: 2805–2810.

Kolkman, JM, Berry ST, Leon AJ, Slabaugh MB, Tang S, Gao W, Shintani DK, Burke JM, Knapp SJ (2007) Single nucleotide polymorphisms and linkage disequilibrium in sunflower. Genetics 177: 457–468.

Kortt AA, Caldwell JB, Lilley GG, Higgins TJV (1991) Amino acid and cDNA sequences of a methionine-rich 2S protein from sunflower seed (*Helianthus annuus* L.). Eur J Biochem 195: 329–334.

Kusterer B, Rozynek1 B, Brahm1 L, Prüfe1 M, Tzigos1 S, Horn R, Friedt W (2004) Construction of a genetic map and localization of major traits in sunflower (*Helianthus annuus* L.) Helia 27(40): 15–24.

Lai Z, Livingstone K, Zou Y, Church SA, Knapp SJ, Andrews J, Rieseberg LH (2005) Identification and mapping of SNPs from ESTs in sunflower. Theor Appl Genet 111: 1532–1544.

Lander ES, Green P, Abrahamson J, Barlow A, Daly MJ, Lincoln SE, Newburg L (1987) MAPMAKER: An interactive computer package for constructing primary genetic linkage maps of experimental and natural populations. Genomics 1: 174–181.

Langar K, Lorieux M, Desmarais E, Griveau Y, Gentzbittel I, Bervillé A (2003) Combined mapping of DALP and AFLP markers in cultivated sunflower using F9 recombinant inbred lines. Theor Appl Genet 106: 1068–1074.

Lawson WR, Goulter KC, Henry RJ, Kong GA, Kochman JK (1998) Marker-assisted selection for two rust resistance genes in sunflower. Mol Breed 4: 227–234.

Lay C, Gerdes J, Kahler A, Whalen R (1988) Inheritance and linkage of nine electrophoretic variants in *Helianthus annuus* L. In: Proc 12th Int Sunflower Conf, Novi Sad, Yugoslavia, pp 411–445.

Leclercq, P (1966) Une stérilité male utilisable pour la production d'hybrides simples de tournesol. (In French). Ann Amelior Plant 16: 135–144

Lister C, Dean C (1993) Recombinant inbred lines for mapping RFLP and phenotypic markers in *Arabidopsis thaliana*. Plant J 4: 745–750.

Lu YH, Melero-Vara JM, García-Tejada JA (2000) Development of SCAR markers linked to the gene *Or5* conferring resistance to broomrape (*Orobanche cumana* Wallr.) in sunflower. Theor Appl Genet 100: 625–632.

Mestries E, Gentzbittel L, Tourvieille de Labrouhe D, Nicolas P, Vear F (1998) Analysis of quantitative trait loci associated with resistance to *Sclerotinia sclerotiorum* in sunflowers (*Helianthus annuus* L.) using molecular markers. Mol Breed 4: 215–226.

Micic Z, Hahn V, Bauer E, Schon CC, Knapp SJ, Tan, S, Melchinger AE (2004) QTL mapping of Sclerotinia midstalk-rot resistance in sunflower. Theor Appl Genet 109: 1474–1484.

Miesfeld R, Krystal M, Arnheim N (1981) A member of a new repeated sequence family which is conserved throughout eucaryotic evolution is found between the human delta and beta globin genes. Nucl Acids Res 9: 5931–5947.

Miller JF, Fick GN (1997) The genetics of sunflower. In: AA Schneiter (ed) Sunflower Technology and Production. CSSA, Madison, WI, USA, pp 441–449.

Mokrani L, Gentzbittel L, Azanza F, Al-Chaarani G, Sarrafi A (2002) Mapping and analysis of quantitative trait loci for grain oil content and agronomic traits using AFLP and SSR in sunflower (*Helianthus annuus* L.). Theor Appl Genet 106: 149–156.

Morgante M, Olivieri AM (1993) PCR-amplified microsatellites as markers in plant genetics. Plant J 3: 175–182.

Mouzeyar S, Roeckel-Drevet P, Gentzbittel L, Philippon J, Tourvieille de Labrouhe D, Vear F, Nicolas P (1995) RFLP and RAPD mapping of the sunflower *Pl1* locus for resistance to *Plasmopara halstedii* race 1. Theor Appl Genet 91: 733–737.

Peerbolte RP and Peleman J (1996) The CARTISOL sunflower RFLP map (146 loci) extended with 291 AFLP® markers. In: Proc 18th Sunflower Res Forum, Jan 1996, Fargo, ND, USA, Natl Sunflower Assoc, Bismarck, ND, USA, pp 174–178.

Perez-Vich B, Fernández-Martínez JM, Grondona M, Knapp SJ, Berry ST (2002) Stearoyl-ACP and oleoyl-PC desaturase genes cosegregate with quantitative trait loci underlying high stearic and high oleic acid mutant phenotypes in sunflower. Theor Appl Genet 104: 338–349.

Pérez-Vich B, Akhtouch B, Knapp SJ, Leon AJ, Velasco L, Fernández-Martínez JM, Berry ST (2004) Quantitative trait loci for brommrape (*Orobanche cumana* Wallr.) resistance in sunflower. Theor Appl Genet 109: 92–102.

Pérez-Vich B, Berry ST, Velasco L, Fernández-Martínez JM, Gandhi S, Freeman C, Heesacker A, Knapp SJ, Leon AJ (2005) Molecular mapping of nuclear male sterility genes in sunflower. Crop Sci 54: 1851–1857.

Poormohammad Kiani S, Grieu P, Maury P, Hewezi T, Gentzbittel L, Sarrafi A (2007) Genetic variability for physiological traits under drought conditions and differential expression of water stress-associated genes in sunflower (*Helianthus annuus* L.). Theor Appl Genet 114: 193–207.

Quillet MC, Madjidian N, Griveau Y, Serieys H, Tersac M, Lorieux M, Bervillé A (1995) Mapping genetic factors controlling pollen viability in an interspecific cross in *Helianthus* sect. *Helianthus*. Theor Appl Genet 91: 1195–1202.

Reiter RS, Williams JGK, Feldmann K, Rafalski JA, Tingey SV, Scolnik PA (1992) Global and local genome mapping in *Arabidopsis thaliana* by using recombinant inbred lines and random amplified polymorphic DNAs. Proc Natl Acad Sci USA 89: 1477–1481.

Rick CM, Fobes JF (1974) Association of an allozyme with nematode resistance. Rep Tomato Genet Coop No 24: 25.

Rieseberg LH (1996) Homology among RAPD fragments in interspecific comparisons. Mol Ecol 5: 99–105.

Rieseberg L (1991) Phylogenetic and systematic inferences from chloroplast DNA and isozyme variation in *Helianthus* sect. *Helianthus* (*Asteraceae*). Sys Bot 16: 50–76.

Rieseberg L, Seiler G (1990) Molecular evidence and the origin and development of the domesticated sunflower (*Helianthus annuus, Asteraceae*). Econ Bot 44: 79–91.

Rieseberg L, Baird SJE, Desrochers AM (1998) Patterns of mating in wild sunflower hybrid zones. Evolution 52: 713–726.

Roath WW, Miller JF, Gulya TJ (1981) Registration of RHA 801 sunflower germplasm. Crop Sci 21: 479.

Robertson DS (1961) Linkage studies of mutants in maize with pigment deficiencies in endosperm and seedling. Genetics 46: 649–662.

Roeckel-Drevet P, Gagne G, Mouzeyar S, Gentzbittel L, Philippon J, Nicolas P, Tourvieille de Labrouhe D, Vear F (1996) Colocation of downy mildew (*Plasmopara halstedii*) resistance genes in sunflower (*Helianthus annuus* L.). Euphytica 91: 225–228.

Rojas-Barros P, Hu J, Jan CC (2008) Molecular mapping of an apical branching gene of cultivated sunflower (*Helianthus annuus* L.). Theor Appl Genet 117: 19–28.

Schwartz D (1971) Genetic control of alcohol dehydrogenase—a competition model for regulation of gene action. Genetics 67: 411–425.

Senior ML, Heun M (1993) Mapping maize microsatellites and polymerase chain reaction confirmation of the targeted repeats using a CT primer. Genome 36: 884–889.

Serre M, Feingold S, Salaberry T, Leon A, Berry S (2001) The genetic map position of the locus encoding the 2S albumin seed storage proteins in cultivated sunflower (*Helianthus annuus* L.). Euphytica 121: 273–278.

Slabaugh, MB, Yu JK, Tang S, Heesacker A, Hu , Lu G, Bidney D, Han F, Knapp SJ (2003) Haplotyping and mapping a large cluster of downy mildew resistance gene candidates in sunflower using multilocus intron fragment length polymorphisms. Plant Biotechnol J 1: 167–185.

Smith S, Helentjaris T (1996) DNA fingerprinting and plant variety protection. In: AH Paterson (ed) Genome Mapping in Plants. RG Landes, New York, USA, pp 95–110.

Stam P (1993) Construction of integrated genetic linkage maps by means of a new computer package: JoinMap. Plant J 3: 739–744.

Stam P (1995) JoinMap 2.0 deals with all types of plant mapping populations. In: Plant & Anim Genom III Conf, San Diego, CA, January 1995. http://www.intl-pag.org/3/abstracts/47pg3.html. Accessed on October 22, 2009.

Tang S, Yu JK, Slabaugh MB, Shintani DK, Knapp SJ (2002) Simple sequence repeat map of the sunflower genome. Theor Appl Genet 105: 1124–1136.

Tanksley SD, Rick CM (1980) Isozymic gene linkage map of the tomato: Applications in genetics and breeding. Theor Appl Genet 57: 161–170.

Thiel T, Michalek W, Varshney RK, Graner A (2003) Exploiting EST databases for the development and characterization of gene-derived SSR-markers in barley (*Hordeum vulgare* L.). Theor Appl Genet 106: 411–422.

Torres AM (1974a) Sunflower alcohol dehydrogenase: ADH_1 genetics and dissociation-recombination Biochem Genet 11: 17–24.

Torres AM (1974b) An intergenic alcohol dehydrogenase isozyme in sunflowers. Biochem Genet 11: 301–308.

Torres AM (1974c) Genetics of sunflower alcohol dehydrogenase: *Adh$_2$*, Nonlinkage to *Adh$_1$* and *Adh$_1$* early alleles Biochem Genet 12: 385–392.

Torres AM, Diedenhofen U (1976) The genetic control of sunflower seed acid phosphatase. Can J Genet Cytol 18: 709–716.

Van Ooijen JW, Voorrips RE (2001) Joinmap version 3.0: Software for the calculation of genetic linkage maps. CPRO-DLO, Wageningen, The Netherlands.

Vear F, Gentzbittel L, Philippon J, Mouzeyar S, Mestries E, Roeckel-Drevet P, Tourvieille de Labrouhe D, Nicolas P (1997) The genetics of resistance to five races of downy mildew (*Plasmopara halstedii*) in sunflower (*Helianthus annuus* L.). Theor Appl Genet 95: 584–589.

Vonder Haar RA, Allen RD, Cohen EA, Nessler CL, Thomas TL (1988) Organization of the sunflower 11S storage protein gene family. Gene 74: 433–443.

Vos P, Hogers R, Bleeker M, Reijans M, van de Lee T, Hornes M, Frijters A, Pot J, Peleman J Kuiper M, Zabeau M (1995) AFLP: A new technique for DNA fingerprinting. Nucl Acids Res 23: 4407–4414.

Weber JL, May PE (1989) Abundant class of human DNA polymorphisms which can be typed using the polymerase chain reaction. Am J Hum Genet 44: 388–396.

Welsh J, McClelland M (1990) Fingerprinting genomes using PCR with consensus tRNA gene primers. Nucl Acids Res 19: 861–866.

Whitton J, Rieseberg LH, Ungerer MC (1997) Microsatellite loci are not conserved across the Asteraceae. Mol Biol Evol 14: 204–209.

Williams JG, Kubelik AR, Livak KJ, Rafalski JA, Tingey SV (1990) DNA polymorphisms amplified by arbitrary primers are useful as genetic markers. Nucl Acids Res 18: 6531–6535.

Wu KS, Tanksley SD (1993) Abundance, polymorphism and genetic mapping of microsatellites in rice. Mol Gen Genet 24: 225–235.

Xiao J, Li J, Yuan L, McCouch SR, Tanksley SD (1996a) Genetic diversity and its relationship to hybrid performance and heterosis in rice as revealed by PCR-based markers. Theor Appl Genet 92: 637–643.

Yue B, Radi SA, Vick BA, Cai X, Tang S, Knapp SJ, Gulya TJ, Miller JF, Hu J (2008a) Identifying quantitative trait loci for resistance to Sclerotinia head rot in two USDA sunflower germplasms. Phytopathology 98: 926–931.

Yue B, Vick BA, Yuan W, Hu J (2008b) Mapping of one of the 2 genes controlling lemon ray flower color in sunflower (*Helianthus annuus* L.). J Hered 99: 564–567.

Zhang LS, Le Clerc V, Li S, Zhang D (2005) Establishment of an effective set of simple sequence repeat markers for sunflower variety identification and diversity assessment. Can J Bot 83: 66–72.

4

Mapping and Tagging of Simply Inherited Traits

Volker Hahn[1]* and *Silke Wieckhorst*[2]

ABSTRACT

This chapter summarizes the studies about the identification of simply inherited traits in sunflower. The understanding of how a trait is inherited, combined with the identification of molecular markers linked to the trait will lead to an improvement of molecular breeding programs of sunflower. As a lot of work was published regarding disease resistances, this chapter emphasizes on simply inherited disease resistances with a focus on downy mildew and broomrape. Other important simply inherited traits are herbicide resistances, male sterility, seed quality traits and morphological traits.

Keywords: gene mapping; single traits; resistance; male sterility; seed quality; morphological traits

4.1 Introduction

The fundamental goal of genetic linkage mapping is gene discovery. Identification of genes conditioning economic traits, their mode of expression and nature of interactions provide breeders the greatest leverage in manipulating the traits to the desired phenotypes. Although agronomically important traits such as yield or drought resistance are influenced by

[1]Universität Hohenheim, Landessaatzuchtanstalt, Fruwirthstr. 21, 70593 Stuttgart, Germany.
[2]Technische Universität München, Wissenschaftszentrum Weihenstephan für Ernährung, Landnutzung und Umwelt, Lehrstuhl für Pflanzenzüchtung, Am Hochanger 4, 85350 Freising, Germany.
*Corresponding author: *vhahn@uni-hohenheim.de*

multiple genes, many traits exist that are simply inherited, at least at first sight. The genetics of sunflower and the inheritance factors of morphological, resistance, and quality traits are summarized in Miller and Fick (1997).

For gene mapping, a crucial prerequisite is a genetic map. Different genetic maps constructed in sunflower are described in the Chapter 3. However, the simple sequence repeat (SSR) marker-based maps published by Tang et al. (2002) and Yu et al. (2003) are considered as public sunflower reference maps. A great help in further characterization of the mapped loci will be the expressed sequence tags (EST) published in the Compositae Genome Project Database (CGPDB, *http://cgpdb.ucdavis.edu*). The database now contains approximately 93,500 cultivated sunflower EST sequences corresponding to approximately 31,600 unique genes. Available sequence data are being already utilized for a variety of purposes.

This chapter summarizes numerous studies investigating mapping and tagging of simply inherited traits in sunflower. Table 4-1 provides information regarding traits, loci, mapping populations, linkage groups (LG), and the corresponding references. Figure 4-1 depicts the approximate positions of the mapped loci in relation to the SSR markers described in Tang et al. (2002).

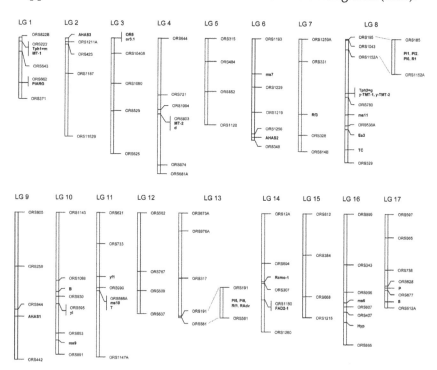

Figure 4-1 The approximate positions of 38 mapped loci governing simply inherited traits in the public sunflower SSR map.

Table 4-1 Mapping of simply inherited traits in sunflower.

Trait	Gene/locus	Linkage group (LG)	Population/s, line/s	References
Resistance to downy mildew	Pl_1	LG 8	GH x RHA266, Ha370 x Ha372	Mouzeyar et al. 1995; Gedil et al. 2001
	Pl_6	LG 8	H52 x HA335	Roeckel-Drevet et al. 1996; Bouzidi et al. 2002
	Pl_2 Pl_2	LG 8	GH x PAC2 CmsHA89 x AS110PL$_2$,	Vear et al. 1997 Brahm et al. 2000
	Pl gene conferring resistance to race 730	LG 8	HA26 S x NIL HA26 R	Pankovic et al. 2007
	Pl_5 Pl_5/Pl_8	LG 13 LG 13	XRQ x PSC8 OCxYSQ, CAY x QIR8	Bert et al. 2001 Radwan et al. 2003, 2004
	Pl_{ARG}	LG 1	CmsHA342 x Arg1575-2	Dussle et al. 2004; Wieckhorst et al. 2008
Resistance gene candidates for resistance to downy mildew	3 RGC loci (NBS-R3)	LG 8	SD x PAC1, CP73 x PAC1, GH x PAC2	Gentzbittel et al. 1998
	Ha-4W2	LG 8	HA370 x HA372	Gedil et al. 2001
	24 HaRGC1 loci	LG 8	Ha370 x Ha372, PHA x PHB, RHA280 x RHA801	Slabaugh et al. 2003
	Ha-NTIR7 Ha-NTIR11, Ha-NTIR2, Ha-NTIR3	LG 13	OC x YSQ, CAY x QIR8	Radwan et al. 2003
	Ha-NTIR11g, Ha-NTIR3A	LG 13	OC x YSQ, CAY x QIR8	Radwan et al. 2004
Resistance to sunflower rust	R_1		CmsHA89 x RHA279	Lawson et al. 1998
	R_1	LG 8	HA370 x HA372, PHA x PHB	Yu et al. 2003
	$R_{Adv,}$		CmsHA89 x RHA279	Lawson et al. 1998
	$R_{Adv,}$	LG 13	HA370 x HA372, PHA x PHB	Yu et al. 2003
Resistance to sunflower chlorotic mottle virus	Rcmo-1	LG 14	L2 x L33	Lenardon et al. 2005
Resistance to Orobanche	Or_5	LG 3	SD x PAC1, RPG01 x AEHC1	Lu et al. 1999
	Or_5		RPG01 x AEHC1, RPG01 x R300V	Lu et al. 2000

Table 4-1 contd....

Table 4-1 contd....

Trait	Gene/locus	Linkage group (LG)	Population/s, line/s	References
	Or_5	LG 3	PHC x PHD	Tang et al. 2003
	Or_5		LR1	Letousey et al. 2007
Herbicide resistance	AHAS3	LG 2	RHA280 x RHA801, NMS373 x ANN1811, (HA425 x HA89) x HA89, IMISUN-2 x ZENB9	Kolkman et al. 2004
	AHAS2	LG 6	see AHAS3	Kolkman et al. 2004
	AHAS1	LG 9	see AHAS3	Kolkman et al. 2004
Pollen fertility restoring genes	Rf1	LG 13	HA89 x RHA266, CX x RHA266, PAC2 x RHA266	Gentzbittel et al. 1995, 1999
	msc1	LG 7	CP73 x PAC1, GH x PAC2	Gentzbittel et al. 1995, 1999
	msc1	LG P	GH x PAC2	Mestries et al. 1998
	Rf1	LG 13	RHA325 x HA342	Horn et al. 2003; Kusterer et al. 2005
	Rf3	LG7	RHA340 x ZENB8	Abratti et al. 2008
Nuclear male sterility	ms6	LG 16	NMS HA89-872 x RHA271	Capatana et al. 2008
	ms7	LG 6	NMS HA89-552 x RHA271	Li et al. 2008
	ms9	LG 10	NMS360 x RHA271	Chen et al. 2006
	ms10	LG 11	NMS801 x 30059	Pérez-Vich et al. 2005
	ms11	LG 8	P21 x P-96, P21 x K-96	Pérez-Vich et al. 2005
	ms11	LG 8	P21 x CAS-14	Pérez-Vich et al. 2006b
Anthocyanin	T	LG 11	NMS373 x ANN1811	Perez-Vich et al. 2005
Branching	b1	LG 10	SD x PAC1, CP73 x PAC1, GH x PAC2, HA89 x RHA266, PAC2 x RHA266	Gentzbittel et al. 1995, 1999
	b1, bbr	LG A	GH x PAC2	Mestries et al. 1998
	Bbr	LG 10	GH x PAC2	Gentzbittel et al. 1999
	B	LG 10	RHA280 x RHA801	Tang et al. 2006a
	b1	LG 10	HA234 x RHA271	Rojas-Barros et al. 2008

Trait	Gene/locus	Linkage group (LG)	Population/s, line/s	References
Lemon ray flower color	*yf1*	LG 11	HA89 x NCCL	Yue et al. 2008a
Chlorophyll deficiency	*yl*	LG 10	YL x NCCL	Yue et al. 2008b
Hull hypodermis pigments	*Hyp*	LG 16	RHA280 x RHA801	Tang et al. 2006a
Hull phytomelanin pigment locus	*P*	LG 17	RHA280 x RHA801	Tang et al. 2006a
Self-incompatibility locus	*S*	LG 17	NMS x ANN1811	Gandhi et al. 2005
High-oleic acid content	delta12 oleate desaturase, delta12 oleate desaturase		HA89, HA341 RHA274, RHA345, BD40713 x BE78079 HA292, HA394	Hongtrakul et al. 1998 Lacombe and Bervillé 2001
	Ol	LG 14	HA89 x CAS3, HAOL9 x CAS3	Perez-Vich et al. 2002
	Ol / FAD2-1	LG 14	diverse high linoleic and high oleic lines	Schuppert et al. 2006
	FAD2-1, FAD2-2, FAD2-3		HA89, HAOL9	Martínez-Rivas et al. 2001
High-stearic acid content	*Es1*	LG 1	HA89 x CAS3, HAOL9 x CAS3	Perez-Vich et al. 2002
	Es3	LG 8	P21 x CAS-14	Pérez-Vich et al. 2006
Fatty acid biosynthesis	Stearate desaturase SAD17 locus A	LG 1	HA89 x CAS20	Pérez-Vich et al. 2004b
	Stearate desaturase SAD6 locus A	LG 11	HA89 x CAS20	S.T. Berry in Pérez-Vich et al. 2004b
	Stearate desaturase SAD6 locus B	LG 1	HA89 x CAS20	Pérez-Vich et al. 2004b
	Stearate desaturase SAD6 locus C	LG 4	HA89 x CAS20	Pérez-Vich et al. 2004b
	FatA thioesterase locus A	LG 1	HA89 x CAS20	Pérez-Vich et al. 2004b
	FatA thioesterase locus B	LG 2	HA89 x CAS20	Pérez-Vich et al. 2004b
	FatB thioesterase locus A	LG 7	HA89 x CAS20	Pérez-Vich et al. 2004b
	Oleate desaturase locus A	LG 1	HA89 x CAS20	Pérez-Vich et al. 2004b
Tocopherol composition	Gamma tocopherol methyltransferase 1 and 2	LG 8	R112 x LG24, RHA280 x RHA801	Hass et al. 2006
	MBPQ/MSBQ-MT	LG 4	R112 x LG24, RHA280 x RHA801	Hass et al. 2006

Table 4-1 contd....

Table 4-1 contd....

Trait	Gene/locus	Linkage group (LG)	Population/s, line/s	References
	Tocopherol cyclase	LG 8	R112 x LG24, RHA280 x RHA801	Hass et al. 2006
	Tocopherol mutant *g*	LG 8	R112 x LG24,	Hass et al. 2006
	Tocopherol mutant *d*	LG 4	R112 x LG24,	Hass et al. 2006
	MBPQ/MSBQ-MT	LG 1	NMS373 x SRA16, RHA280 x RHA801	Tang et al. 2006b
	Tocopherol mutant *m*	LG 1	NMS373 x SRA16	Tang et al. 2006b
	Tph2	LG 8	CAS-12 x IAST-540	Garcia-Moreno et al. 2006
	Tph1	LG 1	CAS12 x T589	Vera-Ruiz et al. 2006

4.2 Disease Resistance

4.2.1 Resistance to Downy Mildew

As resistance to different races of downy mildew belonged to the first-mapped traits in sunflower, mapping of downy mildew resistance genes will be described in more detail. Downy mildew caused by *Plasmopara halstedii* (Farl.) Berl and de Toni is one of the destructive diseases in cultivated sunflower (*Helianthus annuus* L.). Vranceanu and Stoenescu (1970) described the first gene conferring resistance against downy mildew (Pl_1). They identified this gene as dominant and independent. In the following years, further *Pl* genes were discovered from different sources and were incorporated into the breeding material. Some of these *Pl* genes confer resistance against only one race of the pathogen (e.g., Pl_1 against race 100), others are effective against few (e.g., Pl_2) or all known races (e.g., Pl_8 and Pl_{ARG}). In the last few years, new races of *P. halstedii* emerged in the cultivation areas of sunflower (Tourvieille de Labrouhe et al. 2000; Molinero-Ruiz et al. 2002). Moreover, some races developed tolerance to the fungicide metalaxyl (Albourie et al. 1998; Spring et al. 2006), which is the only effective plant protecting agent against this disease. For these reasons, a constant search for new *Pl* genes and the elucidation of the nature of resistance is imperative. The discovery of linkage between resistance genes and molecular markers in order to be able to use these purposefully and durably in plant breeding is of high importance, for example for pyramiding of several resistance genes (Brahm et al. 2000).

Using bulked segregant analysis (BSA; Michelmore et al. 1991), Mouzeyar et al. (1995) mapped Pl_1 as the first *Pl* locus on group 1 of the CARTISOL map (Gentzbittel et al. 1995), which is identical with LG 8 of the public SSR map. They could identify two polymorphic restriction fragment length polymorphism (RFLP) markers (SUN17, SUN124) and screened them against 135 F_2 plants derived from a cross between GH x RHA266 (Pl_1). The

Pl_1 locus was located between SUN017 (5.6 cM) and SUN124 (7.1 cM), thus the first markers were available for marker-assisted selection (MAS), but they were not linked closely enough for molecular cloning of Pl_1. Roeckel-Drevet et al. (1996) and Vear et al. (1997) used BSA in order to map Pl_6 and Pl_2 in the F_2 progeny of H52 x HA335 (Pl_6) and GH x PAC2 (Pl_2). Pl_6 (4.8% recombination) as well as Pl_2 (8.3% recombination) were found to be closely linked with the RFLP marker SUN124, previously mapped closely to Pl_1. Vear et al. (1997) additionally investigated the segregation pattern of F_3 and F_4 progenies of the cross HA335 (Pl_6) x H52 after infection with five downy mildew races (100, 300, 700, 703, 710). The results indicated that Pl_6 could be separated at least into two parts, one giving resistance to race 100 and 300 and the other giving resistance to races 700, 703 and 710. This was the first report that a Pl locus (Pl_6) is not a single independent gene, conferring resistance to many downy mildew races, but rather a cluster of genes, each providing resistance to one, or a few, downy mildew races. Brahm et al. (2000) were in search of markers for the Pl_2 locus and used two sets of near-isogenic lines (AS110/AS110Pl_2, S1358/S1358Pl_2) and bulks of a segregating F_2 population (cmsHA89 x AS110Pl_2) to identify nine random amplified polymorphic DNA (RAPD) and two amplified fragment length polymorphism (AFLP) markers. The most closely linked markers were OPAA14$_{750}$, OPAC20$_{831}$, and E35M48-3 within about 2 cM distance from the Pl_2 locus; all markers were linked in coupling phase. Markers in repulsion phase are also available so that they can be used in combination to select against heterozygous genotypes in breeding programs. The markers for the Pl_2 locus can also be useful for Pl_6, because Pl_2 is a part of the complex Pl_6 locus (Vear et al. 1997; Brahm et al. 2000).

A second linkage group carrying genes controlling resistance to downy mildew was identified with molecular mapping of Pl_5 and segregation studies in testcrosses between different sources of downy mildew resistance (Bert et al. 2001). Pl_5 showed linkage with ten AFLPs, two RFLPs, and $Rf1$, a fertility restorer locus, in the F_2 progeny of XRQ (Pl_5) x PSC8 (Pl_2). The two RFLP markers as well as the $Rf1$ gene were previously mapped on group 6 of the the CARTISOL map (Gentzbittel et al. 1995), which is identical with LG 13 of the public map. In the testcross XRQ (Pl_5) x Ha338 (Pl_7) segregation showed a pattern of two independent genes each conferring resistance to races 100, 710, 703, but for the cross between XRQ (Pl_5) x RHA340 (Pl_8) no segregation occurred. The authors considered that Pl_5 and Pl_8 may be closely linked, but that the two genes are not identical because of their non-identical race resistance patterns (Bert et al. 2001). Radwan et al. (2003) confirmed the position of Pl_5 on group 6 of the RFLP map of Gentzbittel et al. (1999) and mapped Pl_8 in addition on the same linkage group.

Vear et al. (2003) investigated the segregation pattern of progenies, which originated from different crosses between lines with different Pl loci [HA335

(Pl_6), HA338 (Pl_7), RHA340 (Pl_8), YDQ (Pl_6), XRQ (Pl_5), and susceptible line], and RHA419 that obtained its downy mildew resistance from *H. argophyllus* (Arg1575-2). Their results indicate that RHA419 carries a single gene or a cluster of resistance genes to many downy mildew races and is independent of the Pl_6/Pl_7 and Pl_5/Pl_8 clusters. Dussle et al. (2004) confirmed this finding by identifying 12 polymorphic SSRs with BSA in the F_2 progenies of the cross cmsHA342 x Arg1575-2 (Pl_{ARG}). The 12 SSR markers and Pl_{ARG} were mapped on LG 1.

A large number of resistance (*R*) genes from different plant species were cloned and analyzed in the last few years (Gebhardt 1997; Michelmore 2000). The comparison of the sequence revealed that *R*-genes contain conserved domains. Two groups of NBS-LRR (nucleotide-binding site—leucine-rich repeats) resistance gene candidates have been described in plants: the TIR-NBS-LRR subclass (the TIR domain is homologous to the *Drosophila Toll* and mammalian *Interleukin-1* receptor) and the non-TIR-NBS-LRR subclass. The resistance gene analog (RGA, also termed resistance gene candidate—RGC) approach allows the search of new *Pl* resistance genes in wild sunflower species. It also enables the analysis of resistance mechanisms against *P. halstedii* and the development of molecular markers derived from the gene itself or from closely linked sequences as a starting point for map-based cloning or MAS.

Gentzbittel et al. (1998) first used an RGA approach to analyze the resistance of sunflower to downy mildew. Degenerate primers were designed from the conserved NBS domain of the virus resistance gene *N* from tobacco (Whitham et al. 1994) and the rust resistance gene *L6* from flax (Lawrence et al. 1995). After cloning and sequencing, one RGA (NBS-R3), belonging to the TIR-NBS-LRR group, was mapped in three populations in the region of the *Pl* loci on group 1 of the CARTISOL map (Gentzbittel et al. 1995).

Bouzidi et al. (2002) conducted further analysis on the complexity of the Pl_6 locus in order to develop molecular markers suitable for positional cloning of *Pl* genes. To clone a full-length cDNA RGA, the sequence of the RGA product HA-NBS3 (Gentzbittel et al. 1998) was used as a template in RACE-PCR to obtain 5' and 3' ends of the cDNA. By using the sequence of the full-length RGA (GenBANK accession number AF316405) specific primers were designed and tested for polymorphism using BSA. Thirteen dominant sequence tagged site (STS) markers were obtained covering a genetic distance of about 3 cM centered on the Pl_6 locus. All 13 STS markers showed homology to the TIR-NBS-LRR subclass. Pankovic et al. (2007) deployed the three specific primer pairs HAP1, HAP2, HAP3 of Bouzidi et al. (2002) for screening a near-isogenic line (NIL) population [HA26 S x NIL HA26 R (Pl_6)]. The fragment HAP3/S1 (resembling Ha-NBS 12S; Bouzidi et al. 2002) was the only one, which was observed for both resistant and susceptible parental line. Based on this RGA, two co-dominant cleaved amplified polymorphic site (CAPS) markers were developed, which completely co-segregated with the *Pl* gene conferring resistance to race 730.

Hewezi et al. (2006) aimed to clarify the function of the RGA *PLFOR48* using a transgenic loss-of-function approach with expression of antisense cDNA in RHA266 (Pl_1) sunflower line. *PLFOR48* is a full-length RGA obtained with RACE-PCR from the clone HA-NBSR3 of Gentzbittel et al. (1998). The transgenic sunflower lines showed developmental abnormalities and thus could not be used for infection studies with *P. halstedii*. The sequence of *PLFOR48* showed homology to sequences in *Nicotiana tabacum* var. Samsun NN, thus transgenic tobacco was developed with the same antisense cDNA and the susceptibility to *Phytophthora parasitica* was investigated. A higher susceptibility was observed in plants, which showed higher transgene expression. The authors concluded that *PLFOR48* seems to play a role in both pathogen resistance and normal plant development. Genetic transformation of susceptible sunflower lines with the *PLFOR48* sense construct is now being considered as an additional approach towards demonstrating the functional role of this gene in downy mildew resistance.

Gedil et al. (2001) used degenerate oligonucleotide primers (Leister et al. 1996; Yu et al. 1996) targeted to conserved amino acid sequences in known NBS-LRR genes to amplify RGA fragments from sunflower genomic DNA of HA89. PCR products were cloned, sequenced and analyzed by single-strand conformation polymorphism (SSCP) gel electrophoresis to identify redundant clones. Six RGA probes (RFLP) detected fragments that were polymorphic between HA370 (Pl_1) and HA372 (pl_1) and mapped to three linkage groups on the HA370 x HA372 map. Four RGAs belonging to the non-TIR-NBS-LRR subclass were mapped on LG 13 (HR-1W23, HR-1B39, HR-1W41, HR-1W53) and the two others belonging to the TIR-NBS-LRR subclass were mapped on LG 8 (HR-4W2) and 15 (HR-1W22). Gedil et al. (2001) developed a CAPS marker for Ha-4W2 on LG 8, which was linked but did not completely co-segregate with Pl_1. According to Ha-4W2, Slabaugh et al. (2003) demonstrated a multigene family of HaRGC1 (= Ha-4W2; Gedil et al. 2001) in the Pl_1-Pl_2-Pl_6 region, all belonging to the TIR-NBS-LRR subclass and were assessed by multilocus intron fragment length polymorphism (IFLP). Twenty-four HaRGC1s were mapped in three populations spanning 2-4 cM on LG 8 and three SSRs (ORS166, ORS299, ORS1043) were identified segregating within this cluster. This was further evidence that the Pl_1-Pl_2-Pl_6 region is a large cluster of resistance genes. In addition, the method of IFLP fingerprinting is suitable to search for new downy mildew resistance genes from wild sunflowers. Altogether, 63 germplasm accessions (elite inbred lines, partially isogenic inbred lines, open-pollinated populations, Native American land races, wild *H. annuus* populations) were screened with the specific markers for HaRGC1 observing 48 unique markers. Only half of the total HaRGC1 copies were present in domesticated germplasm, demonstrating that wild sunflowers represent a rich source of untapped resistance genes.

Radwan et al. (2003) concentrated on the non-TIR-NBS-LRR RGAs to test whether RGA markers of this subclass of plant resistance genes could be linked to Pl loci, which segregate independently of Pl_6. They used one degenerate (Leister et al. 1996; Yu et al. 1996) and four specific primer pairs, which were designed using the sunflower RGA sequences of the four non-TIR-NBS-LRR on LG 13 (Gedil et al. 2001). The amplification products were cloned, sequenced and used as RFLP probes to group RGAs with identical RFLP profiles in the same class. A total of 16 RGAs were identified and distinguished into six classes of RGA. Two of these classes correspond to the TIR-NBS-LRR type while the remaining four classes correspond to the non-TIR-NBS-LRR type of resistance genes. BSA was used to identify polymorphic RGA fragments (Ha-NTIR7, Ha-NTIR11, Ha-NTIR2, Ha-NTIR3) and those were scored on 150 F_2 plants of each cross. All the non-TIR-LRR RGAs were found to be linked to Pl_5 and Pl_8. Thus, Gentzbittel et al. (1998), Gedil et al. (2001), Bouzidi et al. (2002) and Radwan et al. (2003) provided evidence that TIR-NBS-LRR sequences as well as non-TIR-NBS-LRR sequences exist in the sunflower genome.

In order to develop markers for MAS, Radwan et al. (2004) used the partial sequence of two RGAs, Ha-NTIR11 and Na-NTIR3 (Radwan et al. 2003), to clone and map two full-length RGAs. The two sequences were used to develop 14 STS markers, which are clustered within a genetic distance of about 13.6 cM and 16.7 cM, for the Pl_5 and Pl_8 loci, respectively. Furthermore, Radwan et al. (2005) investigated the transcriptional regulation of the full-length RGA Ha-NTIR11g (Radwan et al. 2004). They compared the expression level of Ha-NTIR11g at different times in resistant (QIR, Pl_8) and susceptible (CAY) lines without infection, after infection with *P. halstedii*, after treatment with signaling molecules such as salicylic acid, methyl-jasmonic acid, hydrogen peroxide, and after wounding. The results show that HaNTI11g is constitutively expressed at a low level in the healthy hypocotyls and cotyledons of the resistant genotype, but not in the susceptible genotype. The expression level of Ha-NTIR11g increased in the resistant genotype only after infection with *P. halstedii*, but not after wounding or treatment with signaling molecules. Thus, a relation between infections with *P. halstedii* and the expression of Ha-NTIR1g1 seems to exist.

Recently, Radwan et al. (2008) identified 630 NBS-LRR homologs in sunflower by database analysis and sequencing of DNA fragments spanning conserved NBS sequences, which were isolated from common and wild sunflower species. They developed DNA markers from 196 unique NBS-LRR sequences and mapped 167 NBS-LRR loci. NBS-LRR loci were distributed in cluster or singleton throughout the sunflower genome. On LG 8, in the region of the downy mildew resistance loci Pl_1, Pl_2, Pl_6, the largest and most complex cluster were identified and on LG 13 (Pl_5, Pl_8) the second largest NBS-LRR cluster with TIR-NBS and non-TIR-NBS-LRR loci were found. Thus, Radwan et al. (2008) allocated a source for identification and cloning of *R*-genes like Pl_8 and R_{ADV} (against rust). Furthermore, they also showed the importance of wild species as a source of *R*-genes in sunflower, because RGAs identified from wild species ESTs were linked to several important *R*-genes like Pl_{ARG}, OR_5, Pl_8, R_1, R_{Adv} (see Sections 4.2.2 and 4.2.4 for details).

4.2.2 Rust Resistance

Sunflower rust caused by the fungus *Puccinia helianthi* Schw. is a major disease of sunflower that has been attributed to causing significant yield losses in oilseed and confectionary crops grown in all regions of the world. As gene pyramiding through traditional plant breeding can be a cumbersome task because the presence of new genes introduced into the common background is masked by genes already present, the use of genetic markers is necessary for pyramiding of genes. Lawson et al. (1998) developed two SCAR markers for the rust resistance genes R_1 and R_{Adv} through conversions of polymorphisms identified by RAPD-PCR and BSA. SCAR marker *SCTO6151* was tightly linked to the R_1 gene for resistance to rust race 1 and was mapped to LG 8 close to a cluster of NBS-LRR resistance gene candidates and the Pl_1 gene for resistance to downy mildew by Yu et al. (2003). SCAR marker *SCX20600* is linked to the R_{Adv} gene and was mapped to LG 13 distal to the fertility restorer locus Rf_1 (Yu et al. 2003).

4.2.3 Resistance to Sunflower Chlorotic Mottle Virus

Sunflower chlorotic mottle virus (SuCMoV), a member of the Potyvirus genus within the Potyviridae family is one of the most widely distributed potyviruses on cultivated and wild sunflowers in Argentina. SuCMoV infection reduces yield parameters such as plant height, stem and capitulum diameter, seed yield, and weight of 1,000 seeds significantly (Lenardon et al. 2001). From 232 accessions tested, the line L33 exhibited the highest level of resistance, characterized by an isolated chlorotic pinpoint which resembles hypersensitive reactions, accompanied by a delay in the beginning of viral replication. Linkage analysis in the F_2 plants from the cross L2 x L33 detected a single locus (*Rcmo-1*) for resistance located on LG 14, flanked by the markers MS0022 (5 cM) and ORS307 (4 cM) (Lenardon et al. 2005).

4.2.4 Broomrape Resistance

Broomrape (*Orobanche cernua* Loelf.) is a parasitic plant that feeds on sunflower roots, causing severe economic damage. The first reports of this parasite attacking sunflower were in Russia, and the attack of broomrape intensified at the beginning of the 20th century when all cultivars grown suffered large scale losses (Miller and Fick 1997). Broomrape is a serious pest of cultivated sunflower in Europe, especially southern Europe, the Balkans, and the Mediterranean (Parker and Riches 1993).

Because *O. cumana* has a broad host range and produces an extraordinarily large number of small, long-lived, facilely dispersed seeds, control through crop management has been difficult. The most economical and effective means of controlling sunflower broomrape is by growing resistant cultivars (Sukno et al. 1999; Lu et al. 2000). While various genetically simple and complex sources of *Orobanche* resistance have been described in sunflower, the most important and widely used are single dominant genes (for references see Tang et al. 2003).

As a first report on molecular markers linked to a broomrape resistance gene, Lu et al. (2000) identified five sequence characterized amplified region (SCAR) markers and one RAPD marker linked to the gene Or_5 which confers resistance to race E of broomrape. Their investigation based on two F_2 populations with the resistant parent line RPG01. As a next step Lu et al. (1999) could integrate their markers in group 17 of the CARTISOL RFLP map published by Gentzbittel et al. (1995). Tang et al. (2003), using BSA, placed the Or_5 gene in a telomeric region of LG 3 with the closest SSR marker (ORS1036) being 6.2 cM proximal to the Or_5 locus. In a QTL study for broomrape resistance Pérez-Vich et al. (2004a) identified one major QTL (*or3.1*) affecting race E resistance that was stable over different experiments carried out in different environments and with different race E broomrape populations. The map position of the QTL *or3.1* detected on top of LG 3 corresponds to the region where Or_5 was mapped by Tang et al. (2003).

A study with molecular, biochemical, proteomic and histochemical approaches to investigate the resistance mechanisms of sunflower inbred line LR1 was published by Letousey et al. (2007). LR1 exhibits strong polygenic resistance occurring at different stages of *O. cumana* development. The resistance of line LR was mainly characterized by *O. cumana* necrosis, which took place soon after attachment. The results of this study suggested that the resistant genotype LR1 reacts by strong induction of salicylic acid and jasmonate pathways and a later reaction was the induction of ethylene pathways. Two pathogenesis-related (PR) genes (*chit.* and *def.*) were also induced during the later stages of the interaction. However, just before the first *O. cumana* necrosis, resistant roots exhibited a transient quasi-total repression of most of the studied genes. Only *HaDef1* showed a strong induction. An explanation of this phenomenon could be the setting up of a mechanism of regulation that organized the overall response towards a more specific and effective pathway (Letousey et al. 2007). *HaDef1*, originally named *def.* by Hu et al. (2003) has been described as a leaf molecular marker of the salicylic pathway (Zélicourt et al. 2007). In the study of Zélicourt et al. (2007), a recombinant peptide corresponding to the Ha-DEF1 defensin domain was produced in an *E. coli* expression system. Ha-DEF1 induced necrotic symptoms on *Orobanche* seedlings that should be correlated with the necrosis occurring in *Orobanche* tubercles attached to the root of a resistant sunflower. Further studies have to show whether there is a link between one of the broomrape resistant genes and *HaDef1*.

4.3 Herbicide Resistance

Sulfonylurea (SU) and imidazolinone (IMI) herbicides have been widely used to control sunflowers in corn, soybean, and other crop rotations and have been selected for herbicide resistance in wild sunflowers (Al-Khatib et al. 1998, 1999; White et al. 2002, 2003). SU and IMI herbicides are specific inhibitors of acetohydroxyacid synthase (AHAS, EC 2.2.1.6). Species differ in herbicide susceptibility and can develop resistance to different classes of AHAS inhibitors. With few exceptions, resistances to AHAS-inhibiting herbicides, in otherwise susceptible species, are caused by point mutations in genes encoding AHAS that reduce the sensitivity of the enzyme to herbicide inhibition. Genes for resistance to AHAS-inhibiting herbicides in sunflower have been introgressed from resistant wild populations into elite inbred lines to develop and deploy herbicide resistant cultivars and hybrids (for references see Kolkmann et al. 2004). By searching the sunflower EST database (*http://compgenomics.ucdavis.edu*, *Kozik et al. 2002*) 11 sunflower *AHAS* ESTs were found. When the DNA sequences of various amplicons were aligned, three paralogous *AHAS* genes were discovered and named *AHAS1*, *AHAS2*, and *AHAS3*. SNP markers were developed for *AHAS1* and *AHAS3* and SSCP markers were developed for a six base-pair INDEL in *AHAS2* and a G/A SNP in *AHAS3*. Additionally, an SSR marker was developed for *AHAS1* based on the poly-Thr repeat in the putative transit peptide of *AHAS1*. *AHAS1* and *AHAS3* were genotyped and genetically mapped in population RHA280 x RHA801 and *AHAS2* was mapped in NMS373 x ANN1811. The three loci mapped to LG 2 (*AHAS3*), LG 6 (*AHAS2*), and LG 9 (*AHAS1*) (Kolkmann et al. 2004) of the public sunflower map. The loci were flanked by mapped SSR or INDEL markers (Tang et al. 2002; Yu et al. 2003). In the study of Kolkman et al. (2004), DNA polymorphisms were not found between herbicide-susceptible and -resistant inbred lines in the *AHAS2* and *AHAS3* coding sequences but two mutations in the sunflower *AHAS1* gene were identified, an Ala205Val mutation and a Pro197Leu mutation, conferring resistance to AHAS-inhibiting herbicides. Pro197 and Ala205 are conserved amino acids in AHAS enzymes in numerous species (Tranel and Wright 2002). The mutation of Pro197 is one of the most common mutations found in plants resistant to AHAS-inhibiting herbicides but mutation of Ala205 in inhibitor-resistant plants had thus far only been reported in cocklebur, *Arabidopsis*, and sunflower (Kolkman et al. 2004).

4.4 Male Sterility

4.4.1 Restoration of Cytoplasmic Male Sterility

Cytoplasmic male sterility (CMS) is characterized by the inability of a plant to produce viable pollen while being female fertile. CMS is caused by a mutation in the mitochondrial genome. In sunflower, the first reliable CMS system (PET1) was reported by Leclercq (1969) in descendants of an interspecific hybrid between *Helianthus petiolaris* Nutt. and the cultivated sunflower. In the meantime, other sources of CMS have been found. Fertility restoration of dominant nuclear genes is essential for hybrid breeding based on CMS to obtain high yields of seeds. For PET1, a large number of fertility restoration lines have been developed by plant breeders. One to four dominant restorer genes have been described depending on the material (Serieys 1996). However, in most of the elite sunflower lines, the two dominant nuclear genes *Rf1* and *Rf2* are responsible for fertility restoration (Leclercq 1984). As *Rf2* is present in nearly all inbred lines, including maintainers of CMS, the *Rf1* gene is most important for sunflower hybrid breeding.

The *Rf1* gene was mapped by Gentzbittel et al. (1995, 1999) on group 6 of their RFLP and consensus map. Horn et al. (2003) constructed a linkage map around the *Rf1* gene using populations from the cross RHA325 x HA342. They identified RAPD and AFLP markers closely linked to the restorer gene *Rf1* and converted two RAPD markers into SCAR markers. Kusterer et al. (2005) saturated the chromosomal region of *Rf1* with PCR marker. They also used SSR markers and mapped marker ORS1030 on this LG. ORS1030 was mapped on LG 13 on the public SSR map of sunflower.

Gentzbittel et al. (1999) mapped another fertility restoration locus of the PET1 cytoplasm (*Msc1*) on group 12 of their map. This linkage group corresponds to LG 7 of the public map. Recently, Abratti et al. (2008) also mapped a novel fertility restoration gene (*Rf3*) for the PET1 cytoplasm on LG 7. *Rf3* was found in the public inbred line RHA340, which was derived from a cross between *H. argophyllus* and HA89 as recurrent parent.

4.4.2 Nuclear Male Sterility

Nuclear male sterility (NMS) in sunflower was first discovered by Kuptsov in 1934 (Gundaev 1971). Since then numeral sources of NMS have been reported (see Miller and Fick 1997). Up to now, 11 nuclear male sterility genes are published (Pérez-Vich et al. 2005; Cheng et al. 2006). *MS6*, *MS7*, *MS9*, *MS10*, and *MS11* have been placed on the genetic map of sunflower.

The lines NMS HA89-872 and NMS HA89-552 each carry a single recessive male sterile gene *ms6* and *ms7*. Capatana et al. (2008) mapped the *ms6* locus to LG 16, which was flanked by the SSR markers ORS807 and ORS996 at a distance of 7.2 and 18.5 cM, respectively. *Ms7* was mapped by Li et al. (2008) to LG 6 of the public SSR genetic map. The NMS mutant NMS360, induced by streptomycin from line HA89, carries the single recessive gene *ms9*. This gene was mapped to LG 10 of the public sunflower SSR linkage map (Chen et al. 2006). *Ms10* and the tightly linked anthocyanin pigment locus *T* were mapped to LG 11 (Pérez-Vich et al. 2005). This male sterility source was first described by Leclercq (1966). Line P21, carrying *ms11* was released by the USDA and the Texas Agricultural Experiment Station in 1970 (Jan 1992). The *ms11* gene was mapped to LG 8 (Pérez-Vich et al. 2005, 2006b).

4.5 Morphological Traits

4.5.1 *Branching*

The inheritance of branching is complex with both dominant and recessive genes controlling this characteristic (see Miller and Fick 1997). For hybrid breeding branching of restorer lines to extend the period of pollen production is desired. As the hybrids have to be single-headed, only recessive branching genes are useful. Inheritance of recessive branching was first reported by Putt (1964) in which a single recessive gene, *b1*, controlled branching. The *b1* locus was mapped by several groups. Gentzbittel et al. (1995, 1999) placed *b1* on group 7 of their consensus map. Mestries et al. (1998) found *b1* to be positioned on linkage group A, and Tang et al. (2006a) mapped *b1* between the markers ORS1088 and ORS930 on LG 10 of the public SSR map. Rojas-Barros et al. (2008) developed target region amplification polymorphism (TRAP) markers TBr4$_{720}$ and TBr8$_{555}$, which were linked to the *b1* gene in the coupling phase at 0.5 cM and markers Tbr20-297 and Tbr20-494 flanked the *b1* locus in the repulsion phase at genetic distances of 7.5 and 2.5 cM, respectively. They mapped *b1* to group 16 of the RFLP map developed by Jan et al. (1998), which also corresponds to LG 10 of the public SSR map. Interestingly, *b1* is linked to several QTLs for seed traits including seed oil concentration, 1,000-seed weight, seed length, and others.

4.5.2 *Leaf and Ray Color*

Nearly 100% hybrids produced with the cytoplasmic male sterility system are used for the commercial sunflower cultivation. Yue et al. (2008a) and Yue et al. (2008b) provide with the genetic characterization and mapping of the traits lemon ray flower color and chlorophyll deficiency, two morphological markers for seed purification identification, and quality control in sunflower hybrid seed production. Yue et al. (2008a) showed that the lemon ray flower color was controlled by two recessive genes and mapped one of the genes *yf1* to LG 11 of the public SSR map. Yue et al. (2008b) identified and mapped a major gene (*yl*) conferring chlorophyll deficiency, which generated a yellow leaf phenotype. TRAP and SSR markers were used to map the recessive gene using BSA to LG 10 of the public SSR map. Furthermore, both traits could be useful to understand the genetic factors, which control the carotenoid biosynthesis and the process of chlorophyll metabolism, respectively.

4.6 Seed Quality

4.6.1 Oil Quality

Sunflower seed oil is a naturally rich source of oleic and linoleic acid, accounting for about 900 g/kg of oil. Linoleic acid contents range from about 600–800 g/kg and oleic acid contents range from 150–350 g/kg (Miller and Fick 1997). Soldatov (1976) developed Pervenets, a high oleic acid sunflower germplasm, by treating seeds with dimethyl sulfate and selecting among progeny with elevated oleic acid contents. In the last years, the cultivation of mid oleic or high oleic sunflower cultivars increased year by year in most countries. The inheritance of the high oleic acid phenotype in sunflower has been widely studied with crosses between different high oleic stocks originating from Pervenets and different low oleic sources. The oleic acid increase in Pervenets was subsequently discovered to be partly caused by an incompletely dominant mutation (*Ol*). *Ol* greatly increases oleic acid concentrations in sunflower seed oils, produces discrete (non-overlapping) or non-discrete (overlapping) oleic acid classes in different genetic backgrounds and is necessary but not always sufficient for producing high oleic acid concentrations (for references see Schuppert et al. 2006). The mutant *Ol* allele identified through phenotypic analyses seems to be a duplicated *FAD2-1* (oleoyl-phosphatidyl choline desaturase) allele identified through molecular analyses (Hongtrakul 1998; Lacombe and Bervillé 2001; Lacombe et al. 2002; Pérez-Vich et al. 2002; Schuppert et al. 2003, 2006). *FAD2* is necessary for the synthesis of linoleic from oleic acid (Mekhedov et al. 2000). *FAD2-1* has been mapped to LG 14 (Lacombe and Berville 2001; Pérez-Vich et al. 2002; Schuppert et al. 2003, 2006).

In addition to high oleic sunflower lines, mutant lines with a high stearic acid content (>25%) in their seed oil have been developed. This high stearic acid oil is demanded for producing a wide range of margarine and spread products (Pérez-Vich et al. 2004b). Two high stearic acid lines have been characterized in more detail: CAS-3 exhibiting a high stearic acid content of 26% (Osorio et al. 1995) and CAS-14 showing a very high stearic acid content of 37% but strongly influenced by the temperature during seed maturation (Fernández-Moya et al. 2002). Genetic studies demonstrated the presence of partially recessive alleles at two independent loci *Es1* and *Es2* in CAS-3 (Pérez-Vich et al. 1999). The high stearic acid trait of CAS-14 was shown to be controlled by a single recessive gene, designated *es3* (Pérez-Vich et al. 2006a). Candidate-gene and QTL analyses demonstrated that *Es1* co-segregated with a stearoyl-acyl carrier protein (ACP) desaturase locus (SAD17A) located on LG 1 of the sunflower genetic map (Perez-Vich et al. 2002). The *es3* gene was mapped on LG 8, closely linked to *Ms11*, a gene determining nuclear male sterility (Perez-Vich et al. 2006b). Fernández-Moya et al. (2003) reported that *es3* might be associated to a regulatory thermosensitive element that determines a lower expression of the stearate desaturase at high temperatures. Further candidate genes putatively involved in stearic acid biosynthesis were mapped by Pérez-Vich et al. (2004b). Two secondary SAD6 loci were mapped to LG 1 and LG 4, the primary locus has been found to map on LG 11. One FatB fragment was mapped to LG 7 and FatA gene probes (FatA-A, FatA-B) were mapped to LG 1, linked to SAD17, and LG 2, respectively. In addition an oleate desaturase-like gene was mapped on LG 1.

4.6.2 *Tocopherol Composition*

As an important nutrient for human beings, vitamin E has been well known for its antioxidative properties (Kamal-Eldin and Appleqvist 1996). It consists of tocopherols and tocotrienols. Alpha-tocopherol exerts the most active biological activity (Traber and Sies 1996) but shows the weakest antioxidant potency in vitro. Beta-, gamma-, and delta-tocopherol possess a lower vitamin E value, but have considerably greater in vitro antioxidant potency than alpha-tocopherol (Pongracz et al. 1995). Sunflower normally produces >90% alpha-tocopherol (Sheppard et al. 1993). However, three loci ($m = Tph1$, $g = Tph2$, and d) are known to disrupt the synthesis of alpha-tocopherol and produce a broad spectrum of off-type tocopherol profiles in sunflower seeds (Demurin 1993; Demurin et al. 1996; Hass et al. 2006; Tang et al. 2006b). The *Tph1* gene, conferring increased beta-tocopherol content to sunflower seeds, has been mapped to the upper end of LG 1 (Tang et al. 2006b) and co-segregated with the SSR markers ORS1093, ORS222 and ORS598 (Vera-Ruiz et al. 2006). *Tph2* was mapped to LG 8 (Hass et al. 2003), about 30 cM down-stream from the upper end of this LG (García Moreno et al. 2006; Hass et al. 2006). This recessive gene affects a high gamma-tocopherol content. The *d* mutation partially disrupts the synthesis of alpha-tocopherol and causes an accumulation of beta-tocopherol. The *d* gene was tightly linked to ORS676 and mapped to LG 4 (Tang et al. 2006b). Further analyses of the genes of the tocopherol biosynthetic pathway and the functions of the *m*, *g*, and *d* mutations have been undertaken (Hass et al. 2006; Tang et al. 2006b), but the results go beyond the scope of this chapter.

References

Abratti G, Bazzalo ME, León A (2008) Mapping a novel fertility restoration gene in sunflower. Proc 17th Int Sunflower Conf, vol 2, Córdoba, Spain, pp 617–621.

Albourie JM, Tourvieille J, Tourvieille de Labrouhe D (1998) Resistance to metalaxyl in isolates of the sunflower pathogen *Plasmopara halstedii*. Eur J Plant Pathol 104: 235–242.

Al-Khatib K, Baumgartner JR, Peterson DE, Currie RS (1998) Imazethapyr resistance in common sunflower (*Helianthus annuus*). Weed Sci 46: 403–407.

Al-Khatib K, Baumgartner JR, Currie RS (1999) Survey of common sunflower (*Helianthus annuus*) resistance to ALS-inhibiting herbicides in northeast Kansas. In: Proc 21st Sunflower Res Workshop, Natl Sunflower Assoc, Bismark, ND, USA, pp 210–215.

Bert PF, Tourvieille de Labrouhe D, Philippon J, Mouzeyar S, Jouan I, Nicolas P, Vear F (2001) Identification of a second linkage group carrying genes controlling resistance to downy mildew (*Plasmopara halstedii*) in sunflower (*Helianthus annuus* L.). Theor Appl Genet 103: 992–997.

Bouzidi MF, Badaoui S, Cambon F, Vear F, Tourvieille de Labrouhe D, Nicolas P, Mouzeyar S (2002) Molecular analysis of a major locus for resistance to downy mildew in sunflower with specific PCR-based markers. Theor Appl Genet 104: 592–600.

Brahm L, Röcher T, Friedt W (2000) PCR-based markers facilitating marker assisted selection in sunflower for resistance to downy mildew. Crop Sci 40: 676–682.

Capatana A, Feng J, Vick BA, Duca M, Jan CC (2008) Molecular mapping of a new induced gene for nuclear male sterility in sunflower (*Helianthus annuus* L.). Proc 17th Int Sunflower Conf, vol 2, Córdoba, Spain, pp 641–644.

Chen J, Hu J, Vick BA, Jan CC (2006) Molecular mapping of a nuclear male-sterility gene in sunflower (*Helianthus annuus* L.) using TRAP and SSR markers. Theor Appl Genet 113:122–127.

Demurin Y (1993) Genetic variability of tocopherol composition in sunflower seeds. Helia 16: 59–62.

Demurin Y, Skoric D, Karlovic D (1996) Genetic variability of tocopherol composition in sunflower seeds as a basis of breeding for improved oil quality. Plant Breed 115: 33–36.

Dussle CM, Hahn V, Knapp SJ, Bauer E (2004) Pl_{Arg} from *Helianthus argophyllus* is unlinked to other known downy mildew resistance genes in sunflower. Theor Appl Genet 109: 1083–1086.

Fernández-Moya V, Martínez-Force E, Garcés R (2002) Temperature effect on a high stearic acid sunflower mutant. Phytochemistry 59: 33–37.

Fernández–Moya V, Martínez-Force E, Garcés R (2003) Temperature-related non-homogeneous fatty acid desaturation in sunflower (*Helianthus annuus* L.) seeds. Planta 216: 834–840.

Gandhi SD, Heesacker AF, Freemann CA, Argyris J, Bradford K, Knapp SJ (2005) The self-incompatibility locus (S) and quantitative trait loci for self-pollination and seed dormancy in sunflower. Theor Appl Genet 111: 619–629.

García-Moreno MJ, Vera-Ruiz EM, Fernández-Martínez JM, Velasco L, Pérez-Vich B (2006) Genetic and molecular analysis of high Gamma-tocopherol content in sunflower. Crop Sci 46: 2015–2021.

Gebhardt C (1997) Plant genes for pathogen resistance—variation on a theme. Trends Plant Sci 2: 243–244.

Gedil MA, Slabaugh MB, Berry S, Johnson R, Michelmore R, Miller J, Gulya T, Knapp SJ (2001) Candidate disease resistance genes in sunflower cloned using conserved nucleotide-binding site motifs: Genetic mapping and linkage to the downy mildew resistance gene *Pl1*. Genome 44: 205–212.

Gentzbittel L, Vear F, Zhang YX, Bervillé A, Nicolas P (1995) Development of a consensus linkage RFLP map of cultivated sunflower (*Helianthus annuus* L.). Theor Appl Genet 90: 1079–1086.

Gentzbittel L, Mouzeyar S, Badaoui S, Mestries E, Vear F, Tourvieille de Labrouhe D, Nicolas P (1998) Cloning of molecular markers for disease resistance in sunflower, *Helianthus annuus* L. Theor Appl Genet 96: 519–525.

Gentzbittel L, Mestries E, Mouzeyar S, Mazeyrat F, Badaoui S, Vear F, Tourvieille de Labrouhe D, Nicolas P (1999) A composite map of expressed sequences and phenotypic traits of the sunflower (*Helianthus annuus* L.) genome. Theor Appl Genet 99: 218–234.

Gundaev AI (1971) Basic principles of sunflower selection. In: Genetic Principles of Plant Selection. (In Russian). Nauka, Moscow, USSR, pp 417–465.

Hass CG, Leonard SW, Miller JF, Slabaugh MB, Traber MG, Knapp SJ (2003) Genetics of tocopherol (Vitamin E) composition of mutants in sunflower. In: Plant & Anim Genom XI Conf, San Diego, CA, USA, 11–15 Jan 2003: *http://www.intl-pag.org/11/abstracts/P7b P821 XI.html*: Accessed September 14, 2009.

Hass CG, Tang S, Leonard S, Traber MG, Miller JF, Knapp SJ (2006) Three non-allelic epistatically interacting methyltransferase mutations produce novel tocopherol (vitamin E) profiles in sunflower. Theor Appl Genet 113: 767–782.

Hewezi T, Mouzeyar S, Thion L, Rickauer M, Alibert G, Nicolas P, Kallerhoff J (2006) Antisense expression of a NBS-LRR sequence in sunflower (*Helianthus annus* L.) and tobacco (*Nicotiana tabacum* L.): Evidence for a dual role in plant development and fungal resistance. Transgen Res 15: 165–180.

Hongtrakul V, Slabaugh MB, Knapp SJ (1998) A seed specific D12 oleate desaturase gene is duplicated, rearranged, and weakly expressed in high oleic acid sunflower lines. Crop Sci 38: 1245–1249.

Horn R, Kusterer B, Lazarescu E, Prüfe M, Friedt W (2003) Molecular mapping of the *Rf1* gene restoring pollen fertility in PET1-based F1 hybrids in sunflower (*Helianthus annuus* L.). Theor Appl Genet 106: 599–606.

Hu X, Bidney DL, Yalpani N, Duvick JP, Crasta O, Folkerts O, Lu G (2003) Overexpression of a gene encoding hydrogen peroxide-generating oxalat oxidase evokes defense responses in sunflower. Plant Physiol 133: 170–181.

Jan CC (1992) Inheritance and allelism of mitomycin C- and streptomycin-induced recessive genes for male sterility in cultivated sunflower. Crop Sci 32: 317–320.

Jan CC, Vick BA, Miller JF, Kahler AL, Butler ET (1998) Construction of an RFLP linkage map for cultivated sunflower. Theor Appl Genet 96: 15–22.

Kamal-Eldin A, Appelqvist LA (1996) The chemistry and antioxidant properties of tocopherols and tocotrienols. Lipids 31: 671–701.

Kolkman JM, Slabaugh MB, Bruniard JM, Berry S, Bushman BS, Olungu C, Maes N, Abratti G, Zambelli A, Miller JF, Leon A, Knapp SJ (2004) Acetohydroxyacid synthase mutations conferring resistance to imidazolinone or sulfonylurea herbicides in sunflower. Theor Appl Genet 109: 1147–1159.

Kozik A, Michelmore RW, Knapp SJ, Matvienko MS, Rieseberg L, Lin H, van Damme M, Lavelle D, Chevalier P, Ziegle J, Ellison P, Kolkman JM, Slabaugh MB, Livingston K, Zhou LZ, Church S, Edberg S, Jackson L, Kesseli R, Bradford K (2002) Sunflower and lettuce ESTs from the compositae genome project: *http://compgenomics.ucdavis.edu*, accessed September 2009.

Kusterer B, Horn R, Friedt W (2005) Molecular mapping of the fertility restoration locus *Rf1* in sunflower and development of diagnostic markers for the restorer gene. Euphytica 143: 35–42.

Lacombe S, Bervillé A (2001) A dominant mutation for high oleic acid content in sunflower (*Helianthus annuus* L.) seed oil is genetically linked to a single oleate-desaturase RFLP locus. Mol Breed 8: 129–137.

Lacombe S, Léger S, Kaan F, Bervillé A (2002) Inheritance of oleic acid content in F2 and a population of recombinant inbred lines segregating for the high oleic trait in sunflower. Helia 25: 85–94.

Lawrence GJ, Finnegan EJ, Ayliffe MA, Ellis JG (1995) The *L6* gene for flax rust resistance is related to the Arabidopsis bacterial resistance gene *RPS2* and the tobacco viral resistance gene *N*. Plant Cell 7: 1195–1206.

Lawson WR, Goulter KC, Henry RJ, Kong GA, Kochman JK (1998) Marker-assisted selection for two rust resistance genes in sunflower. Mol Breed 4: 227–234.

Leclercq P (1966) Une stérilité male utilisable pour la production d'hybrides simples de tournesol. Ann Amélior Plant 16: 135–144.

Leclercq P (1969) Une stérilité mâle chez le tournesol. Ann Amélior Plant 19: 99–106.

Leclercq P (1984) Identification de gènes de restauration de fertilité sur cytoplasms stérilisants chez le tournesol. Agronomie 4: 573–576.

Leister D, Ballvora A, Salamini F, Gebhardt C (1996) A PCR-based approach for isolating pathogen resistance genes from potato with potential for wide application in plants. Nat Genet 14: 421–429.

Lenardon SL, Giolitti F, León AJ, Bazzalo ME, Grondona M (2001) Effects of sunflower chlorotic mottle virus infections on sunflower yield parameters. Helia 24: 55–66.

Lenardon SL, Bazzalo ME, Abratti G, Cimmino CJ, Galella MT, Grondona M, Giolitti F, León AJ (2005) Screening sunflower for resistance to sunflower chlorotic mottle virus and mapping the *Rcmo-1* resistance gene. Crop Sci 45: 735–739.

Letousey P, de Zélicourt A, Vieira Dos Santos C, Thoiron S, Monteau F, Simier P, Thalouarn P, Delavault P (2007) Molecular analysis of resistance mechanisms to *Orobanche cumana* in sunflower. Plant Pathol 56: 536–546.

Li C, Feng J, Ma F, Vick BA, Jan CC (2008) Identification of molecular markers linked to a new nuclear male-sterility gene *ms₇* in sunflower (*Helianthus annuus* L.). Proc 17th Int Sunflower Conf, vol 2, Córdoba, Spain, pp 651–654.

Lu YH, Gagne G, Grezes-Besset B, Blanchard P (1999) Integration of a molecular linkage group containing the broomrape resistance gene *Or5* into an RFLP map in sunflower. Genome 42: 453–456.

Lu YH, Melero-Vara JM, García-Tejada JA (2000) Development of SCAR markers linked to the gene *Or5* conferring resistance to broomrape (*Orobanche cumana* Wallr.) in sunflower. Theor Appl Genet 100: 625–632.

Mekhedov S, Martínez de Ilárduya O, Ohlrogge J (2000) Toward a functional catalog of the plant genome. A survey of genes for lipid biosynthesis. Plant Physiol 122: 389–401.

Mestries E, Gentzbittel L, Tourvieille de Labrouhe D, Nicolas P, Vear F (1998) Analyses of quantitative trait loci associated with resistance to *Sclerotinia sclerotiorum* in sunflower (*Helianthus annuus* L.) using molecular markers. Mol Breed 4: 215–226.

Michelmore R (2000) Genomic approaches to plant disease resistance. Curr Opin Plant Biol 3: 125–131.

Michelmore RW, Paran I, Kesseli RV (1991) Identification of markers linked to disease resistance genes by bulked segregant analysis: a rapid method to detect markers in specific genomic regions by using segregating populations. Proc Natl Acad Sci USA 88: 9828–9832.

Miller JF, Fick GN (1997) The genetics of sunflower. In: AA Schneiter (ed) Sunflower Technology and Production. Agronomy Monographs No 35. ASA, CSSA, SSSA, Madison WI, USA, pp 441–495.

Molinero-Ruiz ML, Dominguez J, Melero-Vara JM (2002) Races of isolates of *Plasmopara halstedii* from Spain and studies on their virulence. Plant Dis 86: 736–740.

Mouzeyar S, Roeckel-Drevet P, Gentzbittel L, Philippon J, Tourvieille de Labroughe, Vear F (1995) RFLP and RAPD mapping of the sunflower *Pl1* locus for resistance to *Plasmopara halstedii* race 1. Theor Appl Genet 91: 733–737.

Osorio J, Fernández-Martínez JM, Mancha M, Garcés R (1995) Mutant sunflowers with high concentration of saturated fatty acids in the oil. Crop Sci 35: 739–742.

Pankovic D, Radovanovic N, Jocic S, Satovic Z, Skoric D (2007) Development of co-dominant amplified polymorphic sequence markers for resistance of sunflower to downy mildew race 730. Plant Breed 126: 440–444.

Parker C, Riches CR (1993) Parasitic weeds of the world: Biology and Control. CAB Int, Wallingford, UK.

Pérez-Vich B, Garcés R, Fernández-Martínez JM (1999) Genetic control of high stearic acid content in the seed oil of the sunflower mutant CAS-3. Theor Appl Genet 99: 663–669.

Pérez-Vich B, Fernández-Martínez JM, Grondona M, Knapp SJ, Berry ST (2002) Stearoyl-ACP and oleoyl-PC desaturase genes cosegregate with quantitative trait loci underlying high stearic and high oleic acid mutant phenotypes in sunflower. Theor Appl Genet 104: 338–349.

Pérez-Vich B, Akhtouch B, Knapp SJ, Leon AJ, Velasco L, Fernández-Martínez JM, Berry ST (2004a) Quantitative trait loci for broomrape (*Orobanche cumana* Wallr.) resistance in sunflower. Theor Appl Genet 109: 92–102.

Pérez-Vich B, Knapp SJ, Leon AJ, Fernández-Martínez JM, Berry ST (2004b) Mapping minor QTL for increased stearic acid content in sunflower seed oil. Mol Breed 13: 313–322.

Pérez-Vich B, Berry TB, Velasco L, Fernández-Martínez JM, Gandhi S, Freeman C, Heesacker A, Knapp SJ, Leon AJ (2005) Molecular mapping of nuclear male sterility genes in sunflower. Crop Sci 45: 1851–1857.

Pérez-Vich B, Velasco L, Muñoz-Ruz J, Fernández-Martínez JM (2006a) Inheritance of high stearic acid content in the sunflower mutant CAS-14. Crop Sci 46: 22–29.

Pérez-Vich B, Leon AJ, Grondona M, Velasco L, Fernández-Martínez JM (2006b) Molecular analysis of the high stearic acid content in sunflower mutant CAS-14. Theor Appl Genet 112: 867–875.

Pongracz G, Weiser H, Matzinger D (1995) Tocopherole. Antioxidantien der Natur. Fat Sci Technol 97: 90–104.

Putt ED (1964) Recessive branching in sunflowers. Crop Sci 4: 444–445.

Radwan O, Bouzidi MF, Vear F, Philippon J, Tourvieille de Labrouhe D, Nicolas P, Mouzeyar S (2003) Identification of non-TIR-NBS-LRR markers linked to the *Pl5/Pl8* locus for resistance to downy mildew in sunflower. Theor Appl Genet 106: 1438–1446.

Radwan O, Bouzidi MF, Nicolas P, Mouzeyar S (2004) Development of PCR markers for the *Pl5/Pl8* locus for resistance to *Plasmopara halstedii* in sunflower, *Helianthus annuus* L. from complete CC-NBS-LRR sequences. Theor Appl Genet 109: 176–185.

Radwan O, Mouzeyar S, Nicolas P, Bouzidi MF (2005) Induction of a sunflower CC-NBS-LRR resistance gene analogue during incompatible interaction with *Plasmopara halstedii*. J Exp Bot 56: 567–575.

Radwan O, Gandhi S, Heesacker A, Whitaker B, Taylor C, Plocik A, Kesseli R, Kozik A, Michelmore RW, Knapp SJ (2008) Genetic diversity and genomic distribution of homologs encoding NBS-LRR disease resistance proteins in sunflower. Mol Genet Genom 280: 111–125.

Roeckel-Drevet P, Gagne G, Mouzeyar S, Gentzbittel L, Phillipon J, Nicolas P, Tourvieille de Labrouhe D, Vear F (1996) Colocation of downy mildew (*Plasmopara halstedii*) resistance genes in sunflower (*Helianthus annuus* L.). Euphytica 91: 225–228.

Rojas-Barros P, Hu J, Jan CC (2008) Molecular mapping of an apical branching gene of cultivated sunflower (*Helianthus annuus* L.). Theor Appl Genet 117: 19–28.

Schuppert GF, Heesacker A, Slabaugh MB, Cole G, Knapp SJ (2003) The high oleic acid phenotype in sunflower is caused by epistatic interactions between oleate desaturase genes. In: Plant & Anim Genom XI Conf, San Diego, California, USA, 11–15 Jan 2003: *http://www.intl-pag.org/11/abstracts/W15_W101_XI.html*; verified September 14, 2009.

Schuppert GF, Tang S, Slabaugh MB, Knapp SJ (2006) The sunflower high-oleic mutant *Ol* carries variable tandem repeats of *FAD2-1*, a seed-specific oleoyl-phosphatidyl choline desaturase. Mol Breed 17: 241–256.

Sheppard AJ, Pennington JAT, Weihrauch JL (1993) Analysis and distribution of vitamin E in vegetable oils and foods. In: L Packer, JFuch (eds) Vitamin E in Health and Disease. Marcel Dekker, New York, USA, pp 9–31.

Serieys H (1996) Identification, study, and utilisation in breeding programs of new CMS sources. FAO Progress Report. Helia 19: 144–158.

Slabaugh MB, Yu J-K, Tang S, Heesacker A, Hu X, Lu G, Bidney D, Han F, Knapp SJ (2003) Haplotyping and mapping a large cluster of downy mildew resistance gene candidates in sunflower using multilocus intron fragment length polymorphisms. Plant Biotechnol J 1: 167–185.

Soldatov KI (1976) Chemical mutagenesis in sunflower breeding. In: Proc 7th Int Sunflower Conf Krasnodar, USSR, pp 352–357.

Spring O, Zipper R, Heller-Dohmen M (2006) First report of metalaxyl resistant isolates of *Plasmopara halstedii* on cultivated sunflower in Germany. J Plant Dis Protect 113: 224.

Sukno S, Melero-Vara JM, Fernández-Martínez JM (1999) Inheritance of resistance to *Orobanche cernua* Loefl. in six sunflower lines. Crop Sci 39: 674–678.

Tang S, Yu JK, Slabaugh MB, Shintani DK, Knapp SJ (2002) Simple sequence repeat map of the sunflower genome. Theor Appl Genet 105: 1124–1136.

Tang S, Heesacker A, Kishore VK, Fernandez A, Sadik ES, Cole G, Knapp SJ (2003) Genetic mapping of the *Or5* gene for resistance to *Orobanche* race E in sunflower. Crop Sci 43: 1021–1028.

Tang S, Leon A, Bridges WC, Knapp SJ (2006a) Quantitative trait loci for genetically correlated seed traits are tightly linked to branching and pericarp pigment loci in sunflower. Crop Sci 46: 721–734.

Tang S, Hass CG, Knapp SJ (2006b) *Ty3/gypsy*-like retrotransposon knockout of a 2-methyl-6-phytyl-1,4-benzoquinone methyltransferase is non-lethal, uncovers a cryptic paralogous mutation, and produces novel tocopherol (vitamin E) profiles in sunflower. Theor Appl Genet 113: 783–799.

Tourvieille de Labrouhe D, Lafon S, Walser P, Raulic I (2000) Une nouvelle race de *Plasmopara halstedii*, agent du mildiou du tournesol. Oleagineux Corps Gras Lipides 7: 404–405.

Traber MG, Sies H (1996) Vitamin E in humans: Demand and delivery. Annu Rev Nutr 16: 321–347.

Tranel PJ, Wright TR (2002) Resistance of weeds to AHAS-inhibiting herbicides: what have we learned? Weed Sci 50: 700–712.

Vera-Ruiz EM, Velasco L, Leon AJ, Fernández-Martínez JM, Pérez-Vich B (2006) Genetic mapping of the *Tph1* gene controlling beta-tocopherol accumulation in sunflower seeds. Mol Breed 17:291–296.

Vear F, Gentzbittel L, Philippon J, Mouzeyar S, Mestries E, Roeckel-Drevet P, Tourvieille de Labrouhe D, Nicolas P (1997) The genetics of resistance to five races of downy mildew (*Plasmopara halstedii*) in sunflower (*Helianthus annuus* L.). Theor Appl Genet 95: 584–589.

Vear F, Tourvieille de Labrouhe D, Miller JF (2003) Inheritance of the wide-range downy mildew resistance in the sunflower line RHA 419. Helia 26: 19–24.

Vranceanu V, Stoenescu F (1970) Immunity to sunflower downy mildew due to a single dominant gene. Problem Agri 22: 34–40.

White AD, Owen MDK, Hartzler RG, Cardina J (2002) Common sunflower resistance to acetolactate-inhibiting herbicides. Weed Sci 50: 432–437.

White AD, Graham MA, Owen MDK (2003) Isolation of acetolactate synthase homologs in common sunflower. Weed Sci 51: 845–853.

Whitham S, Dinesh-Kumar SP, Choi D, Hehl R, Corr C, Baker B (1994) The product of the tobacco mosaic virus resistance gene *N*: Similarity to toll and the interleukin-1 receptor. Cell 78: 1101–1115.

Wieckhorst S, Hahn V, Dußle CM, Knapp SJ, Schön CC, Bauer E (2008) Fine mapping of the downy mildew resistance locus Pl_{ARG} in sunflower. Proc 17th Int Sunflower Conf, vol 2, Córdoba, Spain, pp 645–649.

Yu YG, Buss GR, Saghai Maroof MA (1996) Isolation of a superfamily of candidate disease-resistance genes in soybean based on a conserved nucleotide-binding site. Proc Natl Acad Sci USA 93: 11751–11756.

Yu JK, Tang S, Slabaugh MB, Heesacker A, Cole G, Herring M, Soper J, Han F, Chu WC, Webb DM, Thompson L, Edwards KJ, Berry S, Leon AJ, Grondona M, Olungu C, Maes N, Knapp SJ (2003) Towards a saturated molecular genetic linkage map for cultivated sunflower. Crop Sci 43: 367–387.

Yue B, Vick BA, Yuan W, Hu JG (2008a) Mapping one of the 2 genes controlling lemon ray flower color in sunflower (*Helianthus annuus* L.). J Hered 99: 564–567.

Yue B, Cai X, Vick BA, Hu J (2008b) Genetic characterization and molecular mapping of a chlorophyll deficiency gene in sunflower (*Helianthus annuus* L.). J Plant Physiol, doi: 10.1016/j.jplph.2008.09.008

de Zélicourt A, Letousey P, Thoiron S, Campion C, Simoneau P, Elmorjani K, Marion D, Simier P, Delavault P (2007) Ha-DEF1, a sunflower defensin, induces cell death in *Orobanche* parasitic plants. Planta 226: 591–600.

Molecular Mapping of Complex Traits

Seifollah Poormohammad Kiani[1] and Ahmad Sarrafi[2]*

ABSTRACT

Many agricultural traits are complex quantitative traits, affected by many genes, the environment, and interactions between genes and environments. They do not follow the simple rules of Mendelian genetics. Quantitative trait loci (QTL) mapping is a key tool for studying the genetic architecture of complex traits in plants. In this chapter, we reviewed QTLs identified for complex traits using different germplasm and by different research groups. This would advance our understanding of the genetics of complex traits, including yield, oil content and biotic or abiotic stress tolerance in sunflower.

Keywords: complex traits; quantitative trait loci; oil content; yield; biotic stress; abiotic stress

5.1 Introduction

New technologies have introduced an additional means for improving sunflower yield and quality using quantitative genetics. The aim of molecular genetics in sunflower breeding is to identify, isolate, amplify and modify genes or other sequences of DNA and to combine and express the novel or modified sequences in new genotypes. Despite some limitations,

[1]INRA, Genetics and Plant Breeding (SGAP UR254), 78026, Versailles, France.
[2]INP-ENSAT, IFR 40, Laboratoire de Biotechnologie et Amélioration des Plantes (BAP), 31326, Castanet Tolosan, France.
*Corresponding author: *sarrafi@ensat.fr*

molecular genetics is now producing significant results by using the new technologies to influence basic and applied research in sunflower improvement.

With the advent and development of DNA markers, it became possible to construct saturated genetic maps and to locate quantitative trait loci (QTL) for numerous phenotypes in plants, animals and humans. In sunflower, several molecular genetic linkage maps have been constructed using restriction fragment length polymorphism (RFLP), random amplified polymorphic DNA (RAPD), amplified fragment length polymorphism (AFLP), simple sequence repeat (SSR), insertion/deletion (INDEL), single nucleotide polymorphism (SNP) and target region amplification polymorphism (TRAP) markers (Berry et al. 1995; Gentzbittel et al. 1995, 1998; Jan et al. 1998; Flores-Berrios 2000a; Burke et al. 2002; Mokrani et al. 2002; Bert et al. 2003, 2004; Langar et al. 2003; Yu et al. 2003; Rachid Al-Chaarani et al. 2004; Lai et al. 2005; Hu et al. 2007; Poormohammad Kiani et al. 2007a; Yue et al. 2008a). The advantage of these kinds of maps is that in cases where they have common markers with some other maps, they can be combined in order to construct integrated maps with more markers. These maps are widely used for understanding the genetic basis of complex traits in sunflower.

5.2 Identification of Quantitative Trait Loci

5.2.1 Agronomic Traits

The principal goal of sunflower breeding programs is the development of new cultivars with a high oil yield. Identification of the chromosomal regions, which affect grain yield, oil percentage in grain and other agronomic traits should increase our understanding of the genetic control of the characters and help us to develop marker-assisted selection (MAS) programs.

5.2.1.1 Days from Sowing to Flowering

The genetic and environmental controls of flowering in sunflower are certainly complex and mostly undefined. The flowering phenotype in sunflower has been assessed as days from sowing to flowering (DSF or STF) or days from emergence to flowering (DTF) in most of genetical studies. DSF or DTF is an important trait because cultivars with certain ranges of growth cycle length provide optimum yield in specific environments. Photoperiod and temperature have major effects on STF/DTF and could be important sources of genotype×environment interaction (Leon et al. 2001). Polygenic inheritance patterns have been reported for DSF in most studies (Stoenescu 1974; Machacek 1979; Leon et al. 2000), although there is evidence of genetic factors with major qualitative effects (Jan 1986). Additive gene action has the greatest influence on flowering (Miller et al. 1980; Alvarez et al. 1992), but dominant effects have also been noted (Jan 1986).

Since flowering date is an important agronomic trait for adaptation, several QTL analysis have been carried out on crop species including sunflower. Understanding genetic factors influencing DSF could improve the breeding method and ability to investigate and manipulate other traits in selection programs. An RFLP/isoezymes F_2:F_3 mapping population (162 F_3 plants) from a cross between two inbred lines (GH and PAC2) developed by INRA-France with 82 markers was used for mapping DSF, and three QTLs were detected on linkage group LG A, LG G and LG L (Mestries et al. 1998; Table 5-1). The three QTLs explained 30% of the total phenotypic variation (R^2) and the type of gene action observed was consistent with partial dominance on LG L and overdominance on LG A and LG G. For the QTLs on LG A and LG G, the parent, GH, contributed to positive alleles and for the QTL on LG L, PAC2 alleles increased the trait.

Leon et al. (2000) used 235 F_2:F_3 progeny from the cross between two non-restorer (B) photoperiod sensitive lines ZENB8 (female) and HA89 (male), and mapped QTLs controlling DTF in four different environments (Fargo, ND, and Venado Tuerto, Daireaux and Balcarce in Argentina). An RFLP genetic map from the same cross with 205 markers and 1,380 centiMorgans (cM) genome coverage was used in their studies. Five QTLs were identified on LGs A, B, H, I and L that accounted for 73% of the phenotypic and 89% of the genotypic variation in the mean environment with LOD scores ranging from 2.7% (LG H) to 38.4% (LG B). The authors reported that the genetic variation and parental effects of the QTLs were highly consistent across environments and generations. Two QTLs were environment-specific on LG H and I, and the others were detected in all environments (environment-non-specific). Using the same mapping population, Leon et al. (2001) detected QTLs controlling growing degree days (GDD) to flowering in six different environments with 12.1, 13.1, 14, 15, 15.5 and 16.4 h photoperiods (PP). Six QTLs associated with GDD flowering were detected on LGs A, B, H, I, J and L, contributing 67% and 76% of the phenotypic and genotypic variation in the mean environment. Gene actions were mainly additive and genetic effects for higher values of GDD to flowering were derived from both parents. QTLs with additive effect for higher GDD to flowering were derived from HA89 in the LGs A, F and J, while from ZENB8 in LGs B, I and L. Four of the six QTLs for GDD to flowering (LGs A, B, F and J) had significant QTL × environment interactions. The LOD scores for QTLs in LGs A and B were highly dependent on PP. The LOD scores of QTLs in LG A decreased, while the LOD values of QTLs of LG B increased, as the PP increased from 12.1 to 16.4 h. Moreover, the LOD scores for QTLs in linkage group B were not significant at a PP of 12.1 and

Table 5-1 QTLs detected with different mapping populations and marker systems and their effects on complex traits in sunflower.

Reference / Traits	Abbreviation	Number of QTL	Linkage groups	Total phenotypic variance	Population	Cross	Linkage map length	Average marker interval
Rachid Al-Chaarani et al. 2002								
Downy mildew	dmr	4	1,9,17	54.9	RIL	PAC2×RHA266	2833.7	7.9
Black stem	bsr	7	3,4,8,9,11,15,17	93.1	RIL	PAC2×RHA266	2833.7	7.9
Rachid Al-Chaarani et al. 2004								
Sowing to flowering date	stf	3	4,9,14	52.0	RIL	PAC2×RHA266	2915.9	7.9
Grain weight per plant	gwp	4	4,6,9,21	43.0	RIL	PAC2×RHA266	2915.9	7.9
1000 grain weight	tgw	3	4,6,9	53.0	RIL	PAC2×RHA266	2915.9	7.9
Oil percentage in grain	pog	4	8,11,13,21	39.0	RIL	PAC2×RHA266	2915.9	7.9
Plant height	PH	5	4,11,15	71.0	RIL	PAC2×RHA266	2915.9	7.9
Shoot diameter	SD	5	4,8,14,15	50	RIL	PAC2×RHA266	2915.9	7.9
Head diameter	HD	4	4,9,18,19	25	RIL	PAC2×RHA266	2915.9	7.9
Bert et al. 2002								
Phomopsis mycelium on leaves (year 1998)	pho98	2	4,10	27.7	F2:F3	XRQ×PSC8	2318.0	8.0
Phomopsis mycelium on leaves (year 1999)	pho99	3	4,8,14	59.3	F2:F3	XRQ×PSC8	2318.0	8.0
Phomopsis mycelium on leaves (year 2000)	Pho00	3	4,8,17	47.2	F2:F3	XRQ×PSC8	2318.0	8.0
Phomopsis mycelium on leaves (mean years)	phoMa	4	4,8,10,17	49.2	F2:F3	XRQ×PSC8	2318.0	8.0
Sclerotinia mycelium on leaves (year 1997)	Scl97	3	6,8,13	56.1	F2:F3	XRQ×PSC8	2318.0	8.0
Sclerotinia mycelium on capitulum (year 1997)	Myc97	2	7,8	25.6	F2:F3	XRQ×PSC8	2318.0	8.0
Sclerotinia Percentage attack	Att97	5	5,6,7,8,13	43.3	F2:F3	XRQ×PSC8	2318.0	8.0
Sclerotinia Percentage attack	Att99	3	3,6,10	31.8	F2:F3	XRQ×PSC8	2318.0	8.0
Sclerotinia Latency index (1997)	Lat97	1	7	9.9	F2:F3	XRQ×PSC8	2318.0	8.0
Sclerotinia Latency index (1999)	Lat99	1	7	10.4	F2:F3	XRQ×PSC8	2318.0	8.0
Bert et al. 2003								
Sowing to flowering date (F3 generation)	fd	5	1,4,5,7,11	64.1	F2:F3	XRQ×PSC8	2318.0	8.0
Sowing to flowering date (F3 generation)	fd	6	1,4,5,6,11,18	74.9	F2:F3	XRQ×PSC8	2318.0	8.0
Oil content (F2 generation)	oil	5	2,3,5,7,12	68.1	F2:F3	XRQ×PSC8	2318.0	8.0
Oil content (F3 generation)	oil	5	2,4,5,7,12	70.1	F2:F3	XRQ×PSC8	2318.0	8.0
Seed weight (F2 generation)	sw	2	4,7	16.0	F2:F3	XRQ×PSC8	2318.0	8.0
Seed weight (F3 generation)	sw	2	4,7	25.2	F2:F3	XRQ×PSC8	2318.0	8.0

Plant height (F2 generation)	*ph*	4	1,5,7,8	67.5	F2:F3	XRQ×PSC8	2318.0	8.0
Plant height (F3 generation)	*ph*	4	1,2,5,8	35.4	F2:F3	XRQ×PSC8	2318.0	8.0
Plant lodging	*pl*	3	5,7,19	20.6	F2:F3	XRQ×PSC8	2318.0	8.0
Bert et al. 2004								
P. macdonaldii necrosis on cotyledons	*Bsr*	4	1,3,15,16	39.3	F2:F3	FU×PAZ2	1937.5	9.0
Sclerotinia attack on terminal buds	*Bud*	7	1,7,8,9,10,14,15	50.7	F2:F3	FU×PAZ2	1937.5	9.0
Sclerotinia attack on capitulum	*Att*	3	6,7,9	38.3	F2:F3	FU×PAZ2	1937.5	9.0
Sclerotinia Latency index	*Lat*	4	6,7,9,17	37.4	F2:F3	FU×PAZ2	1937.5	9.0
Darvishzadeh et al. 2007								
Black stem	*bsrMP8*	5	5,9,11,15,17	63.0	RIL	PAC2×RHA266	1824.6	3.7
	bsrMP10	5	1,2,5,15	62.0	RIL	PAC2×RHA266	1824.6	3.7
Duble et al. 2004								
Downy mildew	Pl_{arg}	1	BSA	–	F2	CmsHA342×Arg1575-2	–	–
Ebrahimi et al. 2008								
Oil content (well-watered greenhouse)	1.OC	12	1,2,6,7,10,12, 14,16	>100	RIL	PAC2×RHA266	1824.6	3.7
Oil content (water-stressed greenhouse)	2.OC	8	2,3,6,8,10,11,16	90.0	RIL	PAC2×RHA266	1824.6	3.7
Oil content (well-watered field)	3.OC	6	1,10,12,16,17	66.0	RIL	PAC2×RHA266	1824.6	3.7
Oil content (water-stressed field)	4.OC	3	2,10,16	34.0	RIL	PAC2×RHA266	1824.6	3.7
Palmitic acid (well-watered greenhouse)	1.PA	6	1,2,3,10,15	72.0	RIL	PAC2×RHA266	1824.6	3.7
Palmitic acid (water-stressed greenhouse)	2.PA	15	2,3,5,6,7,8,9,12, 14,16,17	>100	RIL	PAC2×RHA266	1824.6	3.7
Palmitic acid (well-watered field)	3.PA	4	9,12,13,16	42.0	RIL	PAC2×RHA266	1824.6	3.7
Palmitic acid (water-stressed field)	4.PA	5	1,4,7,8	84.0	RIL	PAC2×RHA266	1824.6	3.7
Stearic acid (well-watered greenhouse)	1.SA	5	2,12,14	73.0	RIL	PAC2×RHA266	1824.6	3.7
Stearic acid (water-stressed greenhouse)	2.SA	9	6,7,8,11,12,14	100.0	RIL	PAC2×RHA266	1824.6	3.7
Stearic acid (well-watered field)	3.SA	5	5,10,14,15,16	67.0	RIL	PAC2×RHA266	1824.6	3.7
Stearic acid (water-stressed field)	4.SA	6	1,3,6,10,12,14	70.0	RIL	PAC2×RHA266	1824.6	3.7
Oleic acid (well-watered greenhouse)	1.OA	9	1,3,4,5,7,8,9,15	>100	RIL	PAC2×RHA266	1824.6	3.7
Oleic acid (water-stressed greenhouse)	2.OA	6	3,4,8,9,11,15	82.0	RIL	PAC2×RHA266	1824.6	3.7
Oleic acid (well-watered field)	3.OA	3	3,15,16	31.0	RIL	PAC2×RHA266	1824.6	3.7
Oleic acid (water-stressed field)	4.OA	2	7,13	24.0	RIL	PAC2×RHA266	1824.6	3.7

Table 5-1 contd....

Table 5-1 contd....

Reference Traits	Abbreviation	Number of QTL	Linkage groups	Total phenotypic variance	Population	Cross	Linkage map length	Average marker interval
Linoleic acid (well-watered greenhouse)	1.LA	4	3,4,8,15	58.0	RIL	PAC2×RHA266	1824.6	3.7
Linoleic acid (water-stressed greenhouse)	2.LA	6	3,4,13,15,16	>100	RIL	PAC2×RHA266	1824.6	3.7
Linoleic acid (well-watered field)	3.LA	2	17	26.0	RIL	PAC2×RHA266	1824.6	3.7
Linoleic acid (water-stressed field)	4.LA	2	7,13	30.0	RIL	PAC2×RHA266	1824.6	3.7
Flores-Berrios et al. 2000a								
Shoots per plant	ose	6	2,4,7,9,17	52.0	RIL	PAC2×RHA266	2558.0	10.0
Shoots per regeneration explant	osr	7	2,4,6,7,8,15,17	67.0	RIL	PAC2×RHA266	2558.0	10.0
Flores-Berrios et al. 2000b								
Total embryogenic explant	toe	4	1,3,13,15	48.0	RIL	PAC2×RHA266	2558.0	10.0
Flores-Berrios et al. 2000c								
Total protoplant division	ptd	12	1,7,8,10,13, 14,15,17	72.0	RIL	PAC2×RHA266	2558.0	10.0
Gandhi et al. 2005								
Self-incompatibility	si	1	17	66.2	Back cross	NMS373×ANN1118	1450.0	11.0
Self-pollination	sf	3	6,15,17	63.9	Back cross	NMS373×ANN1118	1450.0	11.0
Seed dormancy	sd	3	3,11,15	38.3	Back cross	NMS373×ANN1118	1450.0	11.0
Hervé et al. 2001								
Chlorophylle concentration	chl	4	5,8,10,18	53.5	RIL	PAC2×RHA266	2558.0	10.0
Net photosynthesis	pho	3	9,14	62.5	RIL	PAC2×RHA266	2558.0	10.0
Stomatal conductance	sco	4	3,8,16,17	61.9	RIL	PAC2×RHA266	2558.0	10.0
Predawn leaf water potential	pot	3	8,10,14	34.1	RIL	PAC2×RHA266	2558.0	10.0
Huang et al. 2007								
Embryogenic explant/100 explant	–	3	5,10,13	38.6	RIL	PAC2×RHA266	2915.9	7.9
Leon et al. 1995								
Seed oil percentage	sop	4	B,C,I,N	57.0	F2/F3	ZENB8×HA89	1380.0	5.9
Leon et al. 2003								
Seed oil percentage	sop	8	B,C,G,I,L,M,N	61.0	F2/F3	ZENB8×HA89	1380.0	5.9
Leon et al. 2000								
Days to flowering	DTF	5	A,B,H,I,L	72.9	F3	ZENB8×HA89	1380.0	5.9

Leon et al. 2001

Trait								
Growing degree days to flowering	gdd	6	A,B,F,I,J,L	76.0	F₃ and F₄	ZENB8×HA89	1380.0	6.7

Mestries et al. 1998

Sowing to flowering date (F3 generation)	–	3	A,L,G	27.0	F2:F3	GH×PAC2	760.0	12.6
Seed weight (F3 generation)	–	2	A,C	28.5	F2:F3	GH×PAC2	760.0	12.6
Oil content (F2 generation)	–	2	A,Q	19.4	F2:F3	GH×PAC2	760.0	12.6
Oil content (F3 generation)	–	3	A,Q,C	53.7	F2:F3	GH×PAC2	760.0	12.6
Oil content (F4 generation)	–	2	A,Q	26.1	F2:F3	GH×PAC2	760.0	12.6
S. sclerotiorum reaction on leaves (F2 generation)	–	1	A	10.5	F2:F3	GH×PAC2	760.0	12.6
S. sclerotiorum reaction on leaves (F3 generation)	–	2	G,P	19.5	F2:F3	GH×PAC2	760.0	12.6
S. sclerotiorum reaction on leaves (F4 generation)	–	2	I,I	20.8	F2:F3	GH×PAC2	760.0	12.6
S. sclerotiorum reaction on capitulum (F2 generation)	–	1	A	9.7	F2:F3	GH×PAC2	760.0	12.6
S. sclerotiorum reaction on capitulum (F3 generation)	–	2	A,M	20.8	F2:F3	GH×PAC2	760.0	12.6
S. sclerotiorum reaction on capitulum (F4 generation)	–	2	A,M	30.6	F2:F3	GH×PAC2	760.0	12.6

Micic et al. 2004

Sclerotinia resistance	Leaf lesion	9	1,4,6,8,9,13,15	55.8	F2	CM625×NDBLOS	961.9	9.6
Sclerotinia resistance	Stem lesion	7	2,3,4,6,8,15,16	71.7	F2	CM625×NDBLOS	961.9	9.6

Micic et al. 2005

Sclerotinia resistance	Leaf lesion	3	4,10,17	38.5	F2	CM625×TUB-5-4243	1005.2	14.0
Sclerotinia resistance	Stem lesion	4	4,10,17	80.5	F3	CM625×TUB-5-4243	1005.2	14.0

Mokrani et al. 2002

Grain weight per plant	gwp	2	9	39.7	F3	L1×L2	2539.0	14.9
1000 grain weight	tgw	1	16	5.4	F2	L1×L2	2539.0	14.9
Oil percentage in grain	pog	7	9,11,12,13	97.4	F2	L1×L2	2539.0	14.9
Sowing to flowering date	stf	2	9,10	73.5	F2	L1×L2	2539.0	14.9

Table 5-1 contd.....

Table 5-1 contd.....

Reference / Traits	Abbreviation	Number of QTL	Linkage groups	Total phenotypic variance	Population	Cross	Linkage map length	Average marker interval
Mouzeyar et al.1995								
Downy mildew	P_{ll}	1	BSA	–	F3	GH×RHA266	–	–
Pérez-Vich et al. 2002								
Stearic acid	C18:0	2	1,7	790.	F2	HA89×CAS3	1807.7	12.0
Stearic acid	C18:0	4	1,3,8,14	84.4	F2	HAOL-9×CAS3	1641.5	12.5
Oleic acid	C18:1	3	1,8,14	58.4	F2	HAOL-9×CAS3	1641.5	12.5
Pérez-Vich et al. 2004								
Broomrape race E	SE-194	1	3	37.0	F3	P21×P91	1144.4	13.3
Broomrape race E	CU-796	1	3	59.4	F3	P21×P91	1144.4	13.3
Broomrape race F	SE-296	2	5,16	40,6	F3	P21×P91	1144.4	13.3
Poormohammad Kiani et al. 2007a								
Relative water content (well-watered)	RWC.WW	6	5,6,10,17	84.0	RIL	PAC2×RHA266	1824.6	3.7
Relative water content (water-stressed)	RWC.WS	6	4,5,7,16,17	74.0	RIL	PAC2×RHA266	1824.6	3.7
Leaf water potential (well-watered)	LWP.WW	3	6,8,12	43.0	RIL	PAC2×RHA266	1824.6	3.7
Leaf water potential (water-stressed)	LWP.WS	6	1,5,7,9,16	79.0	RIL	PAC2×RHA266	1824.6	3.7
Turgor potential (well-watered)	TP.WW	6	2,4,6,8,12	70.0	RIL	PAC2×RHA266	1824.6	3.7
Turgor potential (water-stressed)	TP.WS	6	1,4,7,10,16	73.0	RIL	PAC2×RHA266	1824.6	3.7
Osmotic adjustment	OA	8	2,4,5,12,13	94.0	RIL	PAC2×RHA266	1824.6	3.7
Poormohammad Kiani et al. 2007b								
Relative water content under drought	rwc	2	1,2	–	RIL	PAC2×RHA266	2915.9	7.9
Leaf water potential under drought	Wpot	1	13	–	RIL	PAC2×RHA266	2915.9	7.9
Turgor potential under drought	Tpot	4	7,8,13,14	–	RIL	PAC2×RHA266	2915.9	7.9
Osmotic adjustment	OA	5	1,8,13	–	RIL	PAC2×RHA266	2915.9	7.9
Net photosynthesis under drought	Pho	4	3,10,14	–	RIL	PAC2×RHA266	2915.9	7.9
Poormohammad Kiani et al. 2008								
Potential photochemical efficiency of PSII electron transport (well-watered)	FPW	6	3,5,16,17	87.0	RIL	PAC2×RHA266	1824.6	3.7
Potential photochemical efficiency of PSII electron transport (water-stressed)	FPD	7	2,4,5,6,7,17	65.0	RIL	PAC2×RHA266	1824.6	3.7

Actual efficiency of PSII electron transport (well-watered)	*FPSIIW*	6	1,7,8,9,12,14	82.0	RIL	PAC2×RHA266	1824.6	3.7
Actual efficiency of PSII electron transport (water-stressed)	*FPSIID*	5	4,7,9,12,13	67.0	RIL	PAC2×RHA266	1824.6	3.7
Non-photochemical fluorescence quenching (well-watered)	*NPQW*	8	4,7,8,11,12,13	98.0	RIL	PAC2×RHA266	1824.6	3.7
Non-photochemical fluorescence quenching (water-stressed)	*NPQD*	5	1,3,11,16,17	57.0	RIL	PAC2×RHA266	1824.6	3.7
Proportion of closed PSII traps (well-watered)	*1-qPW*	6	4,7,9,12,13,15	62.0	RIL	PAC2×RHA266	1824.6	3.7
Proportion of closed PSII traps (water-stressed)	*1-qPD*	9	2,3,4,7,10,12,13,15,17	>100	RIL	PAC2×RHA266	1824.6	3.7

Poormohammad Kiani et al. 2009

Sowing to flowering date (well-watered greenhouse)	*DSFW*	7	5,6,9,15,17	80.0	RIL	PAC2×RHA266	1824.6	3.7
Sowing to flowering date (water-stressed greenhouse)	*DSFD*	7	1,7,14,16,17	69.0	RIL	PAC2×RHA266	1824.6	3.7
Sowing to flowering date (irrigated field)	*DSFI*	7	1,7,10,11,14,17	67.0	RIL	PAC2×RHA266	1824.6	3.7
Sowing to flowering date (non-irrigated field)	*DSFN*	6	7,10,11,14	>100	RIL	PAC2×RHA266	1824.6	3.7
Leaf number (well-watered greenhouse)	*LNW*	2	7,12	27.0	RIL	PAC2×RHA266	1824.6	3.7
Leaf number (water-stressed greenhouse)	*LND*	4	3,5,14,15	42.0	RIL	PAC2×RHA266	1824.6	3.7
Leaf number (irrigated field)	*LNI*	5	3,7,14,16	60.0	RIL	PAC2×RHA266	1824.6	3.7
Leaf number (non-irrigated field)	*LNN*	6	3,5,7,14,16	63.0	RIL	PAC2×RHA266	1824.6	3.7
Leaf area at flowering (well-watered greenhouse)	*LAFW*	6	3,7,9,12,13	57.0	RIL	PAC2×RHA266	1824.6	3.7
Leaf area at flowering	*LAFD*	6	1,2,3,9,16,17	71.0	RIL	PAC2×RHA266	1824.6	3.7
Leaf area at flowering (irrigated field)	*LAFI*	3	7,11,12	23.0	RIL	PAC2×RHA266	1824.6	3.7
Leaf area at flowering (non-irrigated field)	*LAFN*	6	3,5,7,11,13	60.0	RIL	PAC2×RHA266	1824.6	3.7
Leaf area duration (well-watered greenhouse)	*LADW*	6	2,3,10,16	82.0	RIL	PAC2×RHA266	1824.6	3.7
Leaf area duration (water-stressed greenhouse)	*LADD*	6	1,9,10,12,13,17	56.0	RIL	PAC2×RHA266	1824.6	3.7
Leaf area duration (irrigated field)	*LADI*	6	1,3,4,6,7,10	69.0	RIL	PAC2×RHA266	1824.6	3.7
Leaf area duration (non-irrigated field) (water-stressed greenhouse)	*LADN*	4	1,7,10,12	38.0	RIL	PAC2×RHA266	1824.6	3.7

Table 5-1 contd....

Table 5-1 contd....

Reference Traits	Abbreviation	Number of QTL	Linkage groups	Total phenotypic variance	Population	Cross	Linkage map length	Average marker interval
Plant height (well-watered greenhouse)	PHW	2	9,15	29.0	RIL	PAC2×RHA266	1824.6	3.7
Plant height (water-stressed greenhouse)	PHD	5	1,2,9,13,14	84.0	RIL	PAC2×RHA266	1824.6	3.7
Plant height (irrigated field)	PHI	5	4,7,11	43.0	RIL	PAC2×RHA266	1824.6	3.7
Plant height (non-irrigated field)	PHN	5	7,11,16	71.0	RIL	PAC2×RHA266	1824.6	3.7
Total dry matter (well-watered greenhouse)	BIOW	6	1,10,12,16	81.0	RIL	PAC2×RHA266	1824.6	3.7
Total dry matter (water-stressed greenhouse)	BIOD	2	1,13	26.0	RIL	PAC2×RHA266	1824.6	3.7
Total dry matter (irrigated field)	BIOI	6	3,7,8,10	49.0	RIL	PAC2×RHA266	1824.6	3.7
Total dry matter (non-irrigated field)	BION	5	1,7,10	47.0	RIL	PAC2×RHA266	1824.6	3.7
Head weight (well-watered greenhouse)	HWW	4	5,6,10,15	58.0	RIL	PAC2×RHA266	1824.6	3.7
Head weight (water-stressed greenhouse)	HWD	6	1,4,6,10,16	94.0	RIL	PAC2×RHA266	1824.6	3.7
Head weight (irrigated field)	HWI	7	3,4,8,12,14	58.0	RIL	PAC2×RHA266	1824.6	3.7
Head weight (non-irrigated field)	HWN	7	2,3,7,10,14	51.0	RIL	PAC2×RHA266	1824.6	3.7
Grain yield per plant (well-watered greenhouse)	GYPW	3	5,10,14	76.0	RIL	PAC2×RHA266	1824.6	3.7
Grain yield per plant (water-stressed greenhouse)	GYPD	5	4,9,13,14,16	89.0	RIL	PAC2×RHA266	1824.6	3.7
Grain yield per plant (irrigated field)	GYPI	6	3,4,5,7,12	70.0	RIL	PAC2×RHA266	1824.6	3.7
Grain yield per plant (non-irrigated field)	GYPN	6	2,3,4,10,14	76.0	RIL	PAC2×RHA266	1824.6	3.7
Tang et al. 2006								
Seed oil concentration	soc	6	1,4,9,10,16,17	55.8	RILs	RHA280×RHA801	–	–
100-seed weight	swt	5	5,9,10,14,1710,16,17	73.4	RILs	RHA280×RHA801	–	–
Yue et al. 2008b								
Disease incidence % test 1	QDi	4	2,7,10,12	65.8	F2:F3	HA441×RHA439	1797.6	8.0
Disease incidence % test 2	QDi	5	8,9,10,12,17	81.0	F2:F3	HA441×RHA439	1797.6	8.0
Disease incidence % test 3	QDi	4	2,3,9,12	60.8	F2:F3	HA441×RHA439	1797.6	8.0
Disease severity test 1	QDs	3	2,4,10	72.2	F2:F3	HA441×RHA439	1797.6	8.0
Disease severity test 2	QDs	4	4,8,12,13	78.7	F2:F3	HA441×RHA439	1797.6	8.0
Disease severity test 3	QDs	2	9,12	42.7	F2:F3	HA441×RHA439	1797.6	8.0
Vear et al. 2008								
Latency index	Latency index	2	1,10	31.0	RIL	XRQ×PSC8	–	–
Percentage attck	Percentage attck	3	1,2,10	29.2	RIL	XRQ×PSC8	–	–

13.1 h. Two QTLs for PP were located at the same genetic positions as QTLs associated with GDD to flowering in LGs A and B, respectively.

Mokrani et al. (2002) identified QTLs for STF using 118 F_3 families derived from a cross between L1 (restorer) and L2 (maintainer) lines from Syngenta seed company sunflower collection (Table 5-1). Using a genetic map constructed with 215 AFLP and 61 SSR markers, two QTLs located on LG 9 (*stf-9-1*) and LG 10 (*stf-10-1*) with LOD scores of 3.6 and 4.3, respectively were detected for STF. The phenotypic variation explained by these QTLs was 2.6% for *stf-9-1* and 70.9% for *stf-10-1*, respectively, and L1 parent contributed to the favorable alleles for both the QTLs.

Bert et al. (2003) also identified the genomic regions involved in STF in a F_2:F_3 mapping population derived from a cross between two inbred lines developed by INRA-France (XRQ×PSC8) in two years 1997 and 1999 (Table 5-1). Five and six QTLs were detected in 1997 and 1999, respectively on different LGs among which, four were common across the two years and three (one on LG 7 in 1997 and two on LG 6 and LG 18 in 1999) were specific to the year of study. The total phenotypic variance explained by five QTLs in 1997 and six QTLs in 1999 was 64.1% and 74.9%, respectively, and both parental lines contributed equally to favorable alleles. The QTLs identified by Bert et al. (2003) for STF were not in the same position as those reported for the same trait by Mestries et al. (1998) on LG 9 and LG 10.

Rachid Al-Chaarani et al. (2004) identified three QTLs controlling STF in a recombinant inbred line (RIL) mapping population derived from the PAC2×RHA266 cross on LGs 4, 9 and 14 with LOD score of 7.5, 9.8 and 10.4, respectively with a sum of the individual phenotypic variation (R^2) of 52% (Table 5-1). RHA266 contributed the positive alleles for two of the three QTLs and PAC2 contributed to only one of them. In any QTL analysis for any quantitative or qualitative trait, it is important to compare the position of QTLs obtained with different mapping populations and also across various environmental conditions. Unfortunately, the comparison of the location of QTLs is not always possible as different marker systems and linkage group nomenclature are used for map construction and QTL detection in different studies.

Recently, QTLs for many agronomical, physiological and developmental traits have been detected under different water treatments in the greenhouse and field conditions using a saturated public AFLP/SSR map of RILs from the cross PAC2×RHA266 with standard linkage group nomenclature (Poormohammad Kiani et al. 2007a, b, 2008, 2009). The results of these research works are presented in 5.2.7.

5.2.1.2 Agro-morphological Traits

Morphological traits are also important in sunflower breeding. Although some of the morphological traits, such as plant height, are not essential characters in breeding for yield, but in case of height for example, it may be related to lodging. The modern varieties show a wide range of heights and, for a given yield, a shorter plant is preferred (Miller and Fick 1997).

Bert et al. (2003) reported QTLs for plant height in two successive years and for plant lodging in an one year experiment using the same F_2:F_3 mapping population (XRQ×PSC8) as mentioned above (see 5.2.1.1). They identified four QTLs in each year (1996 and 1997) for plant height, and three QTLs for plant lodging (Table 5-1). The total phenotypic variance explained for plant height was 67.5% and 36.4% for the year 1996 and 1997, respectively, and it was 20.6% for plant lodging. Three QTLs for plant height were detected in both the years and only one QTL was year-specific. Plant lodging shared one of its QTLs with plant height in each year, which further confirmed the genetic relation between these two traits. A large-effect QTL on LG 8 with R^2 of 39.3% (year 1996) and 17.9% (year 1997) was detected in both the years for plant height with positive alleles from the parent, XRQ.

Rachid Al-Chaarani et al. (2004) identified five QTLs for each of plant height (PH) and shoot diameter (SD), and four QTLs for head diameter (HD) on different linkage groups in the RIL population of the cross PAC2×RHA266 (Table 5-1). LOD scores for these QTLs ranged from 4.35 to 13.66 and total phenotypic variances explained by the QTLs were 71% for PH, 50% for SD and 25% for HD. For one QTL controlling SD (LG 14) and two QTLs governing HD (LG 18 and LG 19), positive alleles came from PAC2, but for the remaining QTLs RHA266 contributed to positive alleles.

More recently, Yue et al. (2008a) developed a new genetic linkage map and detected QTLs controlling four morphological traits on it. They have used 120 F_2 individuals from the cross Lg1×HA379 and four traits, including leaf color (chlorophyll content or greenness degree), plant height, leaf shape, and head shape were investigated in the F_2 and F_3 generations. Their linkage map was constructed using 202 polymorphic bands/markers generated from 54 pairs of TRAP primer combinations and 24 polymorphic SSR primers selected from each of the 17 linkage groups. The SSR markers were used to align the new linkage map with the published sunflower reference SSR map developed by Tang et al. (2002). The linkage map had a total length of 1,597.5 cM with an average distance between the adjacent markers of 7.1 cM. A total of six QTLs were detected on LG 2, LG 3, LG 8 (two loci) and LG 17 (two loci) for plant height. Among them, *ph3*, a major QTL on LG 8 alone explained more than 30% of the phenotypic variation in both the generations. Six QTLs were identified for leaf shape, which explained 7.8–14.5% of the phenotypic variation. None of them were detected in both the generations. Only one QTL was detected for head shape in the F_3 generation that explained 12.2% of the phenotypic variance. Alleles from Lg1 at nine of the QTLs for these three traits had positive effects that coincided with the performance of this parent for these traits. For leaf color-related traits two, three, two and two QTLs were detected for Chl*a*, Chl*b*, Chl*a*/Chl*b* and Chlt in the F_2 generation, respectively. Individual QTLs explained 8.8–49.5% of the phenotypic variance and three QTLs for leaf greenness degree were identified in the F_3 generation with R^2 from 10.8% to 27.2%. The authors concluded that the genetic basis of morphological traits is very complex and each trait is under the control of multiple genes. The authors confirmed also one QTL for plant height on LG 17 (*ph6*) to be linked to an SSR marker ORS811, which had been reported earlier in the same genomic region by Burke et al. (2002).

5.2.1.3 Oil Content and Yield-related Traits

Breeding of sunflower have focused mainly on yield and oil content. Seed yield is known to be dependent on both additive and non-additive gene actions, and genotype × environment interactions are important component of variance leading to low heritability (Fick 1978). Furthermore, yield in sunflower, as in all other crops, depends on many component characters, which have polygenic mode of inheritance (Fick and Miller 1977). A polygenic nature of inheritance has been indicated for oil content with heritability estimates of 65 to 70% (Fick 1975). Oil content is determined by grain yield and oil percentage in the grain. Therefore, both the characters are important to be considered in breeding programs simultaneously. The use of molecular markers provided additional information about the genetic basis of seed-oil concentration.

Leon et al. (1995) used a genetic map of 201 RFLP markers and identified six QTLs associated with 57% of genetic variation for oil content among the F_2 progeny. Two QTLs were identified for seed oil concentration (LG C and LG I), two for seed percentage (LG G and LG J) and the other two for both traits (LG B and LG N); and additive effect was predominant for seed-oil concentration. In a later study, the same mapping population was used for mapping QTLs controlling seed oil concentration across four locations in Argentina (Venado T, Daireaux, Balcarce) and USA (Fargo, ND) and the measurements were recorded on F_3 plants. Eight QTLs on seven LGs (B, C, G, I, L, M and N) were identified for mean values of the trait in the four environments. These eight QTLs accounted for 59% and 86% of phenotypic and genotypic variance, respectively. Alleles for increasing seed oil concentration were all derived from HA89, the parent with higher values for that trait. Gene action was additive for four QTLs on LGs B, C, I and M; dominant for two QTLs on LGs G and N and overdominant for one QTL on LG L. The authors confirmed five QTLs detected from a previous work by Leon et al. (1995) in their study. These were located on LGs B, C, G, I and N.

QTLs for percentage of oil in grain and seed weight were also detected by Mestries et al. (1998) in F_2, F_3 and F_4 generations derived from an F_2

population from the cross GH×PAC2. In F_3 progeny, the authors identified two QTLs for seed weight on LGs A and C, which explained 28.5% of the total phenotypic variance. Favorable alleles for these QTLs were derived from the GH parent. For oil content, two, three and two QTLs were detected in F_2, F_3 and F_4 generations, respectively. These QTLs accounted for 19.4%, 53.8% and 26.1% of phenotypic variance of oil content in F_2, F_3 and F_4 generations, respectively. For three QTLs, the PAC2 alleles increased the trait, and for four QTLs GH alleles increased oil concentration. Two QTLs were detected in all of the three generations (LGs A and Q) and one QTL (LG C) was specific to the F_3 generation.

Mokrani et al. (2002) detected two QTLs for grain yield per plant (LG 9), one QTL for 1,000-grain weight (LG 16) and seven QTLs for oil percentage in grain (LGs 9, 11, 12 and 13), with total phenotypic variation of 50.7%, 22.7% and 90.4%, respectively (Table 5-1). Grain yield per plant and oil percentage in grain shared two QTLs on LG 9.

Bert et al. (2003) identified five QTLs for oil content in each of the F_2 and F_3 progeny from the cross XRQ×PSC8. The total genotypic variation explained by these QTLs was 68.1% in F_2 generation and 70.1% in F_3 generation. Four QTLs for oil content were detected in both the generations and one QTL were specific to each of the generations. XRQ contributed positive alleles for seven loci, but PSC8 contributed for only three QTLs.

Rachid Al-Chaarani et al. (2004) identified QTLs for grain weight per plant, 1,000-grain weight and oil percentage in an RIL population. Four QTLs for grain weight per plant (LGs 4, 6, 9 and 21), three QTLs for 1,000-grain weight (LGs 4, 6 and 9) and four QTLs for oil percentage (LG 8, 11, 13 and 21) were identified. The phenotypic variation explained by these QTLs were 43% for grain weight per plant, 53% for 1,000-grain weight and 39% for oil percentage, respectively. LOD scores ranged from 4.04 to 13.51 and a major QTL was detected for 1,000-grain weight (LG 9), which controlled 37% of the phenotypic variance for this trait. This QTL was also detected for grain weight per plant. Three QTLs were common between grain weight per plant and 1,000-grain weight (LGs 4, 6, and 9), but only one QTL appeared to be common between 1,000-grain weight and oil percentage on pseudo-linkage group 21. These results complement the pioneer works that described molecular markers linked to oil characteristics and quantitative genetics analysis (Leon et al. 1995, 2003).

QTLs involved in seed oil concentration and 100-seed weight were detected in another study using 173 F_7 RILs from a cross between RHA280 (an unbranched, confectionary, fertility restorer line) and RHA801 (a branched, oilseed, fertility restorer line) (Tang et al. 2006). Two hundred and three SSR and insertion-deletion (INDEL) markers and three phenotypic loci were used for map construction. Six QTLs were identified on LG 1, 4, 9, 10, 16, and 17 for seed oil concentration. These QTLs explained 55.7% of the phenotypic variance and QTLs alleles for increased oil were transmitted by the oilseed parent (RHA801). Five QTLs were identified on LG 5, 9, 10, 14, and 17 for 100-seed weight. The QTLs collectively explained 73.4% of the phenotypic variability. The RHA280 allele from four of the five QTLs increased 100-seed weight, whereas the RHA280 allele for one of the five QTLs decreased 100-seed weight. Results showed that three QTLs on LG 9, 10 and 17 control both the traits. An apical branching loci *(B)* and a phytomelanin pigment loci *(P)* were linked to a QTL, which control both the traits on LG 10 and 17, respectively. The *B*-liked QTL, which is located on linkage 10, explained 22.5% and 52.5% of phenotypic variation for seed oil concentration and 100-seed weight, respectively. The effect of *B-linked* QTLs on seed oil concentration and 100-seed weight has been also reported in two high-oil×high-oil mapping populations, GH×PAC2 (Mestries et al. 1998) and XRQ×PSC8 (Bert et al. 2003). However, the effect of *B*-linked QTL on seed oil concentration showed different direction in different genetic backgrounds.

The seed oil of standard cultivated sunflower is composed primarily of the saturated fatty acids palmitic (C16:0) and stearic (C18:0), and the unsaturated oleic (C18:1) and linoleic acids (C18 : 2), with C18:1 and C18:2 accounting for about 90% of the total oil fatty acids (Dorrell and Vick 1997). The quality of standard sunflower oil is defined by the relative content of C18:2, which is high in most of the crop grown throughout the world. The genetic control of the synthesis of stearic acid (C18:0) and oleic acid (C18:1) in the seed oil of sunflower was studied through candidate-gene and QTL analysis (Pérez-Vich et al. 2002). Two F_2 mapping populations were developed using the high C18:0 mutant CAS-3 crossed to either HA-89 (standard, high linoleic fatty acid profile), or HAOL-9 (high C18:1 version of HA-9). The HA-89 × CAS-3 map comprised 154 markers covering 1,808 cM with a mean spacing of 12.0 cM. The HAOL-9×CAS-3 map comprised 134 markers covering 1,642 cM with a mean interval of 12.5 cM between markers. Two and four QTLs controlling stearic acid content were detected in HA-89×CAS-3 (LG 1 and 7) and HAOL-9×CAS-3 (LG 1, 3, 8 and 14), respectively. The two QTLs detected in HA-89×CAS-3 and four QTLs detected in HAOL-9×CAS-3 for stearic acid content accounted for 79.3%, and 84.4% of phenotypic variation, respectively (Table 5-1). One major QTL affecting the C18:0 concentration (peak LOD 48.5) was identified on LG 1 in the HA-89×CAS-3 F_2 population, explaining 78.6% of the phenotypic variation in the C18:0 concentration. The CAS-3 allele at this QTL locus increased the C18:0 content. In HAOL-9×CAS-3 F_2 population, two major C18:0 QTLs were detected. The main C18:0 QTL was detected on LG 1 (peak LOD 50.6) and it explained 81.3% of the phenotypic variance for this trait. A stearoyl-ACP desaturase locus (SAD17A) was found to cosegregate with this locus *(Es1)* controlling the high C18:0. A second major C18:0 QTL was located on

LG 14 (peak LOD 5.1), in the same position as the major QTL for C18:1 in this population. This QTL accounted for 19.3% of the C18:0 phenotypic variation. The CAS-3 allele increased the level of C18:0 at all the putative QTLs, apart from the one on LG 8, where it decreased it.

QTLs affecting C18:1 concentration were identified on LGs 1, 8, and 14 in the HAOL-9×CAS-3 population. These QTLs together explained 58.4% of the phenotypic variance for this trait. The major C18:1 QTLs accounted for 56.5% (LG 14; peak LOD 19.7) and 24.5% (LG 1; peak LOD 8.5) of the phenotypic variance in C18:1. HAOL-9 alleles for the QTLs on LG 1 and LG 14 increased the C18:1, whereas the CAS-3 allele for the QTL on LG 8 increased the C18:1 levels. An oleoyl-PC desaturase locus (OLD7) was found to cosegregate with the gene *Ol* controlling high C18:1 on LG 14.

Sunflower seed composition is largely influenced by environmental factors, such as water availability and temperature. Severe water deficit during the reproductive stages decreases the oil content in sunflower (Nel et al. 2002). Water deficit from sowing to the end of anthesis modifies the fatty acid composition of standard hybrids (Roche et al. 2006). Severe water deficit during seed filling decreases oleic acid (OA) by 10–16%, with a concomitant increase of linoleic acid (LA), in standard hybrids of sunflower (Roche et al. 2006). Ebrahimi et al. (2008) identified QTLs controlling oil content and quality (palmitic, stearic, oleid and linoleic acide content) under different water regimes in RIL population of the cross PAC2×RHA266 (Table 5-1). The public saturated AFLP/SSR map (Poormohammad Kiani et al. 2007a) was used for this study. Oil content, palmitic acid, stearic acid, oleic acid and linoleic acid contents were investigated in the greenhouse and field conditions each under well-watered and water-stressed treatments. A wide range of phenotypic variation and transgressive segregation were observed for all the traits under different water regimes. Oil content was negatively correlated with stearic acid and oleic acid in both well-watered and water-stressed conditions in the greenhouse and field. It was positively correlated with palmitic acid and linoleic acid. Twelve, eight, six and three QTLs were identified for oil content in greenhouse well-watered, greenhouse water-stressed, field well-watered and field-water-stressed conditions, respectively. The total percentage of phenotypic variance explained by these QTLs were >100%, 90%, 66% and 34%, respectively. No explanation was given for the $R^2 > 100$%, but it could be attributed to the limited number of RILs used in their study, which leaded to overestimation of individual R^2. Although the parental lines did not differ for oil content in all four growth conditions but alleles transmitted from the PAC2 parent increased the oil content for 22 QTLs and the parent RHA266 contributed to favorable alleles of only seven QTLs out of 29 QTLs detected for oil content in the four growth conditions.

5.2.2 Disease Resistance

5.2.2.1 Resistance to Downy Mildew

Cultivated sunflower is susceptible to diseases under wetter conditions, such as those encountered in western Europe. Downy mildew and black stem respectively caused by *Plasmopara halstedii* and *Phoma macdonaldii* are considered important diseases in sunflower (Acimovic 1984; Mouzeyar et al. 1994).

Downy mildew is the most studied disease, probably because it follows a simple gene-for-gene resistance model. Several resistance clusters were described (Mouzeyar et al. 1995; Roeckel Drevet et al. 1996; Vear et al. 1997b; Bert et al. 2001) and at least one of them was independently confirmed by Gedil et al. (2001). These results are accompanied by the descriptions of molecular markers tightly linked to the downy mildew resistance, often based on candidate gene approaches, and putatively used as tools for sunflower breeding (Gentzbittel et al. 1998; Brahm et al. 2000; Gedil et al. 2001; Bouzidi et al. 2002). Using a quantitative model, Dußle et al. (2004) mapped the Pl_{Arg} locus, which confers resistance to all the known races of the fungus (Seiler 1991; Gulya 2000), in an F_2 mapping population of a cross between CmsHA342 (susceptible) and Arg1575-2 (resistant) using SSRs markers (Table 5-1). Twelve QTLs were identified using bulked segregant analysis (BSA) on 126 individuals of an F_2 progeny from the cross CmsHA342×Arg1575-2, and all of them were located on LG 1 in a window of 9.3 cM. Since Pl_{Arg} was mapped to a linkage group different from other *Pl* genes previously mapped with SSRs, the authors concluded that Pl_{Arg} provides a new source of resistance against *P. halstedii* in sunflower. Recently, Wieckhorst et al. (2008) explored the structure of the Pl_{Arg} locus on LG 1 using a population of more than 1,000 F_2 plants. The authors showed that, the recombination on LG 1 was suppressed and the backcross-derived resistant inbred line carries *H. agrophyllus* (donor) alleles throughout LG 1 and no alleles from the recurrent parent (HA89) is present. However, further work is in progress to investigate the fine structure of Pl_{Arg}, as reported by the authors.

5.2.2.2 Resistance to Diaporthe helianthi

Phomopsis (*Diaporthe helianthi* Munt-Cvet et al.) was first mentioned and identified in Yugoslavia in 1980 (Muntanola-Cvetkovic et al. 1981). It spread quickly and was rapidly found in the neighboring countries (Vranceanu et al. 1983) and in France (Regnault 1985). It was also reported in the USA (Yang et al. 1984). Since this pathogen cannot always be efficiently controlled by chemicals, breeding for resistance has to be employed to offset the disease. Like the resistance to *S. sclerotiorum*, phomopsis resistance shows continuous variation (Vear et al. 1997a). Also, different genotypes appear to possess resistances to different phases of the disease cycle (Viguié et al. 2000).

Bert et al. (2002) identified the QTLs controlling resistance to mycelium extension on leaves. They placed a mycelial explant on the upper side of the leaf tip and tested the rate of lesion development from this explant. An F_2 mapping population from the cross XRQ×PSC8 was used for this study and the experiment was conducted in three years 1998, 1999 and 2000. Two, three and three QTLs were identified for mycelium extension rate respectively in 1998 (LG 4 and 10 with total R^2 of 27.7%), 1999 (LG 4, 8 and 14 with total R^2 of 59.3%) and 2000 (LG 4, 8 and 17 with total R^2 of 47.2%). For the mean of three years, four QTLs were identified on LG 4, 8, 10 and 17 with total phenotypic variance explained of 49.2%. Results showed that both the parental lines transmitted the positive alleles for the QTLs. Interestingly the linkage groups 4, 8 and 10 contained QTLs for resistance to mycelium extension on leaves detected in 1998, 1999 and 2000 that showed a partially stable QTLs across years.

5.2.2.3 Resistance to Phoma macdonaldii

Black stem, caused by *Phoma macdonaldii*, is one of the most important diseases of sunflower in the world. It causes premature ripening associated with yield losses of 10–30%, and also reduction in oil content and 1,000-seed weight (Carson 1991). Roustaee et al. (2000), using parental genotypes and their F_1 hybrids, showed that the variation among genotypes studied was due to general combining ability and thus most of the variation was attributed to additive effects. Developing partially resistant genotypes using QTL approach is the main challenge to cope with this disease.

Rachid Al-Chaarani et al. (2002) identified seven QTLs for resistance to black stem using an AFLP-based RIL map of the cross PAC2×RHA266 (Table 5-1). A population of 83 RILs was used for inoculation with an agressive monopycniospore isolate of *P. macdonaldii* produced from naturally infected plants in the south-west of France. The seven detected QTLs for black stem resistance jointly explained 92% of the total phenotypic variance and were located on LG 3, 4, 8, 9, 11, 15 and 17. For six of the seven QTLs, the parent RHA266 conferred positive alleles and the parent PAC2 contributed to positive alleles in only one QTL.

In a later study, Bert et al. (2004) used a population of 150 F_2:F_3 plants from the cross between FU (an unbranched maintainer line) and PAZ2 (a male fertility restorer line) for mapping QTLs involved in resistance to *P. macdonaldii* (Table 5-1). A genetic linkage map of 18 LGs (a pseudo LG) with 216 molecular markers spanning 1,937 cM was constructed. Four QTLs were identified for resistance to *P. macdonaldii* necrosis on LGs 1, 3, 15 and 16, which explained all together 39% of the phenotypic variance. The LOD scores ranged from 3.13 to 3.75 and each parental line contributed to positive alleles for two QTLs.

Using the PAC2×RHA266 RIL population and a saturated map, Darvishzadeh et al. (2007) identified 10 QTLs for partial resistance to two *P. macdonaldii* isolates (MP6 and MP8), on LGs 1, 2, 5, 9, 11, 15 and 17. The LGs of this map corresponds to sunflower reference SSR map constructed by Tang et al. (2002), thus the comparison between this map and other maps with standard linkage group nomenclature was feasible. Among the 10 QTLs detected, four were isolate-non-specific and six were isolate-specific. The phenotypic variation explained by these QTLs ranged from 6% to 20% with LOD scores varying from 3.4 to 8.0. The total phenotypic variation explained by the QTLs controlling partial resistance to each isolate was 61% and 62%, and both parental lines contributed equally the positive alleles. Only one QTL controlling partial resistance to black stem isolate MP8 was detected by Rachid Al-Chaarani et al. (2002) and confirmed by Darvishzadeh et al. (2007), which is located on LG 17.

QTL analysis of partial resistance to basal stem and root necrosis caused by *Phoma macdonaldii* was also reported by Abou Alfadil et al. (2007). The RIL population from the cross PAC2×RHA266 and corresponding SSR/AFLP map with standard linkage group nomenclature was used for this study. A total of 27 QTLs were detected for partial resistance to four isolates infecting root and basal stem of sunflower. The percentages of phenotypic variation (R^2) explained by each QTL ranged from 7% to 29%, with LOD scores ranging from 3.1 to 11.2. The authors showed that partial resistances to basal stem and root necrosis are controlled by both isolate-specific and isolate-non-specific QTLs. It is of interest that most of the QTLs common to the different isolates were those with the highest contribution in terms of phenotypic variance.

5.2.2.4 Resistance to Sclerotinia sclerotiorum

Sclerotinia sclerotiorum (Lib) de Bary is an omnivorous and non-specific plant pathogen. In all sunflower growing regions of the world, *S. sclerotiorum* is common and widespread (Gulya et al. 1997). Under severe infection, yield losses in sunflower can reach up to 100% (Sackston 1992) depending on the infected plant parts. The fungus causes three distinct types of disease on sunflower: wilt, midstalk rot, and head rot. Results from the literature are ambiguous concerning the association of susceptibility of sunflower genotypes to *S. sclerotiorum* infection on root, leaf, and head. Therefore, different plant parts were considered for inoculation and evaluation of resistance to *S sclerotiorum*. Chemical measures to control *S. sclerotiorum* in sunflower are ineffective and development of highly resistant sunflower cultivars is desirable under ecological and economical aspects. The genetic mechanisms underlying *S. sclerotiorum* resistance are complex. Genetic studies demonstrated a polygenic inheritance of resistance for all three forms of infection (root, stalk, and head; Robert et al. 1987; Tourvieille and Vear 1990) and no race specificity (Thuault and Tourvieille 1988). Earlier studies suggested additive gene action to be more important than dominance (Robert et al. 1987). Several studies have been undertaken for genetic analysis of resistance to *S. sclerotiorum* using molecular markers (Mestries et al. 1998; Bert et al. 2002, 2004; Micic et al. 2004, 2005) and important sources of resistance have been identified (Micic et al. 2004, 2005).

Mestries et al. (1998) used three generations (F_2, F_3 and F_4) derived from a cross between two inbred lines GH and PAC2, for mycelium test on leaves and capitulum in Clermont-Ferrand in central France. The linkage map was constructed using 82 isoenzymes and RFLP markers, which covered 760 cM of the sunflower genome and the average distance between markers was 12.6 cM. For lesion length on leaves, one (LG A), two (LG G and P) and two (LG I) QTLs were found in F_2, F_3 and F_4 generations, with R^2 of 10.5%, 18.8% and 20.5%, respectively. For capitulum index, as another resistant test, one (LG A), two (LG A and M) and two (LG A and M) QTLs were identified in F_2, F_3 and F_4 generations, with R^2 of 11.3%, 20.8% and 30.6%, respectively. For eight QTLs of the 10 QTLs detected for the two traits in three generations, the PAC2 alleles contributed to positive alleles. Linkage group A and M contained four and two QTLs, respectively controlling either both traits (LG A) or one trait in two generations (LG M).

Bert et al. (2002) identified the QTLs controlling resistance to mycelium extension on leaves and capitulum. An F_2:F_3 mapping population from the cross XRQ×PSC8 was used for this study and the experiment was conducted in two years 1997 and 1999. Mycelium extension on leaves and capitulum were studied in 1997 and 1999, respectively, but percentage of pathogen attack and latency index were studied in both the years in F_3 population. Three and two QTLs were identified for mycelium extension on leave and capitulum in 1997, with total R^2 of 56.1% and 25.6%, respectively. Five (total R^2 of 43.3%) and one (R^2 of 9.9%) QTLs were identified for percentage of attack latency index in 1997, but three (total R^2 of 31.8%) and one (R^2 of 10.4%) QTLs were found for the same traits in 1999 respectively. Both the parental lines contributed the positive alleles to the QTLs, but PSC8 contributed solely the positive alleles for latency index in both the years. QTL co-location was observed for some traits on LG 7, 8 and 13. The position of QTLs from this study when compared with those observed previously by Mestries et al. (1998) showed no co-location of QTLs for the same traits. The authors found that sunflower resistance to *S. sclerotiorum* is controlled by several resistant factors, which differ according to the genotypes tested.

In a later study, Bert et al. (2004) used a population of 150 F_2:F_3 plants from the cross between FU (an unbranched maintainer line) and PAZ2 (a male fertility restorer line) for mapping QTLs involved in resistance to *S. sclerotiorum* (Table 5-1). For details on the parental lines, the population and map construction see the section 5.2.2.3. Seven QTLs were identified for *Sclerotinia* attacks on terminal buds, three were found for attack on capitolum and four were identified for latenecy index. The LOD scores were from 2.85 to 12.79. FU parent contributed the positive alleles for six QTLs of the seven detected for the attack on terminal buds. For attack on capitolum, FU parent contributed the positive alleles for only one of the three QTLs. Both the parental lines contributed equally the positive alleles for latency index QTLs. The R^2 of QTLs was moderate varying from 2.5% to a maximum of 18.7%. The authors showed also further evidence for additive gene effect for the resistance to *S. sclerotiorum* in sunflower.

Micic et al. (2004) used an F_2:F_3 mapping population (354 F_3 families) from the cross between a resistant inbred line (NDBLOS) to stalk rot and a susceptible line (CM625) to identify the QTLs controlling resistance to *S. sclerotiorum* using artificial inoculation. The map was constructed by using 117 SSR markers with a total length of 961.9 cM and average interval length of 9.6 cM. The LGs were numbered as sunflower reference map (Tang et al. 2002). Nine QTLs were identified for the leaf lesion on LG 1, 4, 6, 8, 9, 13 and 15; and seven QTLs were detected for stem lesion on LG 2, 3, 4, 6, 8, 15 and 16, among which those located on LG 4 are common for both the traits. The total phenotypic variance explained by the QTLs for leaf lesion was 55.8%, but it was 71.7% for the QTLs controlling stem lesion (Table 5-1). The LOD scores varied from 2.51 to 36.7 and a major QTL were identified for stem lesion on LG 8, the position 20 cM from the north of LG. The authors showed that a large number of QTLs with small effects are involved in resistance to midstalk, which confirmed the polygenic nature of *S. sclerotiorum* resistance in sunflower reported previously by Mestries et al. (1998). The authors also identified significant QTLs for stem lesion on three of the four LGs reported by Mestries et al. (1998). Cross-reference between two maps showed that three QTLs identified for trait mycelium on leaves on LGs 6, 8, and 13 by Bert et al. (2002), coincided with three QTLs for leaf lesion on LGs 13, 9, and 1, respectively of Micic et al. (2004).

Micic et al. (2005) identified three QTLs for stem lesion, leaf lesion and speed of fungal growth, and validated some of them by selective genotyping in an F_3 mapping population derived from the cross betwnn TUB-5-3234 and CM625, being resistant and susceptible parents, respectively (Table 5-1). Some of the QTLs for different disease resistance traits were overlapped on linkage groups 4, 10 and 17, and two QTLs for stem lesion showed large genetic effects and corroborated earlier findings from another cross NDBLOS×CM625 (Micic et al. 2004). The QTLs explained up to 25% of the phenotypic variance and in a simultaneous fit 75.9% of the genotypic variance was explained for leaf lesion and 84.6% for speed of fungal growth.

The LOD scores varied between 2.50 and 11.27. In sunflower, candidate genes for tolerance to *Sclerotinia* were also described (Mouzeyar et al. 1997; Gentzbittel et al. 1998) and were evaluated in subsequent QTL analyses (Bert et al. 2002).

Gentzbittel et al. (1998) conducted a candidate gene approach to analyze the resistance to *S. sclerotiorum*. They obtained two probes homologous to two plant resistance nucleotide binding site (NBS) and serine-threonine protein kinase (PK) -like genes by homology cloning. They mapped these two clones in three mapping population from the crosses (SD×PAC1; CP73×PAC1 and GH×PAC1). In three crosses, resistance to *S. sclerotiorum* QTLs were mapped at the position of PK locus (LG 1), which controlled up to 50% of the phenotypic variation for this trait.

The new QTLs for resistance to *Sclerotinia* head rot in two USA sunflower germplasms were recently identified in $F_{2:3}$ and $F_{2:4}$ families from a cross between HA441 and RHA439 (Yue et al. 2008b; Table 5-1). The experiment was conducted in three tests, and disease incidence (DI) and disease severity (DS) were studied. Disease incidence (DI) was determined as the percentage of plants infected within a row and disease severity (DS) was estimated by the average disease score of the infected plants within each plot. The data from three tests, Carrington in 2006 (test 1), Carrington in 2007 (test 2), and Fargo in 2007 (test 3) were used for QTL analysis (Yue et al. 2008b). Nine QTLs were detected for DI in the three tests, each explaining 9.6% to 26.4% of the phenotypic variance; whereas seven DS resistance QTLs were identified in three tests, explaining from 8.4% to 34.5% of the phenotypic variance. The LOD scores varied from 2.4 to 11.8 and both the parental lines contributed to the trait variation suggesting quantitative nature of resistance to *Sclerotinia* in sunflower.

Recently, QTLs for capitulum resistance to *Sclerotinia sclerotiorum* have also been identified by Vear et al. (2008). A set of 279 RILs from a cross between two INRA lines, XRQ and PSC8, were used for map construction using 39 RFLPs, 162 SSRs and four Mendelian traits. The RILs were then subjected to *Sclerotinia* ascospore infection at the flowering time and two traits, including latency index (delay from infection to symptom appearance) and mean percentage attack were studied (Table 5-1). Two and three QTLs were detected for latency index and percentage attack, respectively, each explaining 8 to 21% of the phenotypic variance with LOD scores of 4.6 to 13.9. Among the five QTLs detected, two were co-located on LG 1 and LG 10. As the phenotypic variance explained by individual QTLs, had low values, the authors concluded that capitulum resistance to *Sclerotinia* is truly 'polygenic' in sunflower.

5.2.2.5 Resistance to Orobanche cumana

Under semidry conditions (Spain, Israel, Northern Africa), *Orobanche cumana*, a root parasite of sunflower appears to be one of the most important constraints of sunflower production. Different races from A to E of broomrape (*Orobanche cumana*) have been described (Alonso 1998), and a monogenic dominant inheritance of resistance to races A to E was found in most of the genetic studies (Pogorletsky and Geshele 1976; Vranceanu et al. 1980; Ish-Shalom-Gordon et al. 1993; Sukno et al. 1999). However, some reports pointed to a more complex inheritance of the trait, including two dominant genes (Domenguez 1996), one recessive gene (Ramaiah 1987), double recessive epistasis (Kirichenco et al. 1987) or even quantitative inheritance (Pustovoit 1966).

Genetic mapping of several sources of resistance to *Orobanche cumana* were described (Lu et al. 1999, 2000) leading the way to putative breeding for appropriate sunflower inbred lines. In the late 1970s, five pathogenic races named A through E, with a set of sunflower differentials carrying the dominant resistance genes *Or1* through *Or5* have been identified. Pérez-Vich et al. (2004) identified five QTLs for resistance to the race E and six QTLs for resistance to the race F (identified later in the mid 1990s in Spain) of broomrape on seven out of the 17 linkage groups (Table 5-1). Phenotypic variance for resistance to the race E was mainly explained by a major QTL ($R^2 = 59\%$), while the race F resistance was explained by QTLs with small to moderate effects ($R^2 = 15.0$–38.7%). The results suggested that resistance to broomrape in sunflower is controlled by a combination of qualitative, race-specific resistance affecting the presence or absence of broomrape and a quantitative non-race specific resistance affecting their number. However, the nature of the *Or1* to *Or5* genes involved in this interaction is not known. Marquez-Lema et al. (2008) investigated the nature of the *Or5* gene through a candidate gene approach, and identified markers close to the gene in order to facilitate map-based cloning strategies. Using an F_2 population from the cross between a resistant line to races E and F of broomrape (P-21) and a susceptible line (P-96), several SSRs, TRAPs and candidate gene markers were genotyped and the linkage map was constructed. The authors mapped the *Or5* gene on LG 3 close to TRAP marker, TRC27133 and SSR marker, CRT392c. Unfortunately, among the 12 candidate genes tested, three were mapped on LGs 7, 9 and 17, but none of them was located close to the *Or5* gene, which needs further investigation (Marquez-Lema et al. 2008).

Linkage of molecular markers with resistance genes to rust was also described in sunflower (Lawson et al. 1998). Thus, several tools are already available to breeders for the improvement of breeding processes for disease resistance in sunflower.

5.2.3 Self-incompatibility and Seed Dormancy

Self-pollination (SP) and seed dormancy (SD) are genetically complex traits and a number of self-incompatibility (SI) loci has been disputed (Gandhi et al. 2005). Wild populations of common sunflower (*Helianthus annuus* L.) are self-incompatible (Heiser 1954; Heiser et al. 1969) and have high seed dormancy (Heiser 1976; Seiler 1988, 1996, 1998), whereas modern cultivars, inbreds, and hybrids are self-compatible and partially-to-strongly self-pollinated (Luciano et al. 1965; Fick and Zimmer 1976; Fick and Rehder 1977; Fick 1978), and have short-lived seed dormancy (Alissa et al. 1986; Corbineau et al. 1990). Self-incompatibility (SI) and seed dormancy (SD) complicate breeding in elite × wild hybrids (Seiler 1992), and the genetic mechanisms underlying SI, SP, and SD in sunflower are either unknown or superficially known. Gandhi et al. (2005) mapped the QTLs controlling SP, SI and SD in a backcross mapping population from the cross between an elite, self-pollinated, non-dormant inbred line (NMS373) and a wild, self-incompatible, dormant population (ANN1811) (Table 5-1). A population consisting of 212 BC_1 progeny was subsequently produced by backcrossing a single hybrid individual to NMS373. The authors identified one QTL for SI ($R^2 = 66.2\%$) and three QTLs for each SP ($R^2 = 63.9\%$) and SD ($R^2 = 38.3\%$) on different linkage groups. The results differentiated successfully between loci governing SI and SP and identified DNA markers for bypassing SI and SD in elite × wild crosses through marker-assisted selection.

5.2.4 Fertility Restoration and Male Sterility

Hybrid breeding in sunflower is based on a single source of cytoplasmic male sterility (CMS), which is present in all the female but male-sterile plants. To produce the hybrid seeds, the CMS plants are crossed with restorer lines that have the *Rf1* (restorer of fertility) gene to obtain fertile plants. A novel fertility restoration gene has been recently mapped in relation to disease resistance loci, downy mildew and black rust (Abratti et al. 2008). Using an F_2 segregating population from the cross RHA340×ZENB8, the authors reported a new restoration gene on linkage group 7. The ability of molecular markers linked to this locus facilitates the introgression of this gene in different lines of breeding programs for developing new restorer lines. However, map-based cloning of the *Rf* genes may also help to clarify the mechanism behind the expression of the CMS-associated mitochondrial gene.

Two new single recessive nuclear male sterility (NMS) genes 'ms_6' and 'ms_7' were also identified and mapped using molecular marker approaches (Capatana et al. 2008; Li et al. 2008). An F_2 population from the cross between nuclear male sterile mutant NMS HA89-872 and the male fertile line RHA271 was used and "ms_6" and "ms_7" were mapped on linkage groups 16 and 6, respectively. The gene "ms_6" was flanked by ORS807 and ORS996 at distances of 7.9 cM and 18.5 cM, respectively, and "ms_7" was flanked by ORS608 and Tg3r165a-185 at distances of 2.6 cM and 4.7 cM, respectively. These molecular markers can offer a quick and precise detection of genotypes bearing a male sterility allele useful in hybrid production.

5.2.5 *Physiological Traits*

Yield components and oil production are known to be positively correlated with photosynthesis and water status traits in sunflower. Hervé et al. (2001) identified QTLs for photosynthesis and water status traits using RILs from the cross PAC2×RHA 266; and the results are summarized in Table 5-1. Four QTLs detected for chlorophyll concentration accounted for 53.5% of the phenotypic variation for this trait and three QTLs for net photosynthesis accounted for 62.5% of total phenotypic variation. As far as stomatal conductance is concerned, four QTLs with 61.9% of the phenotypic variation were detected, whereas predawn water potential was associated with three QTLs explaining only 34.1% of phenotypic variance.

5.2.6 *In vitro Regeneration*

The ability to regenerate large number of plants is important for the development of biotechnological tools such as genetic transformation in sunflower. In vitro regeneration was investigated in 75 RILs and their parents (PAC2 and RHA266). The results summarized in Table 5-1 showed that the six putative QTLs for the number of shoots per plant and seven for shoots per regenerating explants were detected in cotyledon organogenesis culture (Flores-Berrios et al. 2000a). The same RILs were also used for somatic embryogenesis by epidermic layers and four QTLs were identified for the total number of embryogenic explants, which explained 48% of the phenotypic variation for this trait (Flores-Berrios et al. 2002b). The above-mentioned RILs were also used in another experiment in which 12 QTLs were identified for protoplast division with a total phenotypic variation of 72% (Flores-Berrios et al. 2000c). The results showed that some segments of the linkage groups l, 15 and 17 are likely to contain genes important for organogenesis, somatic embryogenesis and protoplast division. The QTLs identified in these three linkage groups should be involved in cell division in early events associated with cell differentiation. Genomic regions involved in somatic embryogenesis have also been identified in PAC2×RHA266 RIL mapping population, and three QTLs were resolved for the number of embryogenic explants per 100 explants cultured, (Huang et al. 2007). These three QTLs explained 38.6% of the phenotypic variation, with LOD scores ranging from 6.4 to 7.5. The resolution of these QTLs was increased by crosses between two highly embryogenic lines (LR35 and C40) with a poorly embryogenic line (C149) and the construction of two F_2/F_3 linkage maps. The author showed that somatic embryogenesis in sunflower is highly genotype-dependent and quantitative in nature. Reducing the QTL interval into a much shorter region resulted in the identification of four AFLP markers on linkage groups 5 and 13 closely linked to somatic embryogenesis potential.

5.2.7 Abiotic Stress Tolerance

5.2.7.1 Plant Water Status and Osmotic Adjustment

Sunflower offers a rich source of genetic diversity for abiotic stress tolerance, particularly drought and salt. Drought tolerance has been considered as a valid breeding target to partially compensate for the loss in yield. However, this has yet to be exploited. The first QTL analysis of drought tolerance-related traits has been reported by Poormohammad Kiani et al. (2007a, b, 2008; Table 5-1) using a saturated linkage map. QTL mapping of water status traits (relative water content, leaf water potential, osmotic potential and turgor potential) as well as osmotic adjustment was investigated in an RIL population derived from a cross between PAC2 and RHA266 under well-watered and water-stressed conditions. Using a saturated linkage map, three to eight QTLs were detected for each trait explaining from 6% to 29% of the phenotypic variance of the traits; and LOD scores varied from 3.05 to 9.86 (Poormohammad Kiani et al. 2007a). Among the 24 QTLs detected under well-watered conditions, 5 (21%) were also detected under water-stressed conditions (stable QTLs). Eight QTLs were detected for osmotic adjustment on LGs 2, 4, 5, 12 and 13, and four of them were overlapped with the QTLs of water-status traits on LGs 4, 5 and 12 (Poormohammad Kiani et al. 2007a).

The most important chromosomal region for osmotic adjustment is located on LG 5, where the QTL for osmotic adjustment (*OA.5.2*) is overlapped with QTLs for relative water content in both water treatments as well as with the QTLs for leaf water potential under water-stressed conditions. This region, which is linked to an SSR marker (ORS523-1), appears very important in water stress tolerance, as it is involved in the expression of osmotic adjustment as well as in the expression of osmotic potential at full turgor, leaf water potential and relative water content under water-stressed condition. This QTL explains 29% of the phenotypic variance in OA and also 10%, 8% and 18% of phenotypic variance in leaf water potential, relative water content and osmotic potential at full turgor under water-stress, respectively. It explains also 25% of relative water content phenotypic variance under well-watered conditions.

The locations of QTLs identified for plant water status and osmotic adjustment when compared with those controlling agronomic traits under well-watered conditions reported by Rachid Al-Chaarani et al. (2004), showed overlapping for some of the water status and agronomical traits. The overlapping QTLs are located on: LG 2 for turgor potential and stem diameter as well as for osmotic adjustment and sowing-to-flowering period; LG 7 for relative water content and sowing-to-flowering period and for leaf water potential and sowing-to-flowering period; LG 12 for leaf water potential and percentage of oil in the grain and linkage group 17 for osmotic potential and stem diameter (Poormohammad Kiani et al. 2007a). The co-location of QTLs for water status traits and oil percentage as well as for agronomical traits reported above suggests a clear biological link between plant water status and yield-related traits.

5.2.7.2 Yield and Related Traits under Drought Conditions

The relationship between grain yield and related agro-physiological traits was confirmed recently by an experiment conducted under different water treatments using the same RIL population previously used for mapping drought-adaptive traits (Poormohammad Kiani et al. 2007a, 2009; see Table 5-1 and Fig. 5-1). The authors evaluated days from sowing to flowering (DSF), leaf number per plant (LN), leaf area at flowering (LAF), leaf area duration (LAD), plant height (PH), total dry matter (BIO), head weight (HW) and grain yield per plant (GYP) as major yield-related traits in the greenhouse and field, under well-watered and water-stressed conditions. Two to seven QTLs were identified depending on the trait and growth conditions with R^2 ranging from 4% to 40% (Table 5-1). Most of the QTLs were detected under more than one growth conditions and interesting co-locations were found between QTLs controlling grain yield per plant (GYP) and related agro-morphological traits (see Fig. 5-1).

Several QTLs were detected for DSF in four different growth conditions and most of them were common across at least two growth conditions. The most important QTL for DSF was located on LG 7 where several QTLs under different growth conditions were co-localized (Fig. 5-1). The positive alleles for these overlapped QTLs came from RHA266. Seventeen QTLs were identified for LN under four growth conditions among which nine were common across at least two growth conditions and eight were detected under only one condition. Both the parental lines contributed almost equally to the QTL expression. For LAF, 21 QTLs were detected under four growth conditions, their number being from three to six depending on growth conditions. Among 21 QTLs, nine were detected in only one of the growth conditions and 12 were detected in at least two growth conditions. The phenotypic variance explained by each QTL ranged from 5% to 19%, and both the parental lines contributed positive alleles. As far as LAD is concerned, 22 QTLs were identified under four growth conditions, explaining from 4% to 17% of the total variation. Eleven QTLs were specific to single water treatment and 11 were detected in at least two water treatments. The positive alleles for 17 QTLs came from RHA266 and for five QTLs they come from PAC2. A total of 17 QTLs were detected for PH being six unique QTLs and 11 QTLs that were detected in at least two growth conditions. The QTLs explained from 5% to 23% of the phenotypic variance and both parental lines contributed to trait expression. However, RHA266 contributed to positive alleles at 10 QTLs. For total dry matter per plant (BIO), 19 QTLs were identified, explaining from 5% to 23% of variation. The number of QTLs in four growth conditions varied from two to six; 13 were detected in only one of the growth conditions and six were common across different growth conditions. RHA266 contributed positive alleles for 14 of the 19 QTLs. A total of 24 QTL were identified for HW per plant under four growth conditions with the phenotypic variance explained from 4% to 24%. The number of QTLs differed from four to seven depending on the growth condition. Among the 24 QTLs, 16 were detected in only one of the growth conditions and eight were common across different water treatments. PAC2 contributed positive alleles at 10 QTLs and RHA266 contributed for 14 QTLs. For GYP, 20 QTLs were identified under four water treatments with the phenotypic variance explained ranging from 4% to 40%. Nine of the 20 QTLs were identified in only one growth condition and the rest were common across at least two growth conditions. PAC2 and RHA266 contributed equally to the QTLs controlling GYP. Interestingly, only two individual QTLs specific for GYP were identified, and the rests were co-located with agro-morphological traits.

An interesting genomic region was identified on LG7 where QTLs for several developmental, morphological and physiological traits including flowering time, leaf number, leaf area, leaf area duration, plant height and total dry matter were co-located (Fig. 5-1). These region, which is linked to an SSR marker (ORS331) has been characterized in detail by Chapman et al. (2008). The authors showed that the region surrounding ORS331 and ORS143 contains candidate genes significantly linked to QTL clusters for flowering time and leaf number in wild × improved RILs.

Figure 5-1 contd....

Figure 5-1 The AFLP and SSR map of sunflower from the cross between two public inbred lines (PAC2 and RHA266) used for genetic analysis of physiological, morphological and agronomic traits in sunflower (from Poormohammad Kiani et al. 2007a, 2008, 2009). Recombinant inbred lines were used for the construction of the map. The map length is 1,824.6 cM and one marker per each 3.7 cM. The markers are presented at left side and the QTLs are presented at right side of the linkage groups. The co-location of yield with yield-related and drought adaptive traits under different water treatment are shown. The vertical bars show confidence interval of QTLs. (For more detail on QTL abbreviations see Table 5-1).

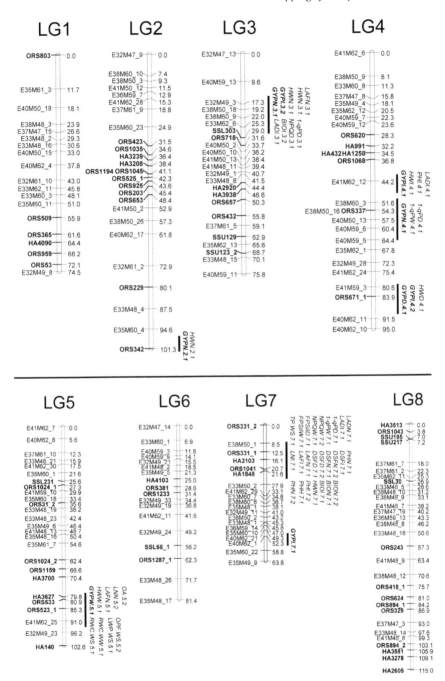

Figure 5-1 contd....

Figure 5-1 contd....

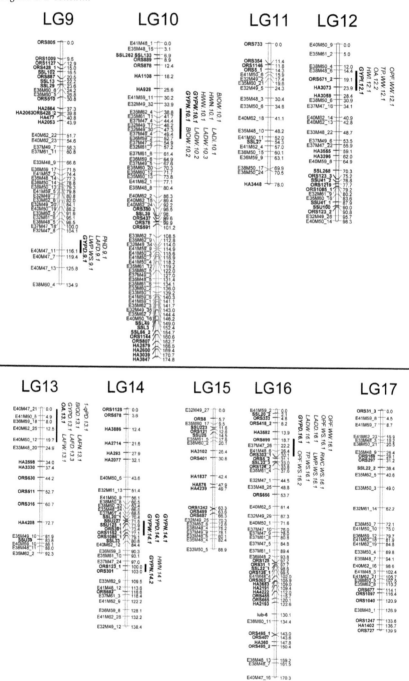

In another experiment conducted by Poormohammad Kiani et al. (2008), QTLs controlling chlorophyll fluorescence parameters, as the functioning of photosynthesis apparatus were mapped under drought stress (Table 5-1). Four chlorophyll fluorescence parameters were studied as: the potential photochemical efficiency of photosystem II (PSII) electron transport (ΦP), the actual efficiency of PSII electron transport ($\Phi PSII$), non-photochemical fluorescence quenching (NPQ) and the proportion of closed PSII traps (1-qP). The number of QTLs varied from 5 to 9, depending on the trait and water treatments, explaining from 5% to 26% of the phenotypic variance of the traits; and LOD scores varied from 3.06 to 12.72. Five QTLs were constitutive and the rests were water treatment-specific (Poormohammad Kiani et al. 2008). Several QTLs controlling photosynthesis-related traits were overlapped with the QTLs controlling plant water status traits on LGs 1, 5, 7, 12, 13, 14, 16 and 17 (Poormohammad Kiani et al. 2007a, 2008), which showed that photosynthesis performance and water status traits are genetically related and should therefore be considered together when used as selection criteria for drought tolerance improvement in sunflower.

Identification of QTLs influencing several traits and yield under different growth conditions, using the same mapping population (PAC2×RHA266) facilitated (1) the identification of consistent genomic regions from those expressed under specific growth conditions for each trait of interest specially for yield and (2) the genetically determination of the trait association by evaluation of overlapping QTLs (Fig. 5-1). This would increase the efficiency of MAS and enhance genetic progress. One of the major goals for sunflower plant breeders is to develop genotypes with high yield potential and the ability to be stable across environments. There are three main ways in which a genotype can achieve yield stability. The first one is the identification of the non-environment-specific QTLs or QTLs with minor interaction with environment (as those located on LG 14 for yield), which should be particularly useful in MAS for yield. The second is the development of widely adapted genotypes by pyramiding different QTLs each controlling adaptation to a different environment (as nine environment-specific QTLs for yield). Finally, the identification of QTLs for yield-related traits which are co-located with yield itself can provide the information about which traits and alleles can increase yield in particular growth condition. The latter helps indirect selection for yield by selecting yield-related traits which share common QTLs with yield. The whole results obtained by Poormohammad Kiani et al. (2007a, b, 2008, 2009) showed major genomic regions controlling plant water status, osmotic adjustment and photosynthesis performance as well as yield-related traits under drought stress, which could be used in MAS for drought tolerance in sunflower.

5.3 Conclusion

The genetics of a broad spectrum of traits, including seed oil concentration, root morphology, branching, seed dormancy, heterosis, male sterility, fertility restoration, flowering time, seed shattering, self-incompatibility, fatty acid and tocopherol composition and concentration, agronomical and physiological traits, drought and diseases resistance have been and are being analyzed in different mapping populations of sunflower. These traits are being mapped relative to candidate genes mined from the expressed sequence tag (EST) databases to provide maps based on transcribed sequences. Recent advances in experimental and statistical procedures, including the simultaneous analysis of QTL that segregate in diverse germplasm, should improved genetic architecture studies of complex traits. However, cost-effective genomic tools as well as accurate phenotypic analysis are also necessary. Improved understanding of the evolution and positive impact on the breeding sunflower should follow advances in the molecular mapping of complex traits. Apart from MAS, the ultimate objective of QTL mapping might be to identify the causal genes, or even the causal sequence changes, the quantitative trait nucleotides (QTN). Most QTN identification has proceeded from whole-genome scans to initially localize QTL to 10–20 cM intervals, followed by high resolution mapping, and finally, detailed genetic and complementation analysis of the sequence variants that co-segregate with the QTL. While this remains a major challenge, the QTL regions presented in this chapter could be an initiation point for resolving the gene level.

References

Abou Alfadil T, Poormohammad Kiani S, Dechamp-Guillaume G, Gentzbittel L, Sarrafi A (2007) QTL mapping of partial resistance to *Phoma* basal stem and root necrosis in sunflower (*Helianthus annuus* L.). Plant Sci 172: 815–823.

Abratti G, Bazzalo ME, Leon A (2008) Mapping a novel fertility restoration gene in sunflower. In: Proc 17th Int Sunflower Conf, 8–12 June 2008, Cordoba, Spain, pp 617–621.

Acimovic M (1984) Sunflower diseases in Europe, the United States and Australia, 1981–1983. Helia 7: 45–54.

Alissa A, Jonard R, Serieys H, Vincourt P (1986) La culture d'embryons isole dans un programme d'amelioration du tournesol. CR Acad Sci Paris 302: 161–164.

Alonso LC (1998) Resistance to *Orobanche* and resistance breeding: a review. In: K Wegmann, LJ Musselman, DM Joel (eds) Current Problems of *Orobanche* Researches. Proc 4th Int Workshop on *Orobanche*, 23–26 Sept 1998, Albena, Bulgaria, pp 233–257.

Alvarez D, Luduena P, Frutos E (1992) Variability and genetic advance in sunflower. In: Proc 13th Int Sunflower Conf, Pisa, Italy, pp 963–968.

Berry ST, Leon AJ, Hanfrey CC, Challis P, Burkolz A, Barnes SR, Rufener GK, Lee M, Caligari PDS (1995) Molecular-marker analysis of *Helianthus annuus* L. 2. Construction of an RFLP map for cultivated sunflower. Theor Appl Genet 91: 195–199.

Bert PF, Tourvieille de Labrouhe D, Philippon J, Mouzeyar S, Jouan I, Nicolas P, Vear F (2001) Identification of a second linkage group carrying genes controlling resistance to downy mildew (*Plasmopara halstedii*) in sunflower (*Helianthus annuus* L.). Theor Appl Genet 103: 992–997.

Bert PF, Jouan I, Tourvieille de Labrouhe D, Serre F, Nicolas P, Vear F (2002) Comparative genetic analysis of quantitative traits in sunflower (*Helianthus annuus* L.) 1. QTL involved in resistance to *Sclerotinia sclerotiorum* and *Diaporthe helianthi*. Theor Appl Genet 105: 985–993.

Bert PF, Jouan I, de Labrouhe DT, Serre F, Philippon J, Nicolas P, Vear F (2003) Comparative genetic analysis of quantitative traits in sunflower (*Helianthus annuus* L.) 2. Characterisation of QTL involved in developmental and agronomic traits Theor Appl Genet 107: 181–189.

Bert PF, Dechamp-Guillaume G, Serre F, Jouan I, de Labrouhe DT, Nicolas P, Vear F (2004) Comparative genetic analysis of quantitative traits in sunflower (*Helianthus annuus* L.) 3. Characterisation of QTL involved in resistance to *Sclerotinia sclerotiorum* and *Phoma macdonaldii*. Theor Appl Genet 109: 865–874.

Bouzidi MF, Badaoui S, Cambon F, Vear F, Tourvieille de Labrouhe D, Nicolas P, Mouzeyar S (2002) Molecular analysis of a major locus for resistance to downy mildew in sunflower with specific PCR-based markers. Theor Appl Genet 4: 592–600.

Brahm L, Rocher T, Friedt W (2000) PCR-based markers facilitating marker assisted selection in sunflower for resistance to downy mildew. Crop Sci 402: 676–692.

Burke JM, Tang S, Knapp SJ, Rieseberg LH (2002) Genetic analysis of sunflower domestication. Genetics 6l: 1257–1267.

Capatana A, Feng J, Vick BA, Duca M, Jan CC (2008) Molecular mapping of a new induced gene for nuclear male sterility in sunflower (*Helianthus annuus* L.). In: Proc 17th Int Sunflower Conf, 8–12 June 2008, Cordoba, Spain, pp 641–644.

Carson ML (1991) Relationship between phoma black stem severity and yield losses in hybrid sunflower. Plant Dis 75: 1150–1153.

Chapman MA, Pashley CH, Wenzler J, Hvala J, Tang S, Knapp SJ, Burke JM (2008) A Genomic scan for selection reveals candidates for genes involved in the evolution of cultivated sunflower (*Helianthus annuus*). Plant Cell 20: 2931–2945.

Corbineau F, Baginol S, Come D (1990) Sunflower (*Helianthus annuus* L.) seed dormancy and its regulation by ethylene. Isr J Bot 39: 313–325.

Darvishzadeh R, Poormohammad Kiani S, Dechamp-Guillaume G, Gentzbittel L, Sarrafi A (2007) Quantitative trait loci associated with isolate specific and isolate nonspecific partial resistance to *Phoma macdonaldii* in sunflower. Plant Pathol 56: 855–861.

Dominguez J (1996) *R-41*, a sunflower restorer inbred line, carrying two genes for resistance against a highly virulent Spanish population of *Orobanche cernua*. Plant Breed 115: 203–204.

Dorrell DG, Vick BA (1997) Properties and processing of oilseed sunflower. In: AA Schneiter (ed) Sunflower Technology and Production. ASA, CSSA and SSSA, Madison, Wisconsin, pp 709–746.

Dußle CM, Hahn V, Knapp SJ, Bauer E (2004) Pl_{Arg} from *Helianthus argophyllus* is unlinked to other known downy mildew resistance genes in sunflower. Theor Appl Genet 109: 1083–1086.

Ebrahimi A, Maury P, Berger M, Poormohammad Kiani S, Nabipour A, Shariati F, Grieu P, Sarrafi A (2008) QTL mapping of seed-quality traits in sunflower recombinant inbred lines under different water regimes. Genome 51: 599–615.

Fick GN (1975) Heritability of oil content in sunflowers. Crop Sci 15: 77–78.

Fick GN (1978) Selection for self-fertility and oil percentage in development of sunflower hybrids. In: Proc 8th Int Sunflower Conf, Int Sunflower Assoc, Minneapolis, USA, pp 418–422.

Fick GN, Zimmer DE (1976) Yield stability of sunflower hybrids and open pollinated varieties. In: Proc 7th Int Sunflower Conf, Jun. 27–Jul. 3, 1976. Krasnadar, USSR, pp 253–258.

Fick GN, Rehder D (1977) Selection criteria in development of high oil sunflower hybrids. In: Proc 2nd Sunflower Forum, Fargo, USA, Natl Sunflower Assoc, Bismarck, ND, USA, pp 26–27.

Fick GN, Miller JF (1997) The genetics of sunflower. In: AA Schneiter (ed) Sunflower Technology and Production. ASA, CSSA and SSSA, Madison, Wisconsin, pp 441–495.

Flores-Berrios E, Gentzbittel L, Kayyal H, Alibert G, Sarrafi A (2000a) AFLP mapping of QTLs for in vitro organogenesis traits using recombinant inbred lines in sunflower (*Hetianthus annuus* L.). Theor Appl Genet 101: 1299–1306.

Flores-Berrios E, Sarrafi A, Fabre F, Alibert G, Gentzbittel L (2000b) Genotypic variation and chromosomal location of QTLs for somatic embryogenesis revealed by epidermal layers culture of recombinant inbred lines in the sunflower (*Helianthus annuus* L.). Theor Appl Genet 101: 1307–1312.

Flores-Berrios E, Gentzbittel L, Mokrani L, Alibert G, Sarrafi A (2000c) Genetic control of early events in protoplast division and regeneration pathways in sunflower. Theor Appl Genet 101: 606–612.

Gandhi SD, Heesacker AF, Freeman CA, Argyris J, Bradford K, Knapp SJ (2005) The self-incompatibility locus (S) and quantitative trait loci for self-pollination and seed dormancy in sunflower. Theor Appl Genet 111: 611–629.

Gedil MA, Slabaugh MB, Berry S, Iohnson R, Michelmore R, Miller I, Gulya T, Knapp SJ (2001) Candidate disease resistance genes in sunflower cloned using conserved nucleotide-binding site motifs: genetic mapping and linkage to the downy mildew resistance gene *Pl1*. Genome 44: 205–212.

Gentzbittel L, Vear F, Zhang YX, Bervillé A, Nicolas P (1995) Development of a consensus linkage RFIP map of cultivated sunflower (*Helianthus annuus* L.). Theor Appl Genet 90: 1079–1086.

Gentzbittel L, Mouzeyar S, Badaoui S, Mestries E, Vear F, Tourvieille de Labrouhe D, Nicolas P (1998) Cloning of molecular markers for disease resistance in sunflower (*Helianthus annuus* L.). Theor Appl Genet 96: 519–525.

Gulya TJ (2000) Metalaxyl resistance in sunflower downy mildew and control through genetics and alternative fungicides. In: Proc 15th Int sunflower Conf, 12–15 June 2000, Toulouse, France, pp G16–G21.

Gulya T, Rashid KY, Masirevic SN (1997) Sunflower diseases. In: Sunflower Technology and Production. ASA, CSSA, SSSA, Madison, Wisconsin, pp 263–379.

Heiser CB (1954) Variation and subspeciation in the common sunflower, *Helianthus annuus*. Am Midl Nat 51: 287–305.

Heiser CB (1976) The sunflower. Univ of Oklahoma Press, Norman, USA.

Heiser CB, Smith DM, Clevenger SB, Martin WC (1969) The North American sunflowers (*Helianthus*). Mem Torr Bot Club 22: 1–218.

Hervé D, Fabre F, Flores Berrios E, Ieroux N, Al Chaarani GR, Planchon C, Sarrafi A, Gentzbittel L (2001) QTL analysis of photosynthesis and water status traits in sunflower (*Helianthus annuus* L.) under greenhouse conditions. J Exp Bot 52: 1857–1864.

Hu J, Yue B, Vick B (2007) Integration of TRAP markers onto a sunflower SSR linkage map constructed from 92 recombinant inbred lines. Helia 30: 25–36.

Huang X, Nabipour A, Gentzbittel L, Sarrafi A (2007) Somatic embryogenesis from thin epidermal layers in sunflower and chromosomal regions controlling the response. Plant Sci 173: 247–252.

Ish-Shalom-Gordon N, Jacobsohn R, Cohen Y (1993) Inheritance of resistance to *Orobanche cumana* in sunflower. Phytopathology 83: 1250–1252.

Jan CC (1986) The inheritance of early maturity and short-stature of *H. annuus* line. In: Proc 9th Sunflower Res Workshop, 10 Dec 1986, Fargo, ND and Natl Sunflower Assoc, Bismarck, ND, USA, p 13.

Jan CC, Vick BA, Miller JF, Kahler AL, Butler ET (1998) construction of an RFLP linkage map for cultivated sunflower. Theor Appl Genet 96: 15–22.

Kirichenco VV, Dolgova EM, Aladina ZK (1987) Virulence of broomrape isolates and the inheritance of resistance. Plant Breed Abstr 57: 1392.

Lai Z, Livingstone K, Zou Y, Church SA, Knapp SJ, Andrews J, Rieseberg LH (2005) Identification and mapping of SNPs from ESTs in sunflower. Theor Appl Genet 111(8): 1532–1544.

Langar K, Lorieux M, Desmarais E, Griveau Y, Gentzbittel L, Bervillé A (2003) A Combined mapping of DALP and AFLP markers in cultivated sunflower using F9 recombinant inbred line. Theor Appl Genet 106: 1068–1074.

Lawson WR, Goulter KC, Henry RJ, Kong GA, Kochrnan JK (1998) Marker-assisted selection for two rust resistance genes in sunflower. Mol Breed 4: 227–234.

Leon AJ, Lee M, Rufener GK, Berry ST, Mowers RP (1995) Use of RFLP markers for genetic linkage analysis of oil percentage in sunflower Crop Sci 35: 558–564.

Leon AJ, Andrade FH, Lee M (2000) Genetic mapping of factors affecting quantitative variation for flowering in sunflower Crop Sci 40: 404–407.

Leon AJ, Lee M, Andrade FH (2001) Quantitative trait loci for growing degree days to flowering and photoperiod response in sunflower (*Helianthus annuus* L.). Theor Appl Genet 102: 497–503.

Leon AJ, Andrade FH, Lee M (2003) Genetic analysis of seed-oil concentration across generations and environments in sunflower. Crop Sci 43: 135–140.

Li C, Feng J, Ma F, Vick BA, Jan CC (2008) Identification of molecular markers linked to a new nuclear male-sterility gene *MS7* in sunflower (*Helianthus annuus* L.). In: Proc 17th Int Sunflower Conf, 8–12 June 2008, Cordoba, Spain, pp 651–654.

Lu YH, Gagne G, Grezes Besset B, Blanchard P, Lu YH (1999) Integration of a molecular linkage group containing the broomrape resistance gene *or5* into an RFLP map in sunflower. Genome 42: 451–456.

Lu YH, Melero Vara JM, Garcia Tejada JA, Blanchard P (2000) Development of SCAR markers linked to the gene *or5* conferring resistance to broomrape (*Orobanche cumana Wallr.*) in sunflower. Theor Appl Genet 100: 625–632.

Luciano A, Kinman ML, Smith JD (1965) Heritability of self-incompatibility in the sunflower (*Helianthus annuus*). Crop Sci 5: 529–532.

Machacek, C (1979) Study of the inheritance of earliness in sunflower (*Helianthus annuus* L.). Genet Slechteni 15: 225–232.

Marquez-Lema A, Delavault P, Letousey P, Hu J, Perez-Vich B (2008) Candidate gene analysis and identification of TRAP and SSR markers linked to the Or5 gene, which confers sunflower resistance to race E of broomrape (*Orobanche cumana* Wallr.). In: Proc 17th Int Sunflower Conf, 8–12 June 2008, Cordoba, Spain, pp 661–666.

Mestries E, Gentzbittel L, Tourvieille de Labrouhe D, Nicolas P, Vear F (1998) Analysis of quantitative trait loci associated with resistance to *Sclerotinia sclerotiorum* in sunflower (*Helianthus annuus* L.) using molecular markers. Mol Breed 4: 215–226.

Micic Z, Hahn V, Bauer E, Schon CC, Knapp SJ, Tang S, Melchinger AE (2004) QTL mapping of *Sclerotinia midstalk*-rot resistance in sunflower. Theor Appl Genet 109: 1474–1484.

Micic Z, Hahn V, Bauer E, Melchinger AE, Knapp SJ, Tang S, Schon CC (2005) Identification and validation of QTL for of *Sclerotinia midstalk*-rot resistance in sunflower by selective genotyping. Theor Appl Genet 111: 233–242.

Miller JF, Fick GN (1997) The genetics of sunflower. In: AA Schneiter (ed.) Sunflower technology and production. Agron Monogr 35. ASA, CSSA, and SSSA, Madison, WI, USA, pp 441–495.

Miller JF, Hammond JJ, Roath WW (1980) Comparison of inbred vs. single-cross testers and estimation of genetic effects in sunflower. Crop Sci 20: 703–706.

Mokrani L, Gentzbittel L, Azanza E, Fitamant L, Al-Chaarani G, Sarrafi A (2002) Mapping and analysis of quantitative trait loci for grain oil and agronomic traits using AFLP and SSR in sunflower (*Helianthus annuus* L.). Theor Appl Genet 106: 149–156.

Mouzeyar S, Tourvieille de Labronhe D, Vear F (1994) Effect of host-race combination on resistance of sunflower to downy mildew. J Phytopathol 141: 249–258.

Mouzeyar S, Roeckel Drevet P, Gentzbittel L, Philippon J, Tourvieille de Labronhe D, Vear F, Nicolas P (1995) RFLP and RAPD mapping of the sunflower *Pl1* locus for resistance to *Plasmopara halstedii* race 1. Theor Appl Genet 91: 733–737.

Mouzeyar S, Gentzbittel L, Badaoui S, Bert-Mestrie E, Perrault A, De conto V, Cock M, Dumas C, Tourvieille de Labrouhe D, Vear F, Nicolas P (1997) Molecular marker for resistance to *Sclerotinia sclerotiorum*. Eur Patent 98917244.0.

Muntanola-Cvetkovic M, Mihaljcevic M, Petrov M (1981) On the identity of the causative agent of a serious Phomopsis Diaporthe disease in sunflower plants. Nova Hedwigia 34: 417–435.

Nel, AA, Loubser, HL, Hammes PS (2002) Development and validation of relationships between sunflower seed quality and easily measurable seed characteristics for grading purposes. S Afr J Plant Soil 19: 201–205.

Pérez-Vich B, Fernandez-Martinez JM, Grondona M, Knapp SJ, Berry ST (2002) Stearoyl-ACP and oleoyl-PC desaturarse genes cosegregate with quantitative trait loci underlying high stearic and high oleic acid mutant phenotypes in sunflower. Theor Appl Genet 104: 338–349.

Pérez-Vich B, Akhtouch S, Knapp SJ, Leon AJ, Velasco L, Fernandez-Martinez JM, Berry ST (2004) Quantitative trait loci for broomrape (*Orobanche cumana* Wallr.) resistance in sunflower. Theor Appl Genet 109: 92–102.

Plaschke J, Ganal MW, Roder MS (1995) Detection of genetic diversity in closely related bread wheat using microsatellites markers. Theor Appl Genet 91: 1001–1007.

Pogorletsky PK, Geshele EE (1976) Sunflower immunity to broomrape and rust. In: Proc 7th Int Sunflower Conf, 27 June–3 July, Krasnodar, Russia, pp 238–243.

Poormohammad Kiani S, Talia P, Maury P, Grieu P, Heinz R, Perrault A, Nishinakamasu V, Hopp E, Gentzbittel L, Paniego N, Sarrafi A (2007a) Genetic analysis of plant water status and osmotic adjustment in recombinant inbred lines of sunflower under two water treatments. Plant Sci 172: 773–787.

Poormohammad Kiani S, Grieu P, Maury P, Hewezi T, Gentzbittel L, Sarrafi A (2007b) Genetic variability for physiological traits under drought conditions and differential expression of water stress-associated genes in sunflower (*Helianthus annuus* L.). Theor Appl Genet 114: 193–207.

Poormohammad Kiani S, Maury P, Sarrafi A, Grieu P (2008) QTL analysis of chlorophyll fluorescence parameters in sunflower under well-watered and water-stressed conditions. Plant Sci 175: 565–573.

Poormohammad Kiani S, Maury P, Nouri L, Ykhlef N, Grieu P, Sarrafi (2009) QTL analysis of yield-related traits in sunflower under different water treatments. Plant Breed: 128: 363–373.

Pustovoit VS (1966) Selection, seed culture and some agrotechnical problems of sunflower. Translated from Russian in 1976 by Indian Natl Sci Document Centre, Delhi, India.

Rachid Al-Chaarani G, Roustae L, Gentzbittel L, Mokrani L, Barrault G, Dechamp-Guillaume G, Sarrafi A (2002) A QTL analysis of sunflower partial resistance to downy mildew (*Plasmopara halstedii*) and black stem (*Phoma macdonaldii*) by the use of recombinant inbred lines (RILs). Theor Appl Genet 104: 490–496.

Rachid Al-Chaarani G, Gentzbittel L, Huang X, Sarrafi A (2004) Genotypic variation and identification of QTLs for agronomic traits using AFLP and SSR in recombinant inbred lines of sunflower (*Helianthus annuus* L). Theor Appl Genet 109: 1353–1360.

Ramaiah KV (1987) Control of Striga and Orobanche species. A review. In: HC Weber, W Forstreuter (eds) Parasitic Flowering Plants. Philipps-Universitt, Marburg, Germany, pp 637–664.

Regnault Y (1985) Premières observations sur le phomopsis du tournesol. Bull Cetiom 92: 13–20.

Robert N, Vear F, Tourvieille de Labrouhe D (1987) L'hérédité de la résistance au *Sclerotinia sclerotiorum* (Lib.) de Bary chez le tournesol. I: Etude des réactions a deux tests mycéliens. Agronomie 7: 423–429.

Roche J, Bouniols A, Mouloungui Z, Barranco T, Cerny M (2006) Management of environmental crop conditions to produce useful sunflower oil components. Eur J Lipid Sci Technol 108: 287–297.

Roeckel Drevet P, Gagne G, Mouzeyar S, Gentzbittel L, Philippon J, Nicolas P, Tourvieille de Labrouhe D, Vear F (1996) Colocation of downy mildew (*Plasmopara halstedii*) resistance genes in sunflower (*Hetianthus annuus* L.). Euphytica 91: 225–228.

Roustaee A, Barrault G, Dechamps-Guillaume G, Lesigne P, Sarrafi A (2000) Inheritance of partial resistance to black stem (*Phoma macdonaldii*) in sunflower. Plant Pathology 49: 396–401.

Seiler G (1988) Influence of pH, storage temperature, and maturity on germination of four wild annual sunflower species (*Helianthus* spp.). In: Proc 12th Int Sunflower Conf, Int Sunflower Assoc, Paris, pp 269–270.

Seiler GJ (1991) Registration of 13 downy mildew tolerant interspecific sunflower germplasm lines derived from wild annual species. Crop Sci 31: 1714–1716.

Seiler GJ (1992) Utilization of wild sunflower species for the improvement of cultivated sunflower. Field Crops Res 30: 195– 230.

Seiler GJ (1996) Dormancy and germination of wild *Helianthus* species. In: DJN Hind , PDS Caligari (eds) Compositae: Biology and Utilitzation. Proc Int Compositae Conf, July 24–Aug 5, 1994, Kew, Royal Botanical Gardens, UK, pp 213–222.

Seiler GJ (1998) Seed maturity, storage time and temperature, and media treatment effects on germination of two wild sunflowers. Agron J 90: 221–226.

Stoenescu F (1974) Genetics. In: AV Vranceanu (eds) Floarea-soarelui. Editura Academiei Republicii Socialiste, Buchrest, Romania, pp 92–125.

Sackston WE (1992) On a treadmill: breeding sunflowers for resistance to disease. Annu Rev Phytopathol 30: 529–551.

Sukno S, Melero-Vara JM, Fernandez-Martinez JM (1999) Inheritance of resistance to *Orobanche cernua* Loefl. in six sunflower lines. Crop Sci 39: 674–678.

Tang S, Yu JK, Slabaugh MB, Shintani DK, Knapp SJ (2002) Simple sequence repeat map of the sunflower genome. Theor Appl Genet 105: 1124–1136.

Tang S, Leon A, Bridges WC, Knapp SJ (2006) Quantitative trait loci for genetically correlated seed traits are tightly linked to branching and pericarp pigment loci in sunflower. Crop Sci 46: 721–734.

Thuault M, Tourvieille D (1988) Etude du pouvoir pathogène de huit isolats de *Sclerotinia* appartenant aux espèces *Sclerotinia sclerotiorum, Sclerotinia minor* et *Sclerotinia trifoliorum* sur tournesol. Inf. Tech. Cetiom 103: 21–27.

Tourvieille D, Vear F (1990) Heredity of resistance to *Sclerotinia sclerotiorum* in sunflower. III. Study of reactions to artificial infections of roots and cotyledons. Agronomie 10: 323–330.

Vear F, Garreyn M, Tourvieille de Labrouhe D (1997a) Inheritance of resistance to phomopsis (*Diaporthe helianthi*) in sunflowers. Plant Breed 116: 277–281.

Vear F, Gentzbittel L, Philippon J, Mouzeyar S, Mestries E, Roeckel Drevet B, Tourvieille de Labrouhe D, Nicolas P (1997b) The genetics of resistance to five races of downy mildew (*Plasmopara halstedii*) in sunflower (*Helianthus annuus* L.). Theor Appl Genet 95: 584–589.

Vear F, Jouan-Dufournel, Bert PF, Serre F, Cambon F, Pont C, Walser P, Roche S, Tourvieille de Labrouhe D, Vincourt P (2008) QTL for resistance to *Sclerotinia sclerotiorum* in sunflower. In: Proc 17th Int Sunflower Conf, 8–12 June 2008, Cordoba, Spain, pp 605–610.

Viguié A, Touvieille de Labrouhe D, Vear F (2000) Inheritance of several sources of resistance to phomopsis stem canker (*Diaporthe helianthi* Munt.-Cvet.) in sunflower (*Helianthus annuus* L.). Euphytica: 116: 167–179.

Vranceanu AV, Tudor VA, Stoenescu FM, Pirvu N (1980) Virulence groups of *Orobanche cumana* Wallr., differential hosts and resistance source genes in sunflower. In: Proc 9th Int Sunflower Conf, 8–9 June 1980, Torremolinos, Spain, pp 74–82.

Vranceanu AV, Chep N, Pirvu N, Stoenescu FM (1983) Genetic variability of sunflower reaction to the attack of *Phomopsis helianthi* Munt. Cvet. et al. Helia 6: 23–25.

Wieckhorst S, Hahn V, Duble CM, Knapp SJ, Schon CC, Bauer E (2008) Fine mapping of the mildew resistance locus Pl_{ARG} in sunflower. In: Proc 17th Int Sunflower Conf, 8–12 June 2008, Cordoba, Spain, pp 645–649.

Yang S, Berry RW, Lutrell ES, Vongkaysone T (1984) A new sunflower disease in Texas caused by *Diaporthe helianthi*. Plant Dis 68: 103–129.

Yu JK, Tang S, Slabaugh MB, Heesacker A, Cole G, Herring M, Soper J, Han F, Chu WC, Webb DM, Thompson L, Edwards KJ, Berry S, Leon AJ, Olungu C, Maes N, Knapp SJ (2003) Towards a saturated molecular genetic linkage map for cultivated sunflower. Crop Sci 43: 367–387.

Yue B, Vick BA, Miller JF, Cai X, Hu J (2008a) Construction of a linkage map with TRAP markers and identification of QTL for four morphological traits in sunflower (*Helianthus annuus* L.). In: Proc 17th Int Sunflower Conf, 8–12 June, Cordoba, Spain, pp 655–660.

Yue B, Radi SA, Vick BA, Cai X, Tang S, Knapp SJ, Gulya TJ, Miller JF, Hu J (2008b) Identifying quantitative trait loci for resistance to *Sclerotinia* head rot in two USA sunflower germplasms. Phytopathology 98: 926–931.

Gene Cloning and Characterization

*Renate Horn** and *Sonia Hamrit*

ABSTRACT

In sunflower, two approaches have been followed to isolate genes: map-based-cloning and candidate gene approaches. Map-based cloning of nuclear male sterility and fertility restoration genes, as well as, genes conferring resistance to downy mildew so far has been unsuccessful. This is primarily due to the large genome size, the high content of repetitive elements, the lack of publicly available large mapping populations and large insert genomic libraries. On the other hand, candidate gene approaches in the areas of developmental biology (e.g., embryogenesis, non-dormancy, pollen and pistil-specific genes), abiotic stress (e.g., heat-shock and drought-related genes), disease resistance and quality traits (e.g., fatty acid biosynthesis or tocopherol biosynthesis) have proven to be very successful. However, in sunflower even if a gene was identified, the lack of a good transformation system hampers verification of the candidate gene. Fortunately, heterologous systems and mutant analyses have proven to be helpful for gene verification and characterization. In spite of these problems, enormous progress has been made on the cloning of genes in sunflower. The current expansion of genomic resources in sunflower promises to accelerate the process of isolation and characterization of agronomically important genes.

Keywords: map-based cloning; molecular marker; BAC libraries; candidate gene approach; mutants

Institute of Biological Sciences, Department of Plant Genetics, University of Rostock, Albert-Einstein-Str. 3, 18051 Rostock, Germany.
*Corresponding author: *renate.horn@uni-rostock.de*

6.1 Introduction

In sunflower, the large size of the genome (3,500 Mb; Baack et al. 2005) and high content of repetitive DNA (e.g., Ty3-*gypsy* and Ty1-*copia*-like DNA; Santini et al. 2002; Natali et al. 2006) pose a challenge for cloning genes of agronomic importance. However, the development of new genomic resources and databases from different genome projects, also for the Asteraceae family, has made the sunflower genome more accessible to molecular analysis. Using different techniques (Chapter 3), several types of molecular markers have been detected that are closely linked to simple inherited traits (Chapter 4) and placed on various genetic maps, constructed for sunflower (Chapter 3). These markers and the positions of the genes on the genetic maps represent the initial steps for isolating genes by map-based cloning approaches. Even though markers can be very efficiently used in marker-assisted breeding programs (Chapter 7), cloning of genes is necessary to understand their function and regulation in different tissues and under various environmental conditions. Map-based cloning or positional cloning has not resulted in any isolation of genes in sunflower so far, but candidate gene approach has been very successful and combination of map-based cloning and candidate gene approaches might in the near future allow successful isolation of additional genes in sunflower. In this chapter, we summarize the attempts made to clone and characterize nuclear-encoded genes in sunflower.

6.2 Map-based Cloning Approaches in Sunflower

Map-based cloning approach is used if the function of the gene of interest is unknown and no corresponding gene has been isolated from another species. Map-based cloning approach is reasonable only under these conditions as it is extremely time-consuming. Map-based cloning requires the identification of markers tightly linked to the gene of interest that can then be used as probes against large genomic insert libraries to identify genomic clones near the gene of interest or containing the gene of interest (Tanksley et al. 1995). So, prerequisites for map-based cloning are: (1) availability of large mapping populations, e.g., F_2, RILs, F_1BC_1 populations to develop high-density genetic maps around the target locus, and (2) large-insert genomic DNA libraries, e.g., BAC libraries. The latter have been and still are representing a limiting factor in map-based cloning in sunflower as neither the genome coverage nor the average insert sizes of the BAC libraries are very satisfactory. This has and will hamper future map-based cloning attempts in sunflower. Table 6-1 provides an overview on the available BAC libraries. So far, all BAC libraries in sunflower except the library constructed by Özdemir et al. (2004) have been made from maintainer lines. Apart from

Table 6-1 Overview on the six BAC libraries in sunflower.

	Gentzbittel et al. 2002	Özdemir et al. 2004	Knapp (CUGI)	Feng et al. 2006	Feng et al. 2006	Bouzidi et al. 2006
Line	HA821	RHA325	HA383	HA89	HA89	YDQ (= HA335)
Average insert size (kb)	80	60	125	140	137	118
Vector	pBelo BAC11	pBelo BAC11	pIndigo-BAC536	PECBAC1	pCLD045 41(binary)	pBelo BAC11
Restriction enzyme	*Hind*III	*Hind*III	*Hind*III	*Bam*HI	*Hind*III	*Hind*III
Genome coverage	4–5-fold	1.9–2-fold	8.3-fold	5-fold	3.9-fold	5-fold
Number of clones	150,000	104,678	202,752	107,136	84,864	147,456

these BAC libraries, Feng et al. (2006) also constructed a BIBAC library. Using a binary vector allows the immediate application of an identified BIBAC clone in transformation experiments using *Agrobacterium* in order to verify candidate genes by complementation of mutants.

With regard to the relationship of genetic maps with physical maps, 1 cM map distance represents about 2 Mb in sunflower genome assuming an estimated haploid genome size (1C) of about 3,000 Mb to 3,500 Mb (Arumuganathan and Earle 1991; Baack et al. 2005) and a map size for the whole genome of about 1,500 cM (961 cM - 1,920 cM; Micic et al. 2004 and Hu et. al. 2007, respectively).

In the ideal case, markers very closely linked to the gene of interest would in a hybridization experiment identify a genomic clone that already contains the target gene ("chromosome landing"; Tanksley et al. 1995). However, a more realistic "chromosome walking" approach will lead to cloning of the gene (Fig. 6-1). Markers from both sides of the target gene would be used to identify BAC clones through colony hybridization or PCR screening of BAC pools. Identified BAC clones need to be arranged into contigs by BAC fingerprinting. BAC end sequences would then be used to develop probes for continuing hybridizations until a closed contig around the gene of interest is formed, which is characterized by overlapping clones coming from both the sides. A relatively large genome size and presence of retroelements pose a major problem for chromosome walking in sunflower. So far, map-based cloning has not led to any successful isolation of a gene in sunflower. However, as mentioned earlier, combination of map-based cloning and candidate gene approaches could hopefully allow isolation of genes in the near future.

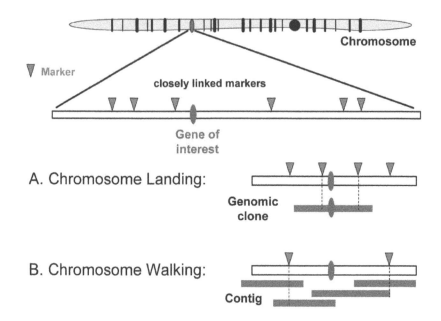

Figure 6-1 Map-based cloning approach. A, Chromosome Landing; B, Chromosome Walking.

6.2.1 *Male Sterility and Fertility Restoration*

Male sterility represents a phenotype in plants, in which the function of the male gametophyte is prevented without affecting the female counterpart. On the basis of inheritance patterns, two types of male sterility occur in sunflower: nuclear male sterility (NMS) and cytoplasmic male sterility (CMS).

NMS is generally a result of a single homozygous recessive locus, and after crossing with any fertile genotype an NMS line produces F_1 progenies that are all male fertile (Jan 1997). Allelic relationships among NMS genes have been reported by Vranceanu (1970). He tested for allelism in 10 NMS sources isolated from a Romanian germplasm and reported the presence of five independent NMS genes, named as *Ms1* to *Ms5*. Jan and Rutger (1988) evaluated the effectiveness of seven NMS mutants induced by mitomycin-C and streptomycin from the cultivated line HA89, which were confirmed and placed in four different allelic groups, each representing a unique NMS gene, designated *Ms6-Ms9* (Jan 1992). NMS genes from two released male sterile lines, P21 and B11A3, have been designated *Ms10* and *Ms11* (Jan 1992), and were subsequently located on the sunflower genetic map by Pérez-Vich et al. (2005). *Ms10* mapped together with the gene *T*, an anthocyanin pigment locus, to the linkage group (LG) 11, while *Ms11* mapped to the LG 8. Four simple sequence repeat (SSR) markers (ORS697, ORS1214,

ORS686, and CRT162) and the phenotypic marker locus *T* cosegregated and were tightly linked to *Ms10* at a genetic distance of less than 1 cM (Pérez-Vich et al. 2005). The availability of tightly linked PCR markers will facilitate marker-assisted breeding programs and provide a basis for physical mapping and map-based-cloning of theses genes. Chen et al. (2006) identified DNA markers linked to the *Ms9* gene in an F_2 population derived from a cross between NMS360 and RHA271 and mapped the *Ms9* gene with SSR markers to LG 10 of the public sunflower map (Yu et al. 2003). The target region amplified polymorphism (TRAP) marker technique, applied by Chen et al. (2006), is a method that uses expressed sequence tag (EST) sequences and bioinformatic tools to generate polymorphic markers around targeted candidate gene sequences (Hu and Vick 2003). Four fixed primers derived from published male sterility inducing genes from wheat (Chen et al. 2005), *Arabidopsis* (Rubinelli et al. 1998; Stintzi and Browse 2000) and maize (Albertsen et al. 1993) were combined with three arbitrary primers. Six of the amplified TRAP markers were linked to the *Ms9* gene, and two of them, Ts4p03-202 and Tt3p09-529, cosegregated with the *Ms9* gene. These two markers will facilitate isolation of the *Ms9* gene by map-based cloning.

CMS, a phenotype in higher plants in which plants are not able to produce or shed any functional pollen, has been associated with chimeric open-reading frames (ORFs) in the mitochondrial genomes (for review see Horn 2006). These ORFs encode proteins that interfere with mitochondrial functions during pollen development. In sunflower, commercial hybrid breeding is based on a single source of male sterility, PET1. This male sterility originated from an interspecific cross between *Helianthus petiolaris* Nutt and *Helianthus annuus* L. (Leclercq 1969). Alterations in the mitochondrial genome of PET1 and fertile lines are limited to a 17-kb region and consist of two mutations: a 12-kb inversion and a 5-kb insertion/deletion, which lead to an altered transcription pattern of the *atp*A gene (Siculella and Palmer 1988). CMS is associated with the expression of a novel ORF, *orfH522* (Köhler et al. 1991) that encodes a 16 kDa polypeptide (Horn et al. 1991; Laver et al. 1991). Nuclear genes known as restorer of fertility (*Rf*) genes have the function to suppress the effect of CMS-associated mitochondrial abnormality on male fertility e.g., reducing the anther-specific level of the cotranscript of *atp*A-*orfH522* in the male florets (Monéger et al. 1994). Several restorer lines are available for PET1; most of them carry the genes *Rf1* and *Rf2* restoring pollen fertility. *Rf2* was described to be present in nearly all inbred lines, including maintainers of PET1 (Miller and Fick 1997) and only *Rf1* gene is introduced by the restorer lines to produce fertile sunflower hybrids. Map-based cloning efforts were started by several groups to isolate the restorer gene *Rf1*. The *Rf1* gene was mapped to LG 6 in the RFLP map and the consensus map (Gentzbittel et al. 1995, 1999) together with the *Pl5* locus conferring resistance to downy mildew (Bert et al. 2001). Horn et al. (2003) developed PCR-based markers for the restorer gene *Rf1* of sunflower hybrids based on PET1 cytoplasm. Using an F$_2$ population (RHA325xHA342), a linkage map around the *Rf1* gene was constructed using AFLP and RAPD markers. Markers tightly linked to the gene *Rf1* were identified, cloned and sequenced (Kusterer et al. 2002, 2004a). The cloned markers were used directly as probes for colony hybridization against high-density filters of two BAC libraries (RHA325 and HA383) or their sequences were used to design overgo probes for hybridizations and primers to screen the BAC library by 3D-PCR of BAC pools. Several BAC clones could be identified. *Rf1* was finally mapped to LG 13 in the consensus map of Tang et al. (2002) using SSR markers (Kusterer et al. 2004b, 2005). Later on, BAC fingerprinting using different restriction enzymes was performed to develop contigs. BAC ends were cloned and sequenced to develop new overgo probes for chromosome walking (Hamrit et al. 2008). However, no closed contig around the restorer gene *Rf1* could be developed so far.

The isolation of fertility restorer genes in petunia (Bentolila et al. 2002), rice (Akagi et al. 2004; Komori et al. 2004; Wang et al. 2006), radish (Brown et al. 2003; Desloire et al. 2003; Imai et al. 2003; Kazama and Toriyama 2003; Koizuka et al. 2003) as well as in sorghum (Klein et al. 2005) have shown that these restorer genes belong to a class of pentatricopeptide repeat (PPR) proteins, which show characteristic tandem repeats of 35 amino acids. These proteins represent a large family of proteins that are assumed to play a role in RNA processing and translation within organelles (Aubourg et al. 2000; Lurin et al. 2004). A combination of candidate gene approach and map-based cloning might allow isolation of the restorer gene *Rf1* in sunflower in the future.

Meanwhile, more than 70 new CMS sources have been described in sunflower (Serieys 1999) that might be used in the future to broaden the base of hybrid production. New restorers have been identified for some of the new CMS sources (Horn and Friedt 1997). However, molecular characterization of 28 new CMS cytoplasms (Horn and Friedt 1999) demonstrated that cytoplasms with very different origins show a considerable similarity, which would not be expected based on their pedigree (Horn 2002). Molecular markers have been identified for isolation of other restorer genes in sunflower by map-based cloning. The restorer gene *Rf_PEF1*, restoring pollen fertility in hybrids based on one of the new CMS sources, PEF1, could be linked to AFLP markers (Schnabel et al. 2008). In addition, SSR marker analyses demonstrated that *Rf_PEF1* is not located on LG 13 as *Rf1*. Feng and Jan (2008) identified a new dominant restorer gene *Rf4*, which restores pollen fertility in the presence of CMS GIG2. The *Rf4* gene was mapped to LG 3 with SSR markers and RFLP-derived STS-markers, and is about 0.9 cM apart from the SSR marker ORS1114 based on a segregating population of 933 individuals.

In the future, isolation of these fertility restorer genes from sunflower will hopefully help to understand the different molecular mechanisms behind cytoplasmic male sterility and its fertility restoration. Comparison of different restorer genes within sunflower will be of special interest.

6.2.2 Resistance to Downy Mildew

Downy mildew caused by *Plasmopara halstedii* (Farl.) Berl. et de Toni, is one of the most widely occurring important pathogens of sunflower (Sackston 1992). It is an obligate parasite and specific to sunflower. A number of physiological races have been reported that can be distinguished by their differential virulence on sunflower genotypes (Gulya et al. 1991a, b). Resistance to downy mildew, which is controlled by several single dominant genes named *Pl*, has been detected in cultivated sunflower (Vranceanu 1970) or introduced from wild *Helianthus* species (Miller and Gulya 1991). Some of them provide resistance to a single race of downy mildew, whereas others impart resistance to two or more races (Miller 1992). So far, 11 *Pl* genes named as Pl_1 to Pl_{11} have been identified conferring resistance to different races of *Plasmapara halstedii* (Zimmer and Kinman 1972; Vranceanu et al. 1981; Miller and Gulya 1991; Rahim et al. 2002).

For marker-assisted selection, several molecular markers have been identified and mapped to the different genomic region containing *Pl* loci (Bert et al. 1991; Gentzbittel et al. 1995; Mouzeyar et al. 1995; Roeckel-Drevet et al. 1996; Brahm et al. 2000; Gedil et al. 2001a, b; Radwan et al. 2003; Radwan et al. 2004). Chapter 4 gives a detailed overview about the marker analyses performed with regard to downy mildew resistance genes. Most of the research groups also intend to use these markers for map-based cloning approaches. Marker analyses were also the first to reveal that Pl_1, Pl_2 and Pl_6 cluster together (Mouzeyar et al. 1995; Roeckel-Drevet et al. 1996; Vear et al. 1997). These three downy mildew resistance gene were located on LG 1 of the RFLP consensus map (Gentzbittel et al. 1995) corresponding to LG 8 of the public SSR map (Yu et al. 2003). In addition, the Pl_5 locus and the Pl_8 locus have been mapped in the same region of LG 6 of the RFLP consensus map (Radwan et al. 2003; Slaubaugh et al. 2003) corresponding to the lower end of LG 13 of the public SSR map (Yu et al. 2003).

Wild species have played a major role in introducing resistance genes for downy mildew into the cultivated sunflower. For example, the downy mildew resistance gene Pl_{Arg} on LG 1, Pl_7 on LG 8 and Pl_8 on LG 13 were introgressed from wild species (Miller et al. 2002; Slabaugh et al. 2003; Dußle et al. 2004). However, introgression from other species can reduce recombination frequencies, which might cause problems in a map-based cloning approach.

Another strategy for cloning resistance genes is the candidate gene approach that has been described in detail in the Section 6.3.4. The number of resistance gene analogs mapped in sunflower or used to develop PCR-based markers had been very low at the beginning (Gentzbittel et al. 1998; Gedil et al. 2001a; Plocik et al. 2004; Radwan et al. 2004). However, mining of the sunflower EST database for NBS-LRR homologs proved to be very efficient (Radwan et al. 2008).

The combination of disease resistance gene analogs (RGAs) and map-based cloning strategies will hopefully allow resolving the structure and the mechanisms behind the resistance to different *Plasmopara halstedii* races.

6.3 Candidate Gene Approach Based on Homology to Genes of Other Species

Using candidate gene approach requires that the gene of interest has been isolated in another species and sequences are available to be used as probes or to design primers for PCR amplification in sunflower. The major problem here is the degree of homology between genes in different species, which either allows to start with specific primers from conserved regions or degenerate primers if the sequence homology proves only moderate between species. Primers are then used to amplify the candidate gene either from genomic DNA or cDNA. The large amount of EST data (Chapter 3) now available for sunflower will be very helpful for employing candidate gene approach in sunflower. Candidate gene approach, although limited to genes of known function, has been very successful in sunflower in the recent years. Table 6-2 gives an overview of the genes that were isolated by a candidate gene approach.

6.3.1 Herbicide Resistance in Sunflower

Wild biotypes of sunflower, native to North America (Rogers et al. 1982), are weeds in corn, soybean and other crops in North America. These weeds are commonly controlled by applying acetohydroxyacid synthase (AHAS)-inhibiting herbicides as sulfonylurea (SU) and imidazoline (IMI). SU and IMI are efficient inhibitors of acetohydroxyacid synthase (AHAS), an enzyme that catalyzes the first step in the biosynthesis of the essential branched-chain amino acids isoleucine, leucine and valine (Duggleby and Pang 2000). To understand the mechanism behind the herbicide resistance, the *AHAS* genes from sunflower were isolated by a candidate gene approach from the two sources of resistance in wild sunflower populations ANN-PUR and ANN-KAN (Kolkman et al. 2004). To design oligonucleotide primers for amplifying *AHAS* gene fragments and isolating the *AHAS* genes from sunflower several sources of DNA sequences were used. First, a cDNA probe

(ZVG437) representing a specific RFLP marker (Berry et al. 1995) was sequenced and through BLASTN and BLASTX analyses was found to be homologous to the 3' end of *AHAS* genes isolated from common cocklebur (*Xanthium stromarium* L., U16279 and U16280; Bernasconi et al. 1995). Three forward primers (p-AHAS-1, p-AHAS-2, and p-AHAS-3) were designed from the 5'end of the cocklebur *AHAS* cDNA and two reverse primers (p-AHAS-4 and p-AHAS-5) were derived from the ZVG437 cDNA probe. In a second strategy, the nucleotide sequences of the cocklebur and lettuce (*Lactuca* sp.) AHAS genes (Mallory-Smith et al. 1990) were aligned. Moderately degenerate forward and reverse primers were then designed based on conserved sequences (Kolkman et al. 2004). These primers were used to amplify the *AHAS* genes from genomic sunflower DNA. The 5' and 3' ends of the coding sequences were completed by genome walking using the Universal Genomewalker kit. Three *AHAS* genes (*AHAS1*, *AHAS2*, and *AHAS3*) were cloned and sequenced from herbicide-resistant (mutant) and susceptible (wild type) genotypes (Kolkman et al. 2004). Two of the three sunflower *AHAS* genes were highly homologous. The nucleotide sequences of *AHAS1* and *AHAS2* were 92% identical, whereas *AHAS3* was 72% identical to *AHAS1* and 73% identical to *AHAS2*. Fourty-eight SNPs were identified in *AHAS1*, a single six-base pair insertion-deletion in *AHAS2*, and a single SNP in *AHAS3*. No DNA polymorphism was found in *AHAS2* among the elite inbred lines. Only six of the 48 SNPs discovered in *AHAS1* would cause amino acid substitutions. With regard to the herbicide resistance, the C-T mutation in codon 205 (Ala to Val) in *AHAS1* confers resistance to IMI and another C-T mutation in codon 197 (Pro to Leu) to SU herbicides. Both mutations have been described for other plant genera as well (Tranel and Wright 2002). White et al. (2003) reported an independent Ala205Val mutation in an *AHAS* gene from common sunflower herbicide resistant biotypes from South Dakota.

In Europe, where wild biotypes of sunflower do not occur, resistance to imidazoline is now used for post-emergence weed control in new sunflower hybrids under the label CLEARFIELD® technology (Pfenning et al. 2008).

6.3.2 Genes Involved in Developmental Aspects in Sunflower

6.3.2.1 Embryogenesis and Plant Development

Higher plant embryogenesis is divided into two major phases: embryo development (or morphogenesis) and seed maturation (West and Harada 1993). In *Arabidopsis thaliana*, the *LEC1* (*LEAFY COTYLEDON1*) gene (*AtLEC1*) and the *LIL1* (*LEAFY COTYLEDON1-LIKE*) gene (*AtLIL*) have distinct functions in embryogenesis (Kwong et al. 2003). Primers recognizing conserved features of the *LEAFY COTYLEDON1-LIKE* (L1L) genes from *A*.

Table 6-2 Overview of genes isolated in sunflower by a candidate gene approach.

Function	Gene	Protein product	Reference
Herbicide resistance			
Herbicide resistance against sulfonylurea and imidazoline	*AHAS-1, AHAS-2* and *AHAS-3*	Acetohydroxyacid synthase 1, 2 and 3	Kolkman et al. 2004
Embryogenesis and Plant Development			
Zygotic and somatic embryogenesis	*HaL1L*	*Helianthus annuus* LEAFY COTYLEDON1 _LIKE protein	Fambrini et al. 2006
Maintenance of meristematic cell identity	*HtKNOT1*	*Helianthus tuberosus* knotted-1-like protein	Chiappetta et al. 2006, Michelotti et al. 2007
Seed development, dessication	*Ha ds10 G1*	*Helianthus annuus* dry seeds 10 G1	Prieto-Dapena et al. 1999
Transcription factor for *trans*-activation of *Ha hsp17.6 G1* and *Ha hsp17.7 G4*	*HaHSFA9*	*Helianthus annuus* heat shock factor A9	Almoguera et al. 2002
Embryo specific heat shock proteins	*Ha hsp17.6 G1*	*Helianthus annuus* heat shock protein 17.6 G1	Almoguera et al. 2002
Zygotic embryogenesis and heat shock response in vegetative tissues	*Ha hsp17.7 G4*	*Helianthus annuus* heat shock protein 17.7 G4	Coca et al. 1996, Almoguera et al. 2002
Transcription factor	*HaDREB2*	*Helianthus annuus* drought-responsive element binding factor 2	Diaz-Martin et al. 2005
Carotinoid biosynthesis	*HaPsy*	*Helianthus annuus* phytoene synthase	Salvini et al. 2005
Early development	*Hahr1*	*Helianthus annuus* homeobox from roots protein	Valle et al. 1997
Non-dormancy	*Zds*	ζ-carotene desaturase	Conti et al. 2004
Expression in mature pollen, pollen tube growth	*HaPLIM1* and *HaPLIM2*	*Helianthus annuus* Pollen LIM protein 1 and 2	Baltz et al. 1992a, b and Eliasson et al. 2000

Expression in leaves, hypocotyls, stems and ovaries	HaWLIM1	*Helianthus annuus* Widely expressed LIM protein	Eliasson et al. 2000
Expression in pistill and pollen Expression in different tissues	sf21A, sf21B sf21C–sf21F	sunflower 21A–21F protein	Kräuter-Canham et al. 1997, Lazarescu et al. 2006
Abiotic Stress Reponses			
Dessication tolerance	Ha hsp17.9	*Helianthus annuus* heat shock protein 17.9	Coca et al. 1994
Dessication tolerance	Ha hsp17.6 G1	*Helianthus annuus* heat shock protein 17.6	Coca et al. 1994
Response to heat shock	Ha hsp18.6 G2	*Helianthus annuus* heat shock protein 18.6	Coca et al. 1996
Zygotic embryogenesis and heat shock response in vegetative tissues	Ha hsp17.7 G4	*Helianthus annuus* heat shock protein 17.7	Coca et al. 1996
Drought-related Genes			
Induction by dehydration drought tolerance	HaDhn1, HaDhn1a and HaDhn2	*Helianthus annuus* dehydrin1, 1a and 2	Cellier et al. 1998, Natali et al. 2003
Induction by dehydration	HaELip1	*Helianthus annuus* early light induced protein	Cellier et al. 1998
Aquaporins, stomatal movement	SunTIP7, 18, 20 Sun Rb7, SunyTIP	Sunflower tonoplast intrinsic protein	Sarda et al. 1999
ABA-responsive nuclear protein in plant stress	HaABRC5	*Helianthus annuus* ABA responsive gene	Liu and Baird 2004
ABA-mediated responses to water stress	Hahb-4	*Helianthus annuus* homeobox-4 protein	Gago et al. 2002
Quality related Genes			
Desaturation of stearic acid	SAD6 and SAD17	Δ9-stearoyl-acyl carrier protein desaturase	Hongtrakul et al. 1998a
Desaturation of oleic acid	FAD2-1, FAD2-2 and FAD2-3	Oleoyl-phosphatidyl choline desaturase	Martinez-Rivas et al. 2001

Table 6-2 contd.....

Table 6-2 contd....

Function	Gene	Protein product	Reference
Fatty acid composition	*FatA* and *FatB*	Acyl-acyl carrier protein thioesterase	Serrano-Vega et al. 2005
Synthesis of α-tocopherol	MT-1 and MT-2	2-methyl-6-phytyl-1,4-benzoquinone/ 2-methyl-6-solanyl-1,4-benzoquinone	Tang et al. 2006
Synthesis of α-tocopherol	γ-TMT	γ-tocopherol methyltransferase	Hass et al. 2006
Synthesis of γ-tocopherol	TC	tocopherolcyclase	Hass et al. 2006

thaliana (*AtL1L*) and *Phaseolus coccineus* (*PcLIL*) were used to amplify cDNAs derived from immature zygotic embryos of *H. annuus* (Fambrini et al. 2006). The rapid amplification of cDNA ends (RACE) approach was used to isolate the 5′ and 3′ ends of the cDNA. The reconstructed full length cDNA sequence (840 bp) from sunflower contained 642 bp coding sequence, 39 nucleotides of a 5′-untranslated region (UTR) and 159 nucleotides of 3′-UTR. The nucleotide sequence represents the entire region coding for a putative peptide of 214 amino acids. The sunflower *HaL1L* gene encodes a heme-activated protein 3 (HAP3) subunit of the CCAAT box-binding factor. *HaL1L* transcripts are accumulated primarily at the early stages of zygotic embryogenesis. In order to see whether *HaL1L* transcription also played a role in somatic embryogenesis, the mutant EMB-2 was investigated. This mutant represents a somaclonal variant, which had been previously induced in vitro from leaf explants of the tetraploid ($2n = 4x = 68$) interspecific hybrid *Helianthus annuus* x *H. tuberosus* (Fambrini et al. 2000). EMB-2 produces ectopic embryo and shoot-like structures, arranged along the veins in clusters. The epiphyllous proliferation of ectopic embryos on EMB-2 leaves was associated with *HaL1L* mRNA accumulation (Fambrini et al. 2006). In conclusion, transcription of the *HaL1L* gene is maintained both in zygotic and somatic embryogenesis. *HaL1L* could be involved in switching the somatic cell fate towards embryogenic competence (Fambrini et al. 2006).

Knotted1-like homoeobox (*KNOX*) genes are expressed in specific patterns in the plant meristems and play an important role in maintaining meristematic cell identity (Michelotti et al. 2007). For cloning of the *H. tuberosus knotted1*-like gene (*HtKNOT1*), total RNA from *H. tuberosus* was used with the Superscript Preamplification kit (Invitrogen) to produce cDNA (Chiappetta et al. 2006). Reverse transcription was carried out with Superscript II retrotranscriptase in the presence of the adapter primer (AP, 5′-GGCCACGCGTCGACTAGTACTTTTTTTTTTTTTTTTT-3′) provided with the kit. The cDNA was then obtained by PCR with the following degenerate primers: P-1, 5′-CCDGMDYTRGAYCARTTCATGG-3′ (forward) and P-4, 5′-ATRAACCARTTGTTKATYTGYTTC-3′ (reverse). These primers were selected in a region with the highest conservation among members of the class I *knox* genes: *KNAT1* and *KNAT2* from *Arabidopsis* (Lincoln et al. 1994), *SBH1* from soybean (Ma et al. 1994) and *LET6* from tomato (Chen et al. 1997). The PCR reaction yielded a PCR product of 450 bp, which was cloned into the pDrive Cloning Vector (Qiagen). The RACE approach was used to isolate the 5′- and 3′-ends of the *HtKNOT1* cDNA. The full length cDNA sequence (1,398 bp, DDBJ/EMBL/GenBank database accession No. AJ519674) contained 1,089 bp CDS, 54-nucleotides of 5′-untranslated region (UTR), and 255-nucleotides of 3′-UTR. The *HtKNOT1* gene encodes a predicted protein of 362 amino acids with a calculated molecular weight of 40.2 kDa. Expression analyses of *HtKNOT1* in stems, inflorescence meristems, floral meristems and floral organs revealed a complex expression pattern (Michelotti et al. 2007). *HtKNOT1* may play a dual role being required to maintain the meristem initials as well as initiating differentiation and/or conferring new cell identity.

The sequence information from a group 1 late embryogenesis-abundant (Lea) cDNA clone of sunflower was used to isolate the *Ha ds10 G1* gene from genomic DNA (Prieto-Dapena et al. 1999). This gene contained an intron at a conserved position with a size of 1,024 bp. Transcription from the *Ha ds10 G1* promoter was strictly seed-specific and originated from at least two close initiation sites. Almoguera et al. (2002) were the first to report cloning and functional characterization of a heat shock transcription factor that was specially expressed during embryogenesis in the absence of environmental stress in sunflower. To isolate *trans*-acting factors involved in the developmental activation of small heat shock protein (sHSP) gene promoters the yeast one-hybrid cloning approach was used (Li and Herskowitz 1993). A cDNA library specific for sunflower embryos containing 830,000 individual transformants, all with an average cDNA insert of 1.3 kb, was established. For one-hybrid screening the reporter yeast strain was transformed with DNA prepared from the embryo cDNA library, after amplification of 1,660,000 primary clones. Five million yeast transformants were plated on SD-His-Leu + 15 mM 3-aminotriazole. Twenty-four putative positive yeast clones were selected. Four of the cDNAs encoded the same heat shock transcription factor, which was named HaHSFA9 (Almoguera et al. 2002) because it was very similar to AtHSFA9 described previously (Nover et al. 2001). In sunflower, HaHSFA9 proved to be a transcription factor involved in the developmental activation of the two small heat shock protein genes *Ha hsp17.6 G1* and *Ha hsp17.7 G4*. The activation is strictly restricted to the embryo (Almoguera et al. 2002). Using the *Ha hsp17.6 G1* promoter as bait (Díaz-Martín et al. 2005), the same yeast one-hybrid technique was used to clone the corresponding transcription factor, named sunflower drought-responsive element binding factor 2 (HaDREB2). Functional analysis of the interaction of the two transcription factors HaHSFA9 and HaADREB2 demonstrated that both transcription factors synergistically trans-activate the *Ha hsp17.6 G1* promoter in bombarded sunflower embryos (Díaz-Martín et al. 2005).

Carotenoid biosynthesis takes place in the plastid, but all known enzymes in the pathway are nuclear-encoded and post-translationally imported into the organelle (reviewed in Sandmann 2001). For cloning the cDNA of the sunflower phytoene synthase gene (*HaPsy*), amino acid sequences of previously cloned *Psy* genes were aligned (Salvini et al. 2005). Regions with the highest conservation were identified and two degenerate oligonucleotide primers were designed: P-A, 5'-TTCCKGGGASTTTGRGYTTGTTG-3' (forward) and P-B,

5'-GCCMAYACMGGCCATCTRCTAGC-3' (reverse). First strand cDNA was made with total RNA using a Superscript preamplification kit. The cDNA was amplified with P-A and P-B primers. The PCR product of 750 bp was cloned into the pCRII vector. The RACE approach was used to isolate the 5'- and 3'-ends of the *Psy* cDNA. The 1,598 bp long reconstructed full-length cDNA sequence (GenBank accession number AJ304825) contained 1,242 bp coding sequence, 172 nucleotides of 5'-untranslated region (UTR), and 170 nucleotides of 3'-UTR (Salvini et al. 2005). The predicted protein (46.8 kDa) displayed a sequence of 414 amino acid residues with a putative transit sequence for plastid targeting in the N-terminal region. The program CloroP 1.1 (Emanuelsson et al. 1999) predicts a transit peptide cleavage site between residue 67 and 68. Organspecific expression of *HaPsy* was analyzed by relative RT-PCR assays (Salvini et al. 2005). The expression of *HaPsy* is very high in cotyledons, and young and mature leaves, but lower in stem and nearly absent in roots. *HaPsy* is regulated during leaf development. The role of the phytoene synthase gene *HaPSy* in controlling carotenoid biosynthesis is demonstrated by the concurrent increase of *HaPsy* transcript levels with the light-dependent, enhanced carotenoid production in green tissues of sunflower (Salvini et al. 2005).

A homeobox (HD)-containing cDNA was isolated by Valle et al. (1997). As template for PCR, a cDNA library prepared from *H. annuus* root mRNA and cloned in lambda gt10 was used. A degenerate oligonucleotide derived from the conserved peptide sequence WFQNRRA from helix 3 of the HD and an oligonucleotide containing the sequence flanking the cloning site of lambda gt10 were used as the primers. The corresponding gene was named *Hahr1* (*Helianthus annuus* homeobox from roots) that encodes a 682 aa protein with a M_r 76,677. *Hahr1* expression was primarily found in dry seeds, hypocotyls and roots at stages associated with early developmental events (Valle et al. 1997).

6.3.2.2 Non-dormancy

Seed dormancy and germination are regulated by a wide range of plant hormones, including abscisic acid (ABA), ethylene, gibberellin (GA), and brassinosteroids, of which ABA is the primary mediator of seed dormancy (Koorneef et al. 2002). Major enzymes of the carotenoid biosynthesis pathway like phytoene desaturase, ζ-carotene desaturase, carotenoid isomerase and lycopene β-cyclase, are also essential for the biosynthesis of carotenoid precursors of ABA. Impairment of carotenoid biosynthesis causes photo-oxidation and ABA-deficient phenotypes in rice, of which the latter is a major factor controlling the pre-harvest sprouting (PHS) or vivipary trait (Fang et al 2008). In addition, the ratio of ABA/GA is distinctly reduced in *phs4-1* mutant seeds in rice. The increased GA might result from a reduced demand of geranyl-geranyl-pyrophosphat (GGPP) by the carotenoid biosynthesis pathway since GGPP is the common precursor for both GA and carotenoid biosynthesis (Rodriguez-Concepcion et al. 2001). However, the contribution of carotenoid synthesis to the regulation of balance between ABA and GA is still elusive (Fang et al. 2008).

The *non-dormant-1* (*nd-1*) mutant in sunflower, characterized by an albino and viviparous phenotype, was isolated by selfing an in vitro-regenerated plant (Pugliesi et al. 1991). Pigment analysis by HPLC demonstrated the absence of ß-carotene, lutein and violaxanthin (Conti et al. 2004). In addition, *nd-1* seedlings grown under very dim light conditions showed a strong accumulation of ζ-carotene and to a lesser extent, of phytofluene and cis-phytoene. All this suggested that ζ-carotene desaturation was impaired. Therefore, Conti et al. (2004) cloned and characterized the ξ-carotene desaturase (*Zds*) gene by a candidate gene approach. Amino acid sequences of cloned *Zds* genes from *Tagetes erecta* (GenBank AF251013), *Arabidopsis thaliana* (AF121947), *Narcissus pseudonarcissus* (AJ224683), *Solanum lycopersicum* (AF195507), *Capsicum annuum* (X89897) and *Zea mays* (AF047490) were aligned. Regions with the highest conservation were identified and two primers were designed. Total RNA was used for first strand cDNA synthesis using a Superscript preamplification kit. The RACE approach was used to isolate the 5′ and 3′ ends of the *Zds* cDNA. The putative full-length *Zds* cDNA was obtained by PCR using specific primers at the extreme 5′ end. To isolate the full-length exons/introns region of the *Zds* gene, sequence information from cDNA was used to design specific primers for PCR amplification from genomic DNA. The *H. annuus Zds* gene (GenBank accession no AJ514406) consists of 14 exons and 13 introns, 44 nucleotides of 5′-untranslated region (UTR), and 26 nucleotides of the 3′-UTR, covering a total of 5,020 bp. RT-PCR analyses showed that *nd-1* plants did not accumulate *Zds* transcripts. Based on genomic Southern hybridizations, the *nd-1* genome might contain a large deletion at the *Zds* locus (Conti et al. 2004).

6.3.2.3 Pistil- and Pollen-specific Genes

LIM proteins contain one or several LIM domains – double zinc–finger motifs with the sequence $C-X_2-C-X_{17-19}-H-X_2-C-X_2-C-X_{16-24}-C-X_2-(C, H, D)$ – which are often found in the same protein together with another functional domain, such as a homeodomain or a kinase domain (Taira et al. 1995). The acronym LIM is derived from the three homeodomain proteins first recognized as sharing this structural feature (Gill 1995): *lin-11*, which functions in asymmetric division of *Caenorhabditis elegans* (Freyd et al. 1990); *isl-1*, which has a major function in motor neuron development in rat (Karlsson et al. 1990); and *mec-3*, which is essential for proper differentiation of touch receptor neurons in *C. elegans* (Way and Chalfie 1988). LIM proteins act as developmental regulators in eukaryots, participating in a variety of basic cellular processes including gene transcription, cytoskeletal organization, and signaling (Dawid et al. 1995).

The first LIM protein encoded by the cDNA clone *sf3* (subsequently renamed HaPLIM1 for *H. annuus* Pollen LIM protein 1) identified in sunflower was exclusively expressed in mature pollen grains (Baltz et al. 1992a, b). It accumulated together with actin at the germination cones of pollen, suggesting that it participates in the growth of the pollen tube (Baltz et al. 1999). The second protein, which is expressed in most of the sporophytic tissues, localizes to cytoplasmic organelles and vesicle-like structures, and in some tissues to the nucleus as well (Mundel et al. 2000). Genomic Southern hybridizations carried out with *HaPLIM1* cDNA as a probe have shown that in sunflower the protein is encoded by a member of a multicopy gene family (Baltz et al. 1992a). The genomic sequences of the original *HaPLIM1* gene and the second gene *HaWLIM1* (*Helianthus annuus* Widely expressed LIM protein 1) were isolated by hybridization screening of a genomic sunflower DNA library constructed in Charon40, using the full-length cDNA clone in case of HaPLIM1, and in case of HaWLIM1, the complete untranslated 3'region as probe. The *HaPLIM2* genomic sequence was obtained by PCR amplification of genomic DNA from the sunflower line HA300 using specific primers (Eliasson et al 2000). Orthologs of these genes were also isolated from tobacco and *Arabidopsis*.

A novel pistil- and pollen-expressed gene (*sf21*) encoding a 352 amino acid long polypeptide was isolated by differential screening of a floral cDNA library from sunflower (Kräuter-Canham et al. 1997). The sunflower *sf21* gene is split by up to 10 introns. Orthologs of this gene were later discovered in *Arabidopsis*, rice and tomato by sequence comparison. The *sf21* gene is expressed in young and mature florets of an open inflorescence but not at the floral bud stage. In sunflower pistils, expression is limited to the transmitting tissue and the phloem cells. Immunohistological studies showed that the SF21 protein in early pistil development is localized exclusively in the nucleus of the stigma cells, while in later stages it appears progressively in the cytoplasm (Kräuter-Canham et al. 2001). However, later Lazarescu et al. (2006) could show that *sf21A* is member of a gene family and transcripts of this family could be found in all organs of sunflower plants. Screening of a sunflower BAC library (Özdemir et al. 2004) with the *sf21* cDNA probe identified six BAC clones, which represented four additional copies to the two genes *sf21A* and *sf21B* identified earlier (Kräuter-Canham et al. 2001). Only one of the BAC clones, 63K3, contained the full set of exons. RT-PCR using different combination of primers was performed. These expression analyses of the sunflower *sf21* gene family revealed multiple alternative and organ-specific splicing of transcripts (Lazarescu et al. 2006).

6.3.3 Genes Involved in Abiotic Stress Responses

6.3.3.1 Heat Shock Proteins

Heat shock proteins (HSPs) are well represented in plants. They have been divided into six different classes on the basis of their subcellular localization, immunological cross-reactivity and sequence similarity (Scharf et al. 2001). Small heat shock proteins are not only induced after heat shock, but also by other abiotic stimuli, such as cold, water deficit, heavy metals, ozone, UV radiation and γ-radiation (Barcala et al. 2008). In sunflower, three different cDNAs were cloned and sequenced from seed-stored mRNA by Almoguera and Jordano (1992). Sequence similarities and response to heat shock identified one of the cDNAs as low-molecular weight heat shock protein (lmw-HSP or small heat shock proteins, sHSP), whereas the two other clones showed significant sequence homology to late embryogenesis abundant (Lea) proteins from cotton and carrots, called D113 and Emb1, respectively. Applying osmotic stress, mRNAs of lmw-HSPs were induced earlier than Lea mRNAs. The coordinate accumulation of Lea and lmw-HSP transcripts during embryogenesis and in response to stress and ABA suggests the existence of common regulatory elements for Lea and lmw-HSP genes, and supports the assumption that HSPs might have alternative functions in the plant cell (Almoguera and Jordano 1992). Apart from this cloned cDNA *Ha hsp17.6*, a second cDNA was isolated and sequenced coding for the gene *Ha hsp17.9* (Coca et al. 1994). The two lmw-HSPs, HSP17.6 and HSP17.9, belong to two different families of cytoplasmic LMW HSPs. Using specific antibodies, Coca et al. (1994) observed differential expression of both proteins during zygotic embryogenesis under controlled environment, and a remarkable persistence of these LMW HSPs during germination. Tissue-print immunolocalization experiments showed that HSP17.6 and HSP17.9 were homogeneously distributed in every tissue of desiccation-tolerant dry seeds and young seedlings under non-stressed conditions (Coca et al. 1994). In 1996, Coca et al. reported the cloning of two additional sunflower genes coding for the small heat shock proteins (sHSPs) Ha hsp18.6 G2 and Ha hsp17.7 G4. In sunflower, *Ha hsp17.7 G4* mRNAs accumulated during zygotic embryogenesis at 25°C and in vegetative tissues either in response to heat shock (42°C), abscisic acid (ABA), or mild water stress treatments. In contrast, *Ha hsp18.6 G2* transcripts accumulated under heat shock conditions. In transgenic tobacco, using a fusion construct G4::*GUS* the developmental induction of *Ha hsp17.7 G4* during zygotic embryogenesis was confirmed. The proximal sequences (from −83 to +163) conferred most of the developmental regulation to the chimeric gene construct, as well as the responses to ABA and heat shock. However, the water stress response of this gene was not reproduced in transgenic tobacco, which indicates that this can be uncoupled from its regulation during embryogenesis (Coca et al. 1996). Chimeric constructs containing the promoter and upstream sequences of *Ha hsp17.6 G1* in tobacco reproduced its unique seed-specific expression previously reported in sunflower (Carranco et al. 1999). The expression of *sHSP* genes is regulated mainly by *cis*-elements, described as heat shock elements (HSEs) present in their promoters. The HSEs are arranged in clusters of three or more inverted repeats of the sequence 5'-nGAAn-3', to which, heat shock factors (HSFs) bind and control transcription (Carranco et al. 1997; Nover et al. 2001).

6.3.3.2 Drought Related Genes

To identify genes regulated by water stress and abscisic acid in drought tolerant sunflower transcript profiles of a drought tolerant and a drought sensitive genotype were compared under field conditions (Ouvrard et al. 1996). By subtractive hybridization, six different cDNA clones (originally designated *sdi* for sunflower *drought induced*), corresponding to transcripts accumulated in the leaves of the drought tolerant line during adaptive response, were identified. Sequence analysis of the six cDNA clones revealed homology to a non-specific lipid transfer protein (nsLTP), early light induced proteins (ELIP), 1-aminocyclopropane-1-carboxylate oxidase (ACC oxidase) and dehydrins (Ouvrard et al. 1996). Dehydrins are proteins produced during late stages of plant embryo development and following any environmental stimulus involving dehydration. Dehydrins represent an immunologically distinct family of proteins, also known as Lea D11 subgroup of late embryogenesis abundant (Lea) proteins (Dure et al. 1989), which have been described in many plant species (Close 1997). The drought induced genes identified in the drought tolerant sunflower line were named *HaDhn1*, *HaDhn2* and *HaELip1* according to their homology to the dehydrin family and to the early light induced protein (ELIP). In leaves of the drought tolerant plants, the accumulation of *HaDhn1* and *HaDhn2* transcripts, but not *HaELip1*, was correlated with the drought adaptive response (Cellier et al. 1998). *HaDhn2* transcription was upregulated in the tolerant genotype compared to the sensitive after ABA treatment, whereas the transcript level of *HaDhn1* was similar in tolerant and sensitive genotype. A sequence allelic to *HaDhn1*, named *HaDhn1a*, was found to be expressed in the later stages of embryogenesis in sunflower, depending on abscisic acid accumulation (Giordani et al. 1999). Natali et al. (2003) investigated the variability of this dehydrin encoding gene *HaDhn1a* in cultivated and wild sunflower accessions. With respect to the overall sequence, variation in both the coding and non-coding (intron and 3'-UTR) sequences was much larger among the wild accessions than among cultivars. Inclusion of wild populations of *H. annuus* into breeding programs might be a way to introduce a broader variability with regard to dehydrins into the cultivated sunflower. Giordani et al. (2003) analyzed the PCR amplification products of the *HaDhn1a* gene from 16 wild *Helianthus* species or subspecies on sequence level. All the isolated sequences included the typical dehydrin domains (Y, S and K), a portion of the 3'-UTR and an intron, inserted in the same position within the S-domain-encoding region. Using these data for phylogenetic trees, perennial and annual species formed a supported clade and *H. annuus* was separated from this clade. Natali et al. (2007) observed light-induced expression of *HaDhn1* during de-etiolation of sunflower seedlings. However, Western blot analyses suggest that transcription of the dehydrin gene is not followed by translation or that the dehydrin production is too low to be detectable.

Aquaporins are water channels, which can facilitate the passage of water across biological membranes. In plants, aquaporins have been localized to the tonoplast (TIP) and the plasma membrane (PIP); a third subfamily, Nod-MIP, includes proteins homologous to Nod26, a protein located in the peribacteriod membranes of symbiotic soybean root nodules (reviewed in Maurel 1997). Five sunflower cDNAs belonging to the TIP family were isolated by Sarda et al. (1999). Expression of the TIP-like genes was studied in roots during 24 h water deprivation. Due to the changes in their transcript levels, it is proposed that SunTIP aquaporins play a role in the sunflower response to drought (Sarda et al. 1999).

The method of differential display reverse transcription PCR (DDRT-PCR) was used to compare overall differences in gene expression between drought- or salinity-stressed and unstressed (control) sunflower plants (Liu and Baird 2003). Five drought-related cDNAs and 12 salinity-regulated cDNAs were cloned and sequenced. Thirteen of these cDNAs were confirmed to be expressed differentially in response to drought or salinity stress by quantitative RT-PCR. Sequence analysis of these clones identified five of them to show homology to known genes: guanylate kinase (signal transduction), lytB (antibiotic/drug resistance), selenium-binding protein (heavy metal stress), polyprotein (reverse transcriptase), and AC-like transposable element. For one of the cDNA clones, *HaABRC5*, the full-length cDNA sequence was obtained by completing the 5'-end by RACE (Liu and Baird 2004). The genomic DNA sequence upstream of the transcription start site of *HaABRC5* was cloned by RAGE. The full-length cDNA contains an ORF of 423 nucleotides encoding 141 amino acids, including a "bipartite nuclear targeting sequence". The deduced amino acid sequence had no similarity to known genes in the database. Quantitative RT-PCR showed that *HaABRC5* is upregulated by drought, high salinity, and exogenous application of abscisic acid (ABA). Three ABRE (ABA responsive elements) were found within the *HaABRC5* promoter region. Therefore, HaABRC5 is probably an ABA-responsive nuclear protein playing a role in plant stress response (Liu and Baird 2004).

Homeodomain-leucine zipper (Hd-Zip) proteins constitute a family of transcription factors found only in plants. For isolation of partial cDNA clones containing homeobox sequences, a PCR-based strategy was employed on total DNA from a sunflower stem cDNA library constructed in lambda gt10 as previously described (Gonzalez and Chan 1993). One of the clones represented a member of the Hd-Zip family that was named *Hahb-4* (Gago et al. 2002). The full-length cDNA sequence was 674 bp long and contained an ORF of 177 amino acids. In order to investigate the genomic structure of the *Hahb-4* gene, genomic DNA was amplified with several oligonucleotides comprising the entire cDNA. A single intron of 101 bp was detected between nucleotides 381 and 382. To isolate the promoter region of *Hahb-4*, inverse PCR was used on genomic DNA. A 720 bp fragment was isolated that contained the TATA box 24 bp upstream from the transcription initiation and two putative ABRE in the –290 and –165 regions (Gago et al. 2002). The transcription factor Hahb-4 is upregulated by drought and ABA in roots, stems and leaves (Gago et al. 2002).

Recently, a sunflower cDNA microarray containing about 800 clones covering major metabolic and signal transduction pathways was used to study gene expression profiles in leaves and embryos of drought-tolerant and -sensitive genotypes subjected to water deficit under field conditions. In total, 409 genes were proven to be differentially expressed among genotypes, water treatment and organs (Roche et al. 2007).

6.3.4 *Disease Resistance*

The identification of common domains like nucleotide binding site (NBS), leucine rich repeats (LRR) and toll interleukin receptor (TIR) in cloned resistance genes (Mindrinos et al. 1994; Whitham et al. 1994; Lawrence et al. 1995) made candidate gene approach for resistance genes in other species possible (Meyers et al. 1999; Dangl and Jones 2001; Jones and Dangl 2006). Conserved amino acid sequence motifs in the NBS domain have been widely used to isolate and classify NBS-LRR encoding genes (Meyers et al. 1999; Pan et al. 2000).

To date, 11 genes have been postulated to provide resistance to one or more races of downy mildew (*Plasmopara halstedii*) in sunflower (Rahim et al. 2002). Apart from map-based cloning attempts, candidate gene approach has been used to clone the genes corresponding to these resistances. Gentzbittel et al. (1998) used degenerate primers designed from conserved NBS domains of *N* from tobacco (Whitham et al. 1994), *RPS* from *A. thaliana* (Mindrinos et al. 1994) and *L6* from flax (Lawrence et al. 1995) to clone resistance gene analogs (RGAs). The amplification products represented a multigene family. One of the clones was sequenced and mapped close to the Pl_6 gene. Gedil et al. (2001a) placed Pl_1 and six RGCs of the NBS-LRR type on the HA370 x HA372 public sunflower map. An RGC of the TIR-NBS-LRR subclass, originally named *Ha-4W2* and later renamed *HaRGC1*, was linked to Pl_1 on LG 8 of the public map (Gedil et al. 2001a). Mapping of resistance gene candidates produced by degenerate primers pointed to the clustering of disease resistance genes in the Pl_1-Pl_2-Pl_6 region (Vear et al. 1997).

Using degenerate primers (Leister et al. 1996) and specific primers derived from sequences of sunflower RGAs (Genbank Accession No. AF272766, AF272767, AF272768 and AF272769) amplification products were obtained by PCR using sunflower DNA from the parents of the investigated crosses and two bulks of the OC × YSO cross (Radwan et al. 2003). Sixteen different RGAs were identified and based on sequence comparison and Southern hybridization patterns grouped into six classes. Two of these classes correspond to TIR-NBS-LRR sequences while the remaining four classes correspond to non-TIR-NBS-LRR type resistance genes. Genetic mapping of these RGAs on two segregating F_2 populations showed that the non-TIR-NBS-LRR RGAs are clustered and linked to the Pl_5/Pl_8 locus for resistance to downy mildew in sunflower (Radwan et al. 2003). Radwan et al. (2004) cloned full-length cDNA and genomic sequences from two of these RGAs, which were named as *Ha-NTIR11g* (accession no. AY490793) and *Ha-NTIR3a* (accession no. AY490791). The genomic sequence length of *Ha-NTIR1g1* was 6,780 bp corresponding to a cDNA of 5,154 bp. This sequence contains one ORF of 3,848 bp. For *Ha-NTIR3* two different clones of 4,034 bp and 3,986 bp were identified with a putative stop codon at position 3,986 for *Ha-NTIR3A* and 3,861 for *Ha-NTIR3B*. The predicted protein structures of the RGA clones *Ha-NTIR11g* and *Ha-NTIR3A* determined using the PFAM (*http://pfam.wustl.edu*) and SMART (*http://smart.embl-heidelberg.de*) databases were 1,279 and 1,302 amino acids long, respectively, and showed domains of the resistance genes of the CC-NBS-LRR class. The genetically incompatible combination involving the downy mildew race 300 and sunflower line QIR8 carrying the Pl_8 resistance gene is characterized by a hypersensitive-like reaction (Radwan et al. 2005). Semi-quantitative RT-PCR analysis showed that the transcript of *Ha-NTIR11g* was specifically induced during the incompatible reaction (Radwan et al. 2005). The high level of transcriptional expression of this RGA coincided with the transcript accumulation of the *hsr203J* gene, which is a marker of the hypersensitive reaction. However, treatment with salicylic acid and methyl jasmonate did not activate transcription of the *Ha-NTIR11g* gene, indicating that *Ha-NTIR11g* is not regulated by defense signaling pathways triggered by these molecules. In conclusion, a sunflower RGA *Ha-NTIR11g* was isolated that is transcriptionally activated in the incompatible reaction with downy mildew race 300. However, the role of this gene in sunflower resistance to *P. halstedii* needs to be verified by both transforming a susceptible sunflower line and analyzing its complete promoter (Radwan et al. 2005). Primers flanking an intron between the TIR-and NBS encoding region of the *HaRGC1* sequence amplify a large RGC family (Slabaugh et al. 2003). These RGC paralogs showed differences in intron length between 90 to >800 nucleotides. Twenty-four HaRGC1 loci were mapped to a 2–4 cM region of LG 8. Nine haplotypes (based on these 24 RGAs) were identified among elite inbred lines and were correlated to known downy mildew resistance specificities (Slabaugh et al. 2003).

In a large-scale approach, Radwan et al. (2008) mined the sunflower EST database and used comparative genomics to develop a deeper understanding of the diversity and distribution of NBS-LRR homologs in the sunflower genome. Collectively, 630 NBS-LRR homologs were identified, 88 by mining a database of 284,241 sunflower ESTs and 542 by sequencing 1,248 genomic DNA clones isolated from common and wild sunflower species. These genomic sequences were obtained by using degenerate primers against four conserved NBS amino acid sequences. DNA markers were developed from 196 unique NBS-LRR sequences and facilitated the mapping of 167 NBS-LRR loci. The latter were distributed throughout the sunflower genome in 44 clusters and singletons (Radwan et al. 2008). The largest and most complex cluster of NBS-LRR loci discovered so far in sunflower is located on linkage group 8 of the public sunflower map. It spans a segment of 36 cM in three subclusters containing 54 NBS-LRR loci. This region also harbors the Pl_1 and other downy mildew R-genes. The second largest NBS-LRR cluster was identified on the lower end of LG 13, with 27 loci in two subclusters and a few singletons spanning a segment of 31 cM. This is the segment that harbors downy mildew (Pl_5 and Pl_8) and black rust ($RADV$) R-genes (Lawson et al. 1998; Radwan et al. 2003; Slabaugh et al. 2003; Yu et al. 2003). This cluster harbors TIR- and non-TIR-NBS-LRR encoding genes. TIR sequences are identified by the presence of the RNBS-A TIR motif (FLENIRExSKKHGLEHLQKKLLSKLL) and aspartic acid (D) as the last amino acid residue in the Kin-2 motif, whereas non-TIR sequences are identified by the presence of the RNBS-A non-TIR motif (FDLxAWVCVSQxF) and tryptophan (W) as the last amino acid residue in the Kin-2 motif (Meyers et al. 1999; Pan et al. 2000). Using multiple sets of primers, NBS fragments were amplified from genomic DNA of three species of the Asteraceae family (Plocik et al. 2004): *Helianthus annuus* (sunflower), *Lactuca sativa* (lettuce), and *Cichorium intybus* (chicory). Analysis suggest that Asteraceae species share distinct families of R-genes, composed of genes related to both coiled-coiled (CC) and toll-interleukin-receptor (TIR) homology domain containing NBS-LRR resistance genes.

6.3.5 *Quality Traits in Sunflower*

6.3.5.1 *Cloning Genes of Fatty Acid Biosynthesis*

Conventional sunflower seed oil contains about 11% of saturated fatty acids (palmitic and stearic acid), a moderate level of monounsaturated fatty acids (oleic acid), and a high concentration of polyunsaturated fatty acids (linoleic acid), with traces of linolenic acid (Dorell and Vick 1997). The quality of seed oil involves the modification of the fatty acid composition, which determines the specific edible purpose (Pérez-Vich et al. 2000).

Stearic acid (C18:0), which is present in about 5% in sunflower oil, is desaturated to oleic acid (C18:1) by the Δ9-stearoyl-acyl carrier protein desaturase (SAD) (Ohlrogge and Browse 1995). A candidate gene approach was used to isolate the corresponding gene(s) (Hongtrakul et al. 1998a). A cDNA library was made from developing seeds and hybridized with a PCR product obtained by using degenerate primers and genomic sunflower DNA. The degenerate primers were designed from two near-consensus sequences identified after aligning 15 plant SAD cDNA sequences from GenBank. Nineteen SAD cDNA clones from sunflower were partially sequenced and found to belong to two groups. Full-length cDNAs from each group (SAD6 and SAD17) were completely sequenced. SAD17 (U91340) was 1,426 bp long with an ORF from bp 59 to 1,246, a 58 bp 5'-UTR and a 180 bp 3'-UTR with an ATAAAA polyadenylation signal beginning at bp 1,380. SAD6 (U91339) was 1,335 bp long with an ORF from bp 84 to 1,271, an 83 bp 5'-UTR and a 64 bp 3'UTR. The ORF of both cDNAs encoded 396 amino acid proteins. Transit peptides of 32 bp were predicted from the transit peptide cleavage site of a safflower cDNA (Thompson et al. 1991). Both genes are strongly expressed in developing seeds, but only moderately expressed in leaves and flowers, and weakly expressed in cotyledons, roots, and stem.

Oleoyl-phosphatidyl choline desaturase (*FAD2*) is necessary for the synthesis of linoleic (C18:2) from oleic acid (C18:1). Three *FAD2* genes (*FAD2-1*, *FAD2-2*, and *FAD2-3*) have been isolated for the sunflower microsomal oleate desaturase by a candidate gene approach from normal-type HA89 (Martinez-Rivas et al. 2001). OLD-7, a previously isolated FAD2 cDNA from the sunflower cultivar Mammoth (Hongtrakul et al. 1998b), showed a 100% identity to the *FAD2-1*. In high oleic lines, *Ol* represents a chemically induced, incomplete dominant mutation, which greatly increases oleic acid content in developing seeds in sunflower. *FAD2-1* cosegregates with *Ol* and has been mapped to LG 14 (Lacombe and Bervillé 2001; Perez-Vich et al. 2002). Hongtrakul et al. (1998b) developed dominant and codominant RFLP markers and a codominant SSR marker for *FAD2-1*. The codominant RFLP distinguished between *Ol* locus genotypes and can be used to accelerate the development of high oleic lines in sunflower.

The specificity of the acyl-acyl carrier protein (ACP) thioesterases plays an important role in controlling the fatty acid composition of seed oil. These enzymes are encoded by nuclear genes but are targeted to the plastid and are usually categorized into two groups, *FatA* and *FatB*, according to their sequence and acyl-ACP preference (Jones et al. 1995). The *FatA* genes encode thioesterases with a preference for unsaturated acyl-ACPs and with specificity for 18:1-ACP. On the other hand, *FatB* genes encode thioesterases with a higher affinity for saturated acyl-ACPs. During sunflower seed development, both FatA and FatB thioesterases have been shown to be expressed (Martinez-Force et al. 2000). For cloning the sunflower thioesterase gene *FatA*, cDNA from developing sunflower endosperm was prepared (Serrano-Vega et al. 2005). FatA protein sequences from public databases were aligned to identify regions of homology using the CLUSTAL v1.8 program (Thompson et al 1997). A PCR fragment was amplified with two degenerate primers designed from two highly conserved regions of FatA: FatA1 (5'-GARATHTAYARRTAYCCNGC-3') and FatAB (5'-CAYTTCNCKN CKRTANTC-3'). Using these primers, a 459 bp fragment from developing sunflower seed cDNA was amplified by PCR, corresponding to the internal part of a *FatA* mRNA. Subsequently, the full-length *HaFatA1* cDNA clone of 1,095 bp was obtained by RACE using additional primers (Serrano-Vega et al. 2005). The full-length cDNA was predicted to generate a pre-protein of 365 amino acids, with a molecular mass of 41.2 kDa. Functional studies of this gene were performed in *E. coli* (Serrano-Vega et al. 2005).

6.3.5.2 Cloning Genes Involved in Tocopherol Biosynthesis in Sunflower

Tocopherols are the most powerful natural fat-soluble antioxidants. The four naturally occurring tocopherols (α-, β-, γ-, and δ-tocopherol) widely differ in their relative in vivo (Vitamin E) and in vitro antioxidant properties (Sattler et al. 2004). Sunflower seeds (wild type) normally accumulate 92-98% α-tocopherol.

Demurin (1993) postulated two non-allelic unlinked genes, *Tph1* and *Tph2*, controlling tocopherol composition in sunflower seeds. *Tph1* gene controls the ratio of α-, and β-tocopherol, and *Tph2* gene affects that of the α - and γ-homologues. Recessive alleles were found by wide-scale screening and selfing of spontaneous mutations. Vera-Ruiz et al. (2006) mapped the *Tph1* gene by SSR markers (Tang et al. 2002) and developed a linkage map of the *Tph1*-encompassing region. The *Tph1* was mapped to the upper end of the LG 1. The two map-based markers linked to this gene can be used to assist selection for increased β-tocopherol content.

Three mutant loci (m = *Tph1*, g = *Tph2*, and *d*) have been shown to disrupt the synthesis of α-tocopherol (α-T) and produce novel tocopherol (vitamin E) profiles in sunflower seeds (Tang et al. 2006). The m (*Tph1*) mutation partially disrupts the synthesis of α-tocopherol (vitamin E) in sunflower seeds and was predicted to disrupt a methyltransferase activity necessary for synthesis of α- and γ-tocopherol. A candidate gene approach was undertaken to isolate two corresponding genes for sunflower, *MT-1* and *MT-2*, encoding for two 2-methyl-6-phytyl-1,4-benzoquinone/2-methyl-6-solanyl-1,4-benzoquinone methyltransferase paralogs (MPBQ/ MSBQ-

MT). BLAST searches of the Compositae Genome Program Database (CGPdb; *http://cgpdb.ucdavis.edu/*) and National Center for Biotechnology Information (NCBI) GenBank Databases (*http://www.ncbi.nlm.niv.gov/*) were performed using cDNA and amino acid sequences for *Arabidopsis* MPBQ/MSBQ-MT (GenBank accession No. AB0542571 and At3g63410; Cheng et al. 2003; Motohashi et al. 2003) as query templates to identify sunflower MPBQ/MSBQ-MT homologs (Tang et al. 2006). Putative MPBQ/MSBQ-*MT* ESTs identified in the initial search were subsequently used as templates in BLAST searches and identified several additional sunflower ESTs homologous to *Arabidopsis* MPBQ/MSBQ-MT. *MT-1* and *MT-2* cDNA ends were isolated using 5'- and 3'-rapid amplification of cDNA ends (RACE) on developing seed RNAs (Tang et al. 2006). Full-length *MT-1* and *MT-2* cDNA sequences were isolated using RT-PCR. Full-length *MT-1* and *MT-2* genomic DNA sequences were subsequently isolated by sequencing amplicons produced by long-distance PCR (Tang et al. 2006). *MT-1* and *m* cosegregated and mapped to LG 1 (Tang et al. 2006). *MT-1* represents a non-lethal knock-out mutation caused by the insertion of a 5.2 kb Ty3/*gypsy*-like retrotransposon in exon 1 (Tang et al. 2006). *MT-1* is not transcribed in the mutant homozygotes (*mm*). *MT-2* and the cryptic codominant mutation *d* cosegregated with the *m* locus on LG 4 being epistatic to the *d* locus. The *MT-2* mutant allele carried a 30 bp insertion at the start of the 5'-UTR which caused a reduced transcription in seeds and leaves in the homozygous mutant (*dd*) compared to the wild type (Tang et al. 2006).

The *g* (*Tph2*) mutation knocks out a γ-tocopherol methyltransferase (γ-TMT) activity and causes an accumulation of > 90% γ-tocopherol (γ-T) in sunflower seeds (Hass et al. 2006). Using a candidate gene approach as described for the *MT* gene isolation, two γ-TMT paralogs starting with *Arabidopsis* γ-TMT (GeneBank accession nos. AF104220 and At1g64970; Shintani and DellaPenna 1998; Bergmüller et al. 2003) could be identified and sequenced in sunflower. The *g* mutation greatly decreased *γ-TMT-1* transcription, caused alternative splicing of *γ-TMT-1*, disrupted *γ-TMT-2* transcription, and knocked out one of two transcription initiation sites identified in the wild type. Both γ-TMT paralogs (*γ-TMT-1* and *γ-TMT-2*) mapped on LG 8, cosegregate with the *g* locus (Hass et al. 2006).

In addition, the tocopherolcyclase (TC) was cloned and isolated starting from the GenBank accession nos. AL022537 and At4g32770 (Porfirova et al. 2002; Sattler et al. 2003). This gene also mapped to LG 8, but segregated independently from the *g* locus (Hass et al. 2006).

6.4 Verification of Candidate Genes and Functional Analyses of Genes

6.4.1 Attempts of Functional Analyses by Transgenic Approaches in Sunflower

Any approach to use sunflower for transformation experiments to confirm functions of the candidate genes isolated has been severely hampered by the fact that sunflower has proven to be recalcitrant in tissue culture due to its low regeneration potential (Hahne 2001). Efforts have been made to select genotypes from interspecific hybrids with superior regeneration potential as compared to the public lines (Weber et al. 2000). Most of the successful transformation approaches applied to produce transgenic offspring used the shoot apical meristems (Bidney et al. 1992; Knittel et al. 1994; Malone-Schoneberg et al. 1994; Grayburn and Vick 1995; Lucas et al. 2000). However, these approaches were inefficient and unreliable as most of the regenerants proved to be chimeric as transformation events giving rise to regenerated shoots predominantly occurred in preexisting meristems (Burrus et al. 1996). Improved transformation protocols using macerating enzymes (0.1% cellulase and 0.05% pectinase) prior to *Agrobacterium* incubation of apical shoot meristems (Weber et al. 2003) or successive excisions of the apical and axillary shoots and cultivating the split embryonic axes on medium containing 0.1 mg l^{-1} 6-benzylaminopurine (Hewezi et al. 2003) have increased the rates of stable transformation events. Mohamed et al. (2006) also used juvenile split apical meristems from two important high oleic sunflower genotypes (Capella and SWSR2) to obtain transformants in sunflower using *gus* and *gfp* as reporter genes. Transformation was performed via *Agrobacterium* and particle bombardment.

However, laboratories still stick with their own protocols and a few reports have been made on transforming sunflowers for characterization or functional analyses of genes isolated from sunflower. Rousselin et al. (2002) successfully transformed sunflower using an interspecific hybrid line STR1/95 derived from a cross between *Helianthus annuus* and *H. strumosus* and a chimaeric construct consisting of the coding sequence of Δ9-stearoyl-(acyl carrier protein) desaturase from *Ricinus communis* under the control of the sunflower promoter *Ha ds10*. Two independent transformants could be obtained containing three and six copies of the T-DNA. Transcript analysis of the Δ9-stearoyl-(acyl carrier protein) desaturase under the control of the seed-specific promoter *Ha ds10* confirmed the tissue-specific expression in developing embryos and not in leaves. Some of the transgenic lines produced oil with a significantly reduced stearic acid content (Rousselin et al. 2002).

An attempt to use gene technology as an alternative method to protect plants from microbial diseases was made by Sawahel and Hagran (2006) in sunflower. The human lysozyme gene, which had been proven to be a simple and effective approach for genetic engineering disease resistance against phytopathogenic bacteria and fungi, only requires the introduction of a single transgene (Nakajima et al. 1997). Sunflower plants were transformed via cocultivation of previously bombarded hypocotyls explants with *Agrobacterium tumefaciens* harboring the plasmid pNGL that contained the human lysozyme gene (Sawahel and Hagran 2006). Two regenerants from one primary transformant that expressed the human lysozyme gene were confirmed to be resistant against *Sclerotinia sclerotiorum*.

Other researchers used *H. annuus* as well as *Nicotiana tabacum* to test the function of their genes (Almoguera et al. 2002; Hewezi et al. 2006). Using degenerate primers derived from NBS domains an NBS-LRR homologous sequence (PLFOR48) was isolated from sunflower, which cosegregated with a downy mildew disease resistance locus (Bouzidi et al. 2002). The antisense sequence of PLFOR48 under the control of the constitutive CaMV 35S promoter was introduced via transformation into sunflower and tobacco to assess loss of function. However, the construct caused major developmental abnormalities in sunflower as well as in tobacco so that it is assumed that PFLOR48 may be also involved in regulating developmental pathways apart from a role in disease resistance.

Rojas et al. (1999) performed successfully transient expression analyses in sunflower embryos by bombardment of microprojectiles coated with a DNA plasmid mixture, including reference, effector and reporter plasmids. Each plasmid combination was bombarded at least 25 times (five replications in each of five independent experiments). These transient expressions showed that ABI3, a seed-specific transcription factor from *Arabidopsis*, also activated the chimeric gene with the *Ha hsp17.7 G4* promoter. *Trans*-activation with LpHSFA1, a heat shock factor from tomato, reproduced the activation patterns of wild type and mutant promoters observed with ABI3 (Rojas et al. 1999). Using LpHsfA1 and LpHsfA2, another well characterized heat stress transcription factors (Hsfs) from tomato, transcription activation of the *Ha hsp17.6 G1* promoter was analysed by transient expression analyses (Rojas et al. 2002). By the same co-bombardment of sunflower embryos, Díaz-Martín et al. (2005) analyzed the interaction of the two transcription factors HaHSFA9 and HaADREB2 that could be shown to synergistically *trans*-activate the *Ha hsp17.6 G1* promoter.

6.4.2 Use of Heterologous Systems to Analyze Genes Isolated from Sunflower

As sunflower proved to be so difficult in transformation experiments heterologous systems, e.g., *Arabidopsis thaliana* or *Nicotiana tabacum*, have been used to study the in vivo function of genes, which had been isolated from sunflower, assuming that basic processes are conserved in dicotyledon plants. The transcription factor Hahb-4, encoded by a sunflower homeobox-leucine zipper gene, had been isolated from sunflower as it is upregulated by drought and ABA in roots, stems and leaves (Gago et al. 2002). To analyze its function *Arabidopsis* plants were transformed with *Hahb-4* under the control of the CaMV 35S promoter (Dezar et al. 2005a). Overexpression of *Hahb-4* in *Arabidopsis* resulted in transgenic plants that were more tolerant to water stress conditions than the wild type plants. However, the transgenic *Arabidopsis* plants showed additional distinct morphological changes and a delay in development due to the constitutive expression of *Hahb-4*. Plants transformed with constructs using the *Hahb-4* promoter sequences fused to *gusA* show the expression of the reporter gene in defined cell-types and developmental stages and are induced by drought and abscisic acid (Dezar et al. 2005b). Transgenic *Arabidopsis* plants carrying *Hahb-4* under control of its own stress-inducible promoter also exhibit water-stress tolerance though to a lower degree, but without any major changes in the phenotype (Manavella et al. 2006). In addition, these plants show a marked delay in senescence and were less sensitive to ethylene. Using microarrays and quantitative RT-PCR, Manavella et al. (2006) could demonstrate that expression of *Hahb-4* has a major repressive effect on genes related to ethylene synthesis and ethylene signaling. The transcription factor Hahb-4 seems to mediate the cross-talk between ethylene and drought signaling pathways. Several inducible promoters were used to control expression of *Hahb-4* in transgenic *Arabidopsis* plants (Cabello et al. 2007). These plants showed a normal phenotype and an enhanced drought tolerance but not as high as the transgenic plants expressing *Hahb-4* constitutively. Finally, a chimeric construct between *Hahb-4* promoter and the leader intron of the *Arabidopsis Cox5c* gene was used either to express *gus* or *Hahb-4* (Cabello et al. 2007). Transgenic plants with the chimerical construct for *Hahb-4* were indistinguishable from the wild type plants under normal growth conditions but water stress tolerance was achieved at a level as strong as the plants constitutively expressing *Hahb-4*.

Plant homeodomain-leucine zipper (HD-Zip) proteins, unlike many animal homeodomains (HDs), are unable to bind DNA as monomers (Tron et al. 2004). To investigate the molecular basis of their different behavior chimeras between HD of the sunflower HD-Zip protein Hahb-4 and that of *Drosophila* engrailed (EN) were constructed and overexpressed in *E. coli*. Electrophoretic mobility shift assays were used to study the binding ability to DNA (Tron et al. 2004). Another transcription factor of the same homeobox-leucine zipper family, *Hahb-10*, was also investigated for its function in *Arabidopsis*. *Hahb-10*, which is regulated by light, promotes early flowering in transgenic *Arabidopsis* plants (Rueda et al. 2005).

Apart from *Arabidopsis*, tobacco was used to test the function of genes isolated from sunflower. Chimeric constructs containing the promoter and upstream sequences of *Ha hsp17.6 G1*, a small heat shock protein gene, reproduced in transgenic tobacco (*Nicotiana tabacum*) its unique seed-specific expression patterns previously reported in sunflower (Carranco et al. 1999). Seed-specific overexpression of the transcription factor HaHSFA9 (*H. annuus* heat-stress factor A9) in transgenic tobacco was linked with increased seed longevity, and with the overaccumulation of a subset of small heat-shock proteins (sHSPs) in seeds (Prieto-Dapena et al. 2006). In addition, Prieto-Dapena et al. (2008) could demonstrate that the constitutive overexpression of the seed-specific HaHSFA9 transcription factor under control of the CaMV 35S promoter is sufficient to confer tolerance to severe dehydration, outside of the developing seed context, to vegetative tissues of transgenic tobacco.

For studying genes from the fatty acid biosynthesis, expression of these genes in *E. coli* proved to be valuable. To characterize the enzyme activities from the stearoyl-ACP-desaturase and the acyl-ACP thioesterases FatA and FatB from sunflower seeds Serrano-Vega et al. (2003, 2005) cloned, sequenced and overexpressed the recombinant genes in *E. coli*.

6.4.3 *Mutant Analyses*

6.4.3.1 *Mutants in Fatty Acid Synthesis*

With regard to fatty acid synthesis, the most well known mutation in sunflower is the pervenet mutation, obtained by treating a wild type population (VNIIMK83) with dimethylsulfate that resulted in the first high-oleic sunflower cultivar (Soldatov 1976). The oleoyl-phosphatidyl choline desaturase (*FAD2*) is necessary for the synthesis of linoleic acid from oleic acid (Ohlrogge and Browse 1995). Three *FAD2* genes (*FAD2-1, FAD2-2* and *FAD2-3*) have been identified in sunflower (Hongtrakul et al. 1998b; Martinéz-Rivas et al. 2001). Whereas *FAD2-1* is strongly expressed in developing seeds (Hongtrakul et al. 1998b), *FAD2-2* and *FAD2-3* are weakly expressed (Martinéz-Rivas et al. 2001). Isolation and sequencing of the corresponding *FAD2* cDNA clones from the high-oleic variety, designed HA-OL9*FAD2-1*, HA-OL9*FAD2-2*, and HA-OL9*FAD2-3*, showed 100% identity to the corresponding *FAD2* sequences from the normal type variety (Martinéz-Rivas et al. 2001). However, Hongtrakul et al. (1998b) as well as Lacombe and Berville (2001) discovered that *FAD2-1* was duplicated and very weakly expressed in mutants homozygous for the *Ol* mutation. Schuppert et al. (2006) further investigated the duplication of the *FAD2-1*. The upstream repeat (*FAD2-1U*) carries a 1.69 kb intron in the 5'-UTR, whereas the downstream repeat (*FAD2-1D*) is missing the first 1.54 kb of the 5'-UTR and intron. *FAD2-1* maps to LG 14 of the public sunflower map (Schuppert et al. 2006). Novel *FAD2-1* alleles were found in exotic low-oleic genotypes.

Osorio et al. (1995) induced additional variability in the fatty acid pattern for sunflower by mutagenesis using ethylmethane sulfonate, sodium azide and X-rays. Four sunflower mutants with altered seed fatty acids composition were isolated: CAS-5, which has oil with a five-fold increase in palmitic acid (C16:0). CAS-3, CAS-4, and CAS-8 have two- to six-fold stearic acid (C18:0) content compared to the normal content observed in the oil of cultivated sunflower. Osorio et al. (1995) concluded that according to the bimodal distribution in fatty acid composition of CAS-3 and CAS-5 major recessive genes were involved in the control of these traits.

Pérez-Vich et al. (1999) studied the inheritance and the genetic control of the CAS-3 mutants in detail. They found that inheritance of the C18:0 content in F_1, F_2, and BC_1F_1 seeds from the cross of CAS-3 with their parental lines indicated the presence of partially recessive alleles at two independent loci *Es1* and *Es2* controlling the high stearic acid content in the seed oil of this mutant. Candidate gene and quantitative trait loci (QTL) analyses demonstrated that *Es1* cosegregates with a stearoyl-acyl carrier protein (ACP) desaturase locus (*SAD17A*) and underlies the major QTL affecting the concentration of stearic acid in CAS-3. This QTL has been named *st1-SAD17A* and is located on LG 1 of the sunflower genetic map (Pérez-Vich et al. 2002). Later on, Pérez-Vich et al. (2004) mapped three minor QTLs for increased stearic acid content in sunflower seed oil on LG 3, LG 11, and LG 13 of the sunflower genetic map (Tang et al. 2002).

Pérez-Vich et al. (2005, 2006a) also reported mapping of the very high stearic acid gene *Es3* from a mutant CAS-14 and identified PCR-based molecular markers linked to this gene. Genetic characterization of the CAS-14 mutant (Pérez-Vich et al. 2006b) showed that the very high stearic acid content in this line, which shows a gradient within the seed from embryo to distal extreme of the seed, is controlled by a single recessive gene (*es3*), segregating independently of the *Es1* gene with a major effect on stearic acid content in CAS-3. In addition, the mutation in the *Es3* gene did not coincide with any boundary group associated to stearoyl-acyl carrier protein desaturase (SAD) enzymes (Pérez-Vich et al. 2005). However, expression analyses of *SAD6* and *SAD17*, two genes coding for SAD forms expressed at high level in sunflower seeds during oil synthesis, showed a coordinated decrease in transcription corresponding to the decrease in enzyme activity (Salas et al. 2008). The observed downregulation of both genes in CAS-14 demonstrated as the previous genetic studies that CAS-3 and CAS-14 represent distinct different mutations. As the mutation did not map within the boundary groups corresponding to SAD genes this means that an alteration in a *trans*-regulatory element affected the expression of these desaturases. Therefore, the gene responsible for the CAS-14 phenotype probably encodes a transcription factor that regulates the expression of *SAD6* and *SAD17* (Salas et al. 2008). Pérez-Vich et al. (2006b) mapped the

Es3 gene to LG 8 of the sunflower genetic map. The nuclear male sterility gene Ms_{11} gene was also mapped to LG 8. The genetic distance between this gene and *Es3* was found to be 7.4 cM. Reliable codominant markers linked to *Es3* have been identified (Pérez-Vich et al. 2006b). These markers can be used to distinguish the heterozygote *Es3/es3* from the *Es3/Es3* in a backcross program without the need for progeny testing. The *Es3* gene is an important target for marker-assisted breeding programs because its expression is strongly influenced by temperature during seed maturation (Fernández-Moya et al. 2002, 2003).

An additional sunflower mutant, CAS-12, was obtained by X-irradiation of the high-oleic line BDS-2-432, which has both high palmitic (about 30%) and high oleic acid contents, and also substantial amount (about 7%) of palmitoleic acid (Fernández-Martínez et al. 1997). Venegas-Caleron et al. (2008) described a new half-palmitic mutant line, CAS-7, with about 14% palmitic acid content that was obtained by chemical mutagenesis using sodium azide. This mutant is also called *wrinkled* because the cotyledons show a characteristic wrinkled phenotype. The homozygous line is lethal within 10–12 days after germination, so the line can only be maintained as a heterozygous. The triacylglycerol content showed a reduction of 57% compared to the control. Furthermore, this mutant has 40% of trilinolein, the highest content found till today in sunflower seeds. The cause of this mutation that gave multiple changes in the phenotype including poor green color and reduced chloroplast development is still unknown.

6.4.3.2 Mutants in Developmental Processes

Giordani et al. (1999) studied expression of the dehydrin gene, *HaDhn1a* in two sunflower mutants for ABA synthesis and accumulation: *nd-1* tan albino, which is a nondormant and lethal mutant with a very low ABA content and no ABA accumulation in response to stress and *w-1*, a wilty mutant with reduced ABA accumulation during embryo and plantlet development. The results indicated the existence of two regulation pathways of *HaDhn1a* transcript accumulation, an ABA-dependent and an ABA-independent one, which may have cumulative effects (Giordani et al. 1999).

Fambrini et al. (2003) described a spontaneous mutant of sunflower, named *missing flower* (*mf*), which was found among progenies of the inbred line *Double Petal Whorl* (DPW), previously selected from an in vitro-derived progeny (Barotti et al. 1995). The DPW genotype is characterized by a homeotic substitution of the anther whorl with chimeric structures mixed with anther and petal tissue (Barotti et al. 1995). The mutant *missing flower* (*mf*) is characterized by a lack of axillary shoots. The phenotype is the result of a defect in axillary meristem initiation. In addition to shoot branching, the mutation affects floral differentiation. The gene product of *missing flower* seems to be essential to provide or perceive an appropriate signal for the initiation of axillary meristems during both vegetative and reproductive phases.

The isolation of pigment-deficient mutants is crucial to identify genes controlling chloroplast biogenesis (Hennigsen and Stummann 1982; Leister 2003). In *H. annuus*, pigment deficient mutants have occurred spontaneously (Rodriguez et al. 1998; Fambrini et al. 2004) or have been induced by in vitro culture (Barotti et al. 1995). The nuclear mutation *xantha1* (*xan1*) is characterized by many pleiotropic effects (Fambrini et al. 2007). Homozygous *xan1/xan1* seedlings under field conditions die after depletion of cotyledonary reserves and can only be recovered by selfing normal heterozygous plants (*Xan1/xan1*). These mutants show aberrant development of chloroplasts, deficient pigment content and reduced CO_2 assimilation rate.

6.4.4 Site-directed Mutagenesis and Chimeric Constructs

To understand the developmental regulation of plant *sHSP* genes, the promoter and regulatory elements need to be cloned and analyzed (Almoguera et al. 1998). Promoter and 5'-flanking sequences of *Ha hsp17.7 G4* (Coca et al. 1996), a small heat shock gene from sunflower, have been isolated and shown to confer developmental regulation in zygotic embryogenesis using chimeric constructs. However, deletion analyses did not separate heat shock response from developmental regulation. Site-directed mutagenesis resulting in sequential nucleotide substitutions in the heat shock element regions of the *Ha hsp17.7 G4* promoter was performed (Almoguera et al. 1998). The substitutions were introduced by PCR amplification of plasmid DNA using *Pfu* DNA polymerase and different olignucleotides containing the desired substitutions. A two-step PCR (Chen and Przybyla 1994) was used, in which "megaprimers" containing the mutations were obtained after the first round of amplification, and gel-purified for utilization in subsequent reaction, which amplified DNA fragments that were flanked by restriction sites in the plasmid polylinker. Obtained promoter sequences were combined with *gus* for expression analyses in tobacco (Almoguera et al. 1998). Dual regulation of the *Ha hsp17.7 G4* promoter during embryogenesis could be demonstrated. Thus, whereas activation of the chimeric genes during early maturation stages did not require intact heat shock elements (HSE) in the promoter region, expression at later desiccation stages was reduced by mutations in both the proximal (-57 to -89) and distal (-99 to -121) HSE. In contrast, two point mutations in the proximal HSE that did not severely affect gene expression during zygotic embryogenesis eliminated the heat shock response of the same chimeric gene in vegetative organs (Almoguera et al. 1998).

Knotted-1 like genes constitute a family of genes, whose products are transcription factors involved in several aspects of plant development. Tioni et al. (2003) had isolated three *knotted-1* like genes in sunflower and analyzed their expression patterns. One of the genes, *HAKN1*, was highly expressed in leaves and roots, whereas other gene, *HaKN2*, showed preferential expression in stem, especially in fascicular and interfascicular cambium and phloem (Tioni et al. 2003). The interaction of the homeodomain of the sunflower *HAKN1* transcription factor with DNA was studied by site-directed mutagenesis, hydroxyl radical footprinting and missing nucleoside experiments (Tioni et al. 2005). For this, the homeodomain of *HAKN1* was expressed in *E. coli* as a fusion protein with the maltose binding protein using the vector pMALc2. The fusion protein was purified by affinity chromatography in amylose resin and used for DNA-protein interaction studies. A 24-bp nucleotide (HAKN1 binding site, BS1) containing the sequence TGT(G/C)ACA was used as the DNA target. Electrophoretic mobility shift assays were performed with HAKN1 and the oligonucleotide BS1 and its variants containing changes at single positions. Based on the analyses, a model for the interaction of HAKN1 with DNA was proposed (Tioni et al. 2005).

The protein PLIM-1 from sunflower is expressed exclusively in mature, free pollens (Baltz et al. 1996). The pollen protein as well as some of the mutant forms created by site-directed mutagenesis were expressed in *E. coli* to obtain protein for studies of its interactions with nucleic acids. The PLIM1 protein bound DNA and RNA in vitro to form large complexes, while the mutant polypeptides lost the ability (Baltz et al. 1996).

Serrano-Vega et al. (2005) studied the substrate specificity of an acyl-acyl carrier protein (ACP) thioesterase, which determines the fatty acids available for biosynthesis of storage and membrane lipids in seeds. Through the heterologous expression of *HaFatA1* in *E. coli*, the acyl-ACP-thioesterase FatA1 was purified and characterized, showing that sunflower *HaFatA1* cDNA encodes a functional enzyme with preference for monounsaturated acyl-ACPs. The HaFatA1 thioesterase was most efficient (kcat/Km) in catalyzing oleoyl-ACP, both in vivo and in vitro. Finally, using available structure prediction models, a 3D model of plant acyl-ACP thioesterases was proposed. In addition, the model was tested by mutating the residues proposed to interact with the ACP protein in the FatA thioesterase by site-directed mutagenesis (Serrano-Vega et al. 2005).

6.5 Conclusions

In the recent years, cloning and characterization of genes from sunflower have made a huge progress. Although, so far, no gene has been successfully cloned by map-based cloning attempts, candidate gene approach has proved to be very useful and successful in cloning genes in sunflower. Intensive studies have also been performed to understand promoter structure of genes in sunflower and to analyze the transcription pattern in different tissues and under different environmental conditions. Functional analyses to characterize or verify candidate genes by transgenic approaches have been performed in sunflower or even more successful in heterologous systems e.g., tobacco and *Arabidopsis*.

The feasibility of association mapping offers new possibilities to identify genes responsible for a phenotype. In addition, the identification of genes underlying major QTLs for agronomically important traits will present a challenge for sunflower researchers in the coming years. For the future, there is still a lot of work to be done to clone and characterize genes, especially those specific for sunflower.

References

Akagi H, Nakamura A, Yokozeki-Misono Y, Inagaki A, Takahashi H, Mori K, Fujimura T (2004) Positional cloning of the rice *Rf-1* gene, a restorer of BT-type cytoplasmic male sterility that encodes a mitochondria-targeting PPR protein. Theor Appl Genet 108: 1449–1457.

Albertsen MC, Trimnell MR, Fox TW (1993) Tagging, cloning and characterizing a male fertility gene in maize. Am J Bot 80 (Suppl): 16.

Almoguera C, Jordano J (1992) Developmental and environmental concurrent expression of sunflower dry-seed-stored low-molecular weight heat shock protein and Lea messenger mRNA. Plant Mol Biol 19: 781–792.

Almoguera C, Prieto-Dapena P, Jordano J (1998) Dual regulation of a heat shock promoter during embryogenesis: stage-dependent role of heat shock elements. Plant J 13: 437–446.

Almoguera C, Rojas A, Diaz-Martin J, Prieto-Dapena P, Carranco R, Jordano J (2002) A seed-specific heat-shock transcription factor involved in developmental regulating during embryogenesis in sunflower. J Biol Chem 277: 43866–43872.

Arumuganathan K, Earle ED (1991) Nuclear DNA content of some important plant species. Plant Mol Biol Rep 9: 208–219.

Auborg S, Boudet N, Kreis M, Lecharny A (2000) In *Arabidopsis thaliana*, 1% of the genome codes a novel protein family unique to plants. Plant Mol Biol 42: 603–613.

Baack EJ, Whitney KD, Rieseberg LH (2005) Hybridization and genome size evolution: Timing and magnitude of nuclear DNA content increases in *Helianthus* homoploid species. New Phytol 167: 623–630.

Baltz R, Domon C, Pillay DT, Steinmetz A (1992a) Characterization of a pollen-specific cDNA from sunflower encoding a zinc finger protein. Plant J 2: 713–721.

Baltz R, Evrard JL, Domon C, Steinmetz A (1992b) A LIM motif is present in a pollen-specific protein. Plant Cell 4: 1465–1466.

Baltz R, Evrard JL, Bourdon V, Steinmetz A (1996) The pollen-specific LIM protein PLIM1 from sunflower binds nucleic acids in vitro. Sex Plant Reprod 9: 264–268.

Baltz R, Schmit AC, Kohnen M, Hentges F, Steinmetz A (1999) Differential localization of the LIM domain protein PLIM-1 in microspores and mature pollen grains from sunflower. Sex Plant Reprod 12: 60–65.

Barcala M, Garcia A, Cubas P, Alomguera C, Jordano J, Fenoll C, Escobar C (2008) Distinct heat-shock element arrangements that mediate the heat shock, but not the late–embryogenesis induction of small heat-shock proteins, correlated with promoter activation in root-knot nematode feeding cells. Plant Mol Biol 66: 151–164.

Barotti S, Fambrini M, Pugliesi C, Lenzi A (1995) Genetic variability in plants regenerated from in vitro culture of sunflower (*Helianthus annuus* L.). Plant Breed 114: 275–276.

Bentolila S, Alfonso AA, Hanson MR (2002) A pentatricopeptide-repeat-containing gene restores fertility to cytoplasmic male-sterile plants. Proc Natl Acad Sci USA 99: 10887–10892.

Bergmüller E, Profirova S, Dörmann P (2003) Characterization of an *Arabidopsis* mutant deficient in γ-tocopherol methyltransferase. Plant Mol Biol 52: 1181–1190.

Bernasconi P, Woodworth AR, Rosen BA, Subramanian MV, Sichl DL (1995) A natural occurring point mutation confers broad range tolerance to herbicides that target acetolactate synthase. J Biol Chem 270: 17381–1785.

Berry ST, Leon AJ, Hanfrey CC, Challis P, Burkholz A, Barnes SR, Rufener GK, Lee M, Caligari PDS (1995) Molecular marker analysis of *Helianthus annuus* L. 2. Construction of an RFLP linkage map for cultivated sunflower. Theor Appl Genet 91: 195–199.

Bert PF, Tournevieille de Labrouche D, Philippon J, Mouzeyar S, Jouan I, Nicolas P, Vear F (2001) Identification of a second linkage group carrying genes controlling resistance to downy mildew (*Plasmopara halstedii*) in sunflower (*Helianthus annuus* L.). Theor Appl Genet 103: 992–997.

Bidney D, Scelonge C, Martich J, Burrus M, Sims L, Huffmann G (1992) Microprojectile bombardment of plant tissues increases transformation frequency by *Agrobacterium tumefaciens*. Plant Mol Biol 18: 301–313.

Bouzidi MF, Badaoui S, Cambon F, Vear F, Tourvieille De Labrouhe D, Nicolas P, Mouzeyar S (2002) Molecular analysis of a major locus for resistance to downy mildew in sunflower with specific PCR-based markers. Theor Appl Genet 104: 592–600.

Bouzidi MF, Franchel J, Tao Q, Stormo K, Mraz A, Nicolas P, Mouzeyar S (2006) A sunflower BAC library suitable for PCR screening and physical mapping of targeted genomic regions. Theor Appl Genet 113: 81–89.

Brahm L, Röcher T, Friedt W (2000) PCR-based markers facilitating marker assisted selection in sunflower for resistance to downy mildew. Crop Sci 40: 676–682.

Brown GG, Formanova N, Jin H, Wargachuk R, Dendy C, Patil P, Laforest M, Zhang J, Cheung WY, Landry BS (2003) The radish *Rfo* restorer gene of Ogura cytoplasmic male sterility encodes a protein with multiple pentatricopeptide repeats. Plant J 35: 262–272.

Burrus M, Molinier J, Himber C, Hunold R, Bronner R, Rousselin P, Hahne G (1996) *Agrobacterium*-mediated transformation of sunflower (*Helianthus annuus* L.) shoot apices: Transformation patterns. Mol Breed 2: 329–338.

Cabello JV, Dezar CA, Manavella PA, Chan RL (2007) The intron of the *Arabidopsis thaliana* COX5c gene is able to improve the drought tolerance conferred by the sunflower Hahb-4 transcription factor. Planta 226: 1143–1154.

Carranco R, Almoguera C, Jordano J (1997) A plant small heat shock protein expressed during zygotic embryogenesis but noninducible by heat stress. J Biol Chem 272: 27470–27475.

Carranco R, Almoguera C, Jordano J (1999) An imperfect heat shock element and different upstream sequences are required for the seed-specific expression of a small heat shock protein gene. Plant Physiol 121: 723–730.

Cellier F, Conéjéro G, Breitler JC, Casse F (1998) Molecular and physiological responses to water deficit in drought-tolerant and drought-sensitive lines of sunflower. Plant Physiol 116: 319–328.

Chen E, Przybyla AE (1994) An efficient site-directed mutagenesis method based on PCR. Bio/Techniques 17: 657–659.

Chen JJ, Janssen BJ, Williams A, Sinha N (1997) A gene fusion at the homeobox locus: Alterations in leaf shape and implications for morphological evolution. Plant Cell 9: 1289–1304.

Chen J, Ren Z, Kong X, Wu J, Zhou R, Jia J (2005) Isolation, characterization and expression analysis of male sterility gene homology sequence in wheat. Acta Genet Sin 32: 566–570.

Chen J, Hu J, Vick BA, Jan CC (2006) Molecular mapping of a nuclear male-sterility gene in sunflower (*Helianthus annuus* L.) using TRAP and SSR markers. Theor Appl Genet 113: 122–127.

Cheng Z, Sattler S, Maeda H, Sakuragi Y, Bryant DA, DellaPenna D (2003) Highly divergent methyltransferases catalyze a conserved reaction in tocopherol and plastoquinone synthesis in cyanobacteria and photosynthetic eukaryotes. Plant Cell 15: 2343–2356.

Chiappetta A, Michelotti V, Fambrini M, Bruno L, Salvini M, Petrarulo M, Azmi A, Van Onckelen H, Pugliesi C, Bitoni MB (2006) Zeatin accumulation and misexpression of a class I *knox* gene are intimately linked in the epiphyllous response of the interspecific hybrid EMB-2 (*Helianthus annuus* x *H. tuberosus*). Planta 223: 917–931.

Close TJ (1997) Dehydrins: A commonalty in the response of plants to dehydration and low temperature. Physiol Plant 100: 291–296.

Coca MA, Almoguera C, Jordano J (1994) Expression of sunflower low-molecular weight heat-shock proteins during embryogenesis and persistence after germination—localization and possible functional implications. Plant Mol Biol 25: 479–492.

Coca MA, Almoguera C, Thomas TL, Jordano J (1996) Differential regulation of small heat shock protein genes in plants: analysis of water-stress-inducible and developmentally activated sunflower promoter. Plant Mol Biol 31: 863–876.

Conti A, Pancaldi S, Fambrini M, Michelotti V, Bonora A, Salvini M, Pugliesi C (2004) A deficiency for ξ-carotene desaturase characterizes the sunflower *non dormant-1* mutant. Plant Cell Physiol 45: 445–455.

Dangle JL, Jones JDG (2001) Plant pathogens and integrated defense responses to infection. Nature 411: 826–833.

Dawid IB, Toyama R, Taira M (1995) LIM domain proteins. CR Acad Sci III 318: 295–306.

Demurin YN (1993) Genetic variability of tocopherol composition in sunflower seeds. Helia 16(18): 59–62.

Desloire S, Gherbi H, Laloui W, Marhadour S, Clouet V, Cattolico L, Falentin C, Giancola S, Renard M, Budar F, Small I, Caboch M, Delourme R, Bendahmane A (2003) Identification of the fertility restoration locus, *Rfo*, in radish, as a member of the pentatricopeptide-repeat protein family. EMBO Rep 4(6): 588–594.

Dezar CA, Gago GM, Gonzalez DH, Chan RL (2005a) *Hahb-4*, a sunflower homeobox-leucine zipper gene, is a developmental regulator and confers drought tolerance to *Arabidopsis thaliana* plants. Transgen Res 14: 429–440.

Dezar CA, Fedrigo GV, Chan RL (2005b) The promoter of the sunflower HD-Zip protein gene *Hahb4* directs tissue-specific expression and is inducible by water stress, high salt concentrations and ABA. Plant Sci 169: 447–459.

Díaz-Martín J, Almoguera C, Prieto-Dapena P, Espinosa JM, Jordano J (2005) Functional interaction between two transcription factors involved in the development regulation of a small heat stress protein gene promoter. Plant Physiol 139: 1483–1494.

Dorell DG, Vick BA (1997) Properties and processing of oilseed sunflower. In: AA Schneiter (ed) Sunflower Technology and Production. CSSA, Madison, Wisconsin, USA, pp 709–745.

Duggleby RG, Pang SS (2000) Acetohydroxyacid synthase. J Biochem Mol Biol 33: 1–36.

Dure L III, Crouch M, Harada J, Ho T-HD, Mundy J, Quantrano R, Thomas T, Sung ZR (1989) Common amino acid sequence domains among LEA proteins of higher plants. Plant Mol Biol 12: 475–486.

Dußle CM, Hahn V, Knapp SJ, Bauer E (2004) *Pl*~*Arg*~ from *Helianthus argophyllus* is unlinked to other known downy mildew resistance genes in sunflower. Theor Appl Genet 109: 1083–1086.

Eliasson A, Gass N, Mundel C, Baltz R, Kräuter R, Evrard JL, Steinmetz A (2000) Molecular and expression analysis of a LIM protein gene family from flowering plants. Mol Gen Genet 264: 257–267.

Emanuelsson O, Nielsen H, von Heijne G (1999) ChloroP, a neural network-based method for prediciting chloroplast transit peptides and their cleavage sites. Protein Sci 8: 978–984.

Fambrini M, Cioni G, Bianchi R, Pugliesi C (2000) Epiphylly in a variant of induced by in vitro tissue culture. Int J Plant Sci 161: 13–22.

Fambrini M, Cioni G, Bertini D, Michelotti V, Conti A, Puglieso C (2003) *Missing Flowers* gene controls axillary meristems initiation in sunflower. Genesis 36: 25–33.

Fambrini M, Castagna A, Vecchia FD, Degl'Innocenti E, Ranieri A, Vernieri P, Pardossi A, Guidi L, Rascio N, Pugliesi C (2004) Characterization of a pigment-deficient mutant of sunflower (*Helianthus annuus* L.) with abnormal chloroplast biogenesis, reduced PSII activity and low endogenous level of abscisic acid. Plant Sci 167: 79–89.

Fambrini M, Durante C, Cioni G, Geri C, Gioretti L, Michelotti V, Salvini M, Pugliesi C (2006) Characterization of LEAFY COTYLEDON1-LIKE gene in *Helianthus annuus* and its relationship with zygotic and somatic embryogenesis. Dev Gen Evol 216: 253–264.

Fambrini M, Cioni G, Rascio N, Dalla Vecchia F, Pugliesi C (2007) Early seedling growth and morphogenesis in the *xantha1* mutant of sunflower with alteration of the chloroplast biogenesis. Photosynthetica 45(3): 400–408.

Fang J, Chai C, Qian Q, Li C, Tang J, Sun L, Huang Z, Guo X, Sun C, Liu M, Zhang Y, Lu Q, Wang Y, Lu C, Han B, Chen F, Cheng Z, Chu C (2008) Mutations of genes in synthesis of the carotenoid precursors of ABA lead to preharvest sprouting and photo-oxidation in rice. Plant J 54: 177–189.

Feng J, Jan CC (2008) Introgression and molecular tagging of *Rf4*, a new male fertility restoration gene from wild sunflower *Helianthus maximiliani* L. Theor Appl Genet 117: 241–249.

Feng J, Vick BA, Lee MK, Zhang HB, Jan CC (2006) Construction of BAC and BIBAC libraries from sunflower and identification of linkage group-specific clones by overgo hybridization. Theor Appl Genet 113: 23–32.

Fernández-Martinez JM, Mancha M, Osorio J, Garcés R (1997) Sunflower mutant containing high levels of palmitic acid in high oleic background. Euphytica 97: 113–116.

Fernández-Moya V, Martinez-Force E, Garcés R (2002) Temperature effect on a high stearic acid sunflower mutant. Photochemistry 59: 33–37.

Fernández-Moya V, Martinez-Force E, Garcés R (2003) Temperature-related non-homogenous fatty acid desaturation in sunflower (*Helianthus annuus* L.) seeds. Planta 216: 834–840.

Freyd G, Kim SK, Horvitz HR (1990) Novel cysteine-rich motif and homeodomain in the product of *Caenorhabditis elegans* cell lineage gene *lin-11*. Nature 344: 876–879.

Gago GM, Almoguera C, Jordano J, Gonzalez HD, Chan RL (2002) *Hahb-4*, a homeobox-leucine zipper gene potentially involved in abscisic acid-dependent response to water stress in sunflower. Plant, Cell Environ 25: 633–640.

Gedil MA, Slabaugh MB, Berry S, Johnson R, Michelmore R, Miller J, Gulya T, Knapp SJ (2001a) Candidate disease resistance genes in sunflower cloned using conserved nucleotide-binding site motifs: Genetic mapping and linkage to the downy mildew resistance gene *Pl1*. Genome 44: 205–212.

Gedil MA, Wye C, Berry S, Segers B, Peleman J, Jones R, Leon A, Slabaugh MB, Knapp SJ (2001b) An integrated restriction fragment length polymorphism—amplified fragment length polymorphism linkage map for cultivated sunflower. Genome 44: 213–221.

Gentzbittel L, Vear F, Zhang YX, Bervillé A, Nicolas P (1995) Development of a consensus linkage RFLP map of cultivated sunflower (*Helianthus annuus* L.). Theor Appl Genet 90: 1079–1086.

Gentzbittel L, Mouzeyar S, Badaoui S, Mestries E, Vear F, Tourvieille De Labrouche D, Nicolas P (1998) Cloning of molecular markers for disease resistance in sunflower, *Helianthus annuus* L. Theor Appl Genet 96: 519–525.

Gentzbittel L, Mestries E, Mouzeyar S, Mazeyrat F, Badaoui S, Vear F, Tourvieille De Labrouhe D, Nicolas P (1999) A composite map of expressed sequences and phenotypic traits of the sunflower (*Helianthus annuus* L.) genome. Theor Appl Genet 99: 218–234.

Gentzbittel L, Abbott A, Galaud JP, Georgi L, Fabre F, Liboz T, Alibert G (2002) A bacterial artificial chromosome (BAC) library for sunflower, and identification of clones containing genes for putative transmembrane receptors. Mol Genet Genom 266: 979–987.

Gill GN (1995) The enigma of LIM domains. Structure 3: 1285–1289.

Giordani T, Natali L, Cavallini A (2003) Analysis of a dehydrin encoding gene and its phylogenetic utility in *Helianthus*. Theor Appl Genet 107: 316–325.

Giordani T, Natali L, D'Ercole A, Pugliesi C, Fambrini M, Vernieri P, Vitagliano C, Cavallini A (1999) Expression of a dehydrin gene during embryo development and drought stress in ABA-deficient mutants of sunflower (*Helianthus annuus* L.). Plant Mol Biol 39: 739–748.

Gonzalez DH, Chan RL (1993) Screening cDNA libraries by PCR using lambda sequencing primers and degenerate oligonucleotides. Trends Genet 9: 231–232.

Grayburn WS, Vick BA (1995) Transformation of sunflower (*Helianthus annuus* L.) following wounding with glass beads. Plant Cell Rep 14: 285–289.

Gulya TJ, Miller JF, Viranyi F, Sackson WE (1991a) Proposed internationally standardized methods for race identification of *Plasmopara halstedii*. Helia 14: 11–20.

Gulya TJ, Sackson WE, Viranyi F, Masirevic S, Rashid KY, (1991b) New races of the sunflower downy mildew pathogen (*Plasmopara halstedii*) in Europe and North and South America. J Phytopathol 132: 303–311.

Hahne G (2001) Sunflower. In: YH Hui, GG Khatchtourians, A McHughen, WK Nip, R Scorza (eds) Handbook of Transgenic Plants. Marcel Dekker, New York, USA, pp 813–833.

Hamrit S, Kusterer B, Wolfgang Friedt, Horn R (2008) Verification of positive BAC clones near the *Rf1* gene restoring pollen fertility in the presence of the PET1 cytoplasm in sunflower (*Helianthus annuus* L.) and direct isolation of BAC ends. In: Proc 17th Int Sunflower Conf, vol 2, Córdoba, Spain, pp 623–628.

Hass CG, Tang S, Leonard S, Traber MG, Miller JF, Knapp SJ (2006) Three non-allelic epistatically interacting methyltransferase mutations produce novel tocopherol (vitamin E) profiles in sunflower. Theor Appl Genet 113: 767–782.

Henningsen KW, Stummann BM (1982) Use of mutants in the study of chloroplast biogenesis. In: B Parthier, D Boulter (eds) Encyclopedia of Plant Physiology, New Series, vol 14B. Springer-Verlag, Berlin, Germany, pp 597–644.

Hewezi T, Jardinaud F, Alibert G, Kallerhoff J (2003) A new approach for efficient regeneration of a recalcitrant genotype of sunflower (*Helianthus annuus*) by organogenesis induction on split embryonic axes. Plant Cell, Tiss Org Cult 73: 81–86.

Hewezi T, Mouzeyar S, Thion L, Rickauer M, Alibert G, Nicolas P, Kallerhoff J (2006) Antisense expression of a NBS-LRR sequence in sunflower (*Helianthus annuus* L.) and tobacco (*Nicotiana tabacum* L.): Evidence for a dual role in plant development and fungal resistance. Transgen Res 15: 165–180.

Hongtrakul V, Slabaugh MB, Knapp SJ (1998a) DFLP, SSCP, and SSR markers for Δ9-stearoyl-acyl carrier protein desaturases strongly expressed in developing seeds of sunflower: Intron lengths are polymorphic among elite inbred lines. Mol Breed 4: 195–203.

Hongtrakul V, Slabaugh MB, Knapp SJ (1998b) A seed specific Δ-12 oleate desaturase gene is duplicated, rearranged, and weakly expressed in high oleic acid sunflower lines. Crop Sci 38: 1245–1249.

Horn R (2002) Molecular diversity of male sterility inducing and male-fertile cytoplasm in the genus *Helianthus*. Theor Appl Genet 104: 562–570.

Horn R (2006) Cytoplasmic male sterility and fertility restoration in higher plants. Progr Bot 67: 31–52.

Horn R, Friedt W (1997) Fertility restoration of new CMS sources in sunflower. Plant Breed 116: 317–322.

Horn R, Friedt W (1999) CMS sources in sunflower: Different origin but same mechanism. Theor Appl Genet 98: 195–201.

Horn R, Köhler RH, Zetsche K (1991) A mitochondrial 16 kDa protein is associated with cytoplasmic male sterility in sunflower. Plant Mol Biol 7: 29–36.

Horn R, Kusterer B, Lazarescu E, Prüfe M, Friedt W (2003) Molecular mapping of the *Rf1* gene restoring pollen fertility in PET1-based F_1 hybrids in sunflower (*Helianthus annuus* L.). Theor Appl Genet 106: 599–606.

Hu J, Vick BA (2003) Target region amplified polymorphism, a novel marker technique for plant genotyping. Plant Mol Biol Rep 21: 289–294.

Hu J, Yue B, Vick BA (2007) Integration of TRAP markers onto a sunflower SSR marker linkage map constructed from 92 recombinant inbred lines. Helia 30(46): 25–36.

Imai R, Koizuka N, Fujimoto H, Hayakawa T, Sakai T, Imamura J (2003) Delimitation of the fertility restorer locus *Rfk1* to a 43-kb contig in Kosena radish (*Raphanus sativus* L.). Mol Genet Genom 269(3): 388–394.

Jan CC (1992) Inheritance and allelism of mitomycin C- and streptomycin-induced recessive genes for male sterility in cultivated sunflower. Crop Sci 32: 317–320.

Jan CC (1997) Cytology and interspecific hybridization. In: AA Schneiter (ed) Sunflower Technology and Production. CSSA, Madison, Wisconsin, USA, pp 497–558.

Jan CC, Rutger JN (1988) Mitomycin C- and streptomycin-induced male sterility in cultivated sunflower. Crop Sci 28: 792–795.

Jones JDG, Dangle JL (2006) The plant immune system. Nature 444: 323–329.

Jones A, Davies HM, Voelker TA (1995) Palmitoyl-acyl carrier protein (ACP) thioesterase and the evolutionary origin of plant acyl-ACP thioesterases. Plant Cell 7: 359–371.

Karlsson O, Thor S, Njorberg T, Ohlsson H, Edlund T (1990) Insulin gene enhancer binding protein Isl-1 is a member of a novel class of proteins containing both homeo- and a Cys-His domain. Nature 344: 879–882.

Kazama T, Toriyama K (2003) A pentatricopeptide repeat-containing gene that promotes the processing of aberrant *atp6* RNA of cytoplasmic male-sterile rice. FEBS Lett 544: 99–102.

Klein RR, Klein PE, Mullet JE, Minx P, Rooney WL, Schertz KF (2005) Fertility restorer locus *Rf1* of sorghum (*Sorghum bicolor* L.) encodes a pentatricopeptide repeat protein not present in the collinear region of rice chromosome 12. Theor Appl Genet 111: 994–1012.

Knittel N, Gruber V, Hahne G, Lenee P (1994) Transformation of sunflower (*Helianthus annuus* L.): a reliable protocol. Plant Cell Rep 14: 81–86.

Köhler RH, Horn R, Lössl A, Zetsche K (1991) Cytoplasmic male sterility in sunflower is correlated with the co-transcription of a new open reading frame with the *atpA* gene. Mol Gen Genet 227: 369–376.

Koizuka N, Imai R, Fujimoto H, Hayakawa T, Kimura Y, Kohno-Murase J, Sakai T, Kawasaki S, Imamura J (2003) Genetic characterization of a pentatricopeptide repeat protein gene, *orf687*, that restores fertility in the cytoplasmic male-sterile Kosena radish. Plant J 34: 407–415.

Kolkman J, Slabaugh MB, Bruniard J, Berr S, Bushman SB, Olungu C, Maes N, Abratti G, Zambelli A, Miller JF, Leon A, Knapp SJ (2004) Acetohydroxyacid synthase mutations conferring resistance to imidazolinone or sulfonylurea herbicides in sunflower. Theor Appl Genet 109: 1147–1155.

Komori T, Ohta S, Murai N, Takakura Y, Kuraya Y, Suzuki S, Hiei Y, Imaseki H, Nitta N (2004) Map-based cloning of a fertility restorer gene, *Rf-1*, in rice (*Oryza sativa* L.). Plant J 37: 315–325.

Koorneef M, Bentsink L, Hilhorst H (2002) Seed dormancy and germination. Curr Opin Plant Biol 5: 33–36.

Kräuter-Canham R, Bronner R, Evrard JL, Hahn V, Friedt W, Steinmetz A (1997) A transmitting tissue- and pollen-expressed protein from sunflower with sequence similarity to the human RTP protein. Plant Sci 129: 191–202.

Kräuter-Canham R, Bronner R, Steinmetz A (2001) SF21 is a protein which exhibits a dual nuclear and cytoplasmic localization in developing pistillar tissue. Ann Bot 87: 241–249.

Kusterer B, Prüfe M, Lazarescu E, Özdemir N, Friedt W, Horn R (2002) Mapping of the restorer gene *Rf1* in sunflower (*Helianthus annuus* L.). Helia 25: 41–46.

Kusterer B, Friedt W, Lazarescu L, Prüfe M, Özdemir N, Tzigos S, Horn R (2004a) Map-based cloning strategy for isolating the restorer gene *Rf1* of the PET1 cytoplasm in sunflower (*Helianthus annuus* L). Helia 27: 1–14.

Kusterer B, Rozynek B, Brahm L, Prüfe M, Tzigos S, Horn R, Friedt W (2004b) Construction of a genetic map and localization of major traits in sunflower (*Helianthus annuus* L.). Helia 27: 15–23.

Kusterer B, Horn R, Friedt W (2005) Molecular mapping of the fertility restoration locus *Rf1* in sunflower and development of diagnostic markers for the restorer gene. Euphytica 143: 35–43.

Kwong RM, Bui AQ, Lee H, Kwong LW, Fischer RL, Goldberg RB, Harada JJ (2003) LEAFY COTYLEDON1-LIKE defines a class of regulators essential for embryo development. Plant Cell 15: 5–18.

Lacombe S, Bervillé A (2001) A dominant mutation for high oleic acid content in sunflower (*Helianthus annuus* L.) seed oil is genetically linked to a single oleate-desaturase RFLP locus. Mol Breed 8: 129–137

Laver HK, Reynolds SJ, Monéger F, Leaver CJ (1991) Mitochondrial genome organization and expression associated with cytoplasmic male sterility in sunflower (*Helianthus annuus* L.). Plant J 1(2): 185–193.

Lawrence GJ, Finnegan EJ, Ayliffe MA, Ellis JG (1995) The *L6* gene for flax rust resistance is related to the *Arabidopsis* bacterial resistance gene *RPS2* and the tobacco viral resistance gene *N*. Plant Cell 7: 1195–1206.

Lawson WR, Goulter KC, Henry RJ, Kong GA, Kochman JK (1998) Marker-assisted selection for two rust resistance genes in sunflower. Mol Breed 4: 227–234.

Lazarescu E, Friedt W, Horn R, Steinmetz A (2006) Expression analysis of the sunflower *SF21* gene family reveals multiple alternative and organ-specific splicing of transcripts. Gene 374: 77–86.

Leclercq P (1969) Une stérilité mâle chez le tournesol. Ann Amélior Plant 19: 99–106.

Leister D (2003) Chloroplast research in the genomic age. Trends Genet 19: 47–56.

Leister D, Ballvora A, Salamini F, Gebhardt C (1996) A PCR-based approach for isolating pathogen resistance genes from potato with potential for wide application in plants. Nat Genet 14: 421–429.

Li JJ, Herskowitz I (1993) Isolation of ORC6, a component of the yeast origin recognition complex by a one-hybrid system. Science 262: 1870–1874.

Lincoln C, Long J, Yamaguchi J, Serikawa K, Hake S (1994) A *knotted1*-like homeobox gene in *Arabidopsis* is expressed in the vegetative meristems and dramatically alters leaf morphology when overexpressed in transgenic plants. Plant Cell 6: 1859–1876.

Liu X, Baird WV (2003) Differential expression of genes regulated in response to drought or salinity stress in sunflower. Crop Sci 43: 678–687.

Liu X, Baird WV (2004) Identification of a novel gene, *HaABRC5*, from *Helianthus annuus* (Asteraceae) that is upregulated in response to drought, salinity and abscisic acid. Am J Bot 91: 184–191.

Lucas O, Kallerhoff J, Alibert G (2000) Production of stable transgenic sunflower (*Helianthus annuus* L.) from wounded immature embryos by particle bombardment and co-cultivation with *Agrobacterium tumefaciens*. Mol Breed 6: 479–487.

Lurin C, Andres C, Aubourg S, Bellaoui M, Bitton F, Bruyere C, Caboche M, Debast C, Gualberto J, Hoffmann B, Lecharny A, Le Ret M, Martin-Magniette ML, Mireau H, Peeters N, Renou JP, Szurek B, Taconnat L, Small I (2004) Genome-wide analysis of *Arabidopsis* pentatricopeptide repeat proteins reveals their essential role in organelle biogenesis. Plant Cell 16: 2089–2103.

Ma H, McMullen MD, Finer JJ (1994) Identification of a homeobox-containing gene with enhanced expression during soybean (*Glycine max* L.) somatic embryo development. Plant Mol Biol 24: 465–473.

Mallory-Smith CA, Thill DC, Dial MJ (1990) Identification of sulfonylurea herbicide-resistant prickly lettuce (*Lactuca serriola*). Weed Technol 4: 163–168.

Malone-Schoneberg J, Scelonge CJ, Burrus M, Bidney DL (1994) Stable transformation of sunflower using *Agrobacterium* and split embryogenic axis explants. Plant Sci 103: 199–207.

Manavella PA, Arce AL, Dezar CA, Bitton F, Renou JP, Crespi M, Chan RL (2006) Cross-talk between ethylene and drought signalling pathway is mediated by the sunflower Hahb-4 transcription factor. Plant J 48: 125–137.

Martinez-Force E, Cantisan S, Serrano-Vega MJ, Garces R (2000) Acyl-acyl carrier protein thioesterase activity from sunflower (*Helianthus annuus* L.) seeds. Planta 211: 673–678.

Martinez-Rivas JM, Sperling P, Lühs W, Heinz E (2001) Spatial and temporal regulation of three different microsomal oleate desaturase genes (*FAD2*) from normal-type and high-oleic varieties of sunflower (*Helianthus annuus* L.) Mol Breed 8: 159–168.

Maurel C (1997) Aquaporins and water permeability of plant membranes. Annu Rev Plant Physiol Plant Mol Biol 48: 399–429.

Meyers BC, Dickermann AW, Michelmore RW, Sivaramakrishnan S, Sobral BW, Young ND (1999) Plant disease resistance genes encode members of an ancient and diverse protein family within the nucleotide-binding superfamily. Plant J 20: 317–332.

Michelotti V, Giorgetti L, Geri C, Cionini G, Pugliesi C, Fambrini M (2007) Expression of the *HtKNOT1*, a class I *KNOX* gene, overlaps cell layers and development compartments of differentiating cells in stems and flowers of *Helianthus tuberosus*. Cell Biol Int 31: 1280–1287.

Micic Z, Hahn V, Bauer E, Schön CC, Knapp SJ, Tang S, Melchinger AE (2004) QTL mapping of *Sclerotinia* midstalk-rot resistance in sunflower. Theor Appl Genet 109: 1474–1484.

Miller JF (1992) Update on inheritance of sunflower characteristics. In: Proc 13th Int Sunflower Conf, 7–11 Sept 1992, Pisa, Italy, pp 905–945.

Miller JF, Fick GN (1997) The genetics of Sunflower. In: Schneiter AA (ed) Sunflower Technology and Production. CSSA, Madison, Wisconsin, USA, pp 441–495.

Miller JF, Gulya TJ (1991) Inheritance of resistance to race 4 of downy mildew derived from interspecific crosses in sunflower. Crop Sci 31: 40–43.

Miller JF, Gulya TJ, Seiler GJ (2002) Registration of five fertility restorer sunflower germplasms. Crop Sci 42: 989.

Mindrinos M, Katagiri F, Yu GL, Ausubel FM (1994) The *A. thaliana* disease-resistance gene *RPS2* encodes a protein containing a nucleotide binding site and leucine-rich repeats. Cell 78: 1089–1099.

Mohamed Sh, Boehm R, Schnabl H (2006) Stable genetic transformation of high oleic *Helianthus annuus* L. genotypes with high efficiency. Plant Sci 171: 546–554.

Monéger F, Smart CJ, Leaver CJ (1994) Nuclear restoration of cytoplasmic male sterility in sunflower is associated with the tissue-specific regulation of a novel mitochondrial gene. EMBO J 13: 8–17.

Motohashi R, Ito T, Kobayashi M, Taji T, Nagata N, Asami T, Yoshida S, Yamaguchi-Shinozaki K, Shinozaki K (2003) Functional analysis of the 37 kDa inner envelope membrane polypeptide in chloroplast biogenesis using *Ds*-tagged *Arabidopsis* pale-green mutant. Plant J 34: 719–731.

Mouzeyar S, Roeckel-Drevet P, Gentzbittel L, Tourvielle de Labrouhe D, Vear F, Nicolas P (1995) RFLP and RAPD mapping of the sunflower *Pl1* locus for resistance to *Plasmopara halstedii* race 1. Theor Appl Genet 91: 733–737.

Mundel C, Baltz R, Eliasson A, Bronner R, Gass N, Kräuter R, Evrard JL, Steinmetz A (2000) A LIM-domain protein from sunflower localizes to the cytoplasm and/or nucleus in a wide range of tissues and associates with the phragmoplast in dividing cells. Plant Mol Biol 42: 291–302.

Nakajima H, Muranaka T, Ishige F, Akutsu K, Oeda K (1997) Fungal and bacterial disease resistance in transgenic plants expressing human lysozyme. Plant Cell Rep 16: 674–679.

Natali L, Giordani T, Cavallini A (2003) Sequence variability of a dehydrin gene within *Helianthus annuus*. Theor Appl Genet 106: 811–818.

Natali L, Santini S, Giordani T, Minelli S, Maestrini P, Cionini PG, Cavallini A (2006) Distribution of Ty3-*gypsy*- and Ty1-*copia*-like DNA sequences in the genus *Helianthus* and other Asteraceae. Genome 49: 64–72.

Natali L, Giordani T, Lercari B, Maestrini P, Cozza R, Pangaro T, Vernieri P, Martinelli F, Cavallini A (2007) Light induced expression of a dehydrin-encoded gene during seedling de-etiolation in sunflower (*Helianthus annuus* L.). J Plant Physiol 164: 263–273.

Nover L, Bharti K, Döring P, Mishra SK, Ganguli A, Scharf KD (2001) *Arabidopsis* and the heat stress transcription factor world: How many heat stress transcription factors do we need? Cell Stress Chaperones 6: 177–189.

Ohlrogge J, Browse J (1995) Lipid biosynthesis. Plant Cell 7: 957–970.

Osorio J, Fernández-Martinez J, Mancha M, Garcés R (1995) Mutant sunflowers with high concentration of saturated fatty acids in the oil. Crop Sci 35: 739–742.

Ouvrard O, Cellier F, Ferrare K, Tousch D, Lamaze T, Dupuis JM, Casse-Delbart F (1996) Identification and expression of water stress- and abscisic acid-regulated genes in a drought-tolerant sunflower genotype. Plant Mol Biol 31: 819–829.

Özdemir N, Horn R, Friedt W (2004) Construction and characterization of a BAC library for sunflower (*Helianthus annuus* L.). Euphytica 138: 177–183.

Pan Q, Liu YS, Budai-Hadrian O, Sela M, Carmel-Goren L, Zamir D, Fluhr R (2000) Comparative genetics of nucleotide binding site-leucine rich repeat resistance gene homologues in the genomes of two dicotyledons: Tomato and *Arabidopsis*. Genetics 155: 309–322.

Pérez-Vich B, Garcés R, Fernández-Martinez JM (1999) Genetic control of high stearic acid content in the seed oil of the sunflower mutant CAS-3. Theor Appl Genet 99: 663–669.

Pérez-Vich B, Garcés R, Fernández-Martinez JM (2000) Genetic relationships between loci controlling the high stearic and the high oleic acid traits in sunflower. Crop Sci 40: 990-995.

Pérez-Vich B, Fernández-Martinez JM, Grondona M, Knapp SJ, Berry ST (2002) Stearoyl-ACP and oleol-PC desaturase genes cosegregate with quantitative trait loci underlying stearic and high oleic acid mutant phenotypes in sunflower. Theor Appl Genet 104: 338–349.

Pérez-Vich B, Knapp SJ, Leon AJ, Fernández-Martinez JM, Berry ST (2004) Mapping minor QTL for increased stearic acid content in sunflower seed oil. Mol Breed 13: 313–322.

Pérez-Vich B, Berry ST, Velasco L, Fernández-Martinez JM, Gandhi S, Freeman C, Heesacker A, Knapp SJ, Leon AJ (2005) Molecular mapping of nuclear male sterility genes in sunflower. Crop Sci 45: 1851–1857.

Pérez-Vich B, Leon AJ, Grondona M, Velasco L, Fernández-Martinez JM (2006a) Molecular analysis of the high stearic acid content in sunflower mutant CAS-14. Theor Appl Genet 112: 867–875.

Pérez-Vich B, Velasco L, MuZoz-Ruz J, Fernández-Martinez JM (2006b) Inheritance of high stearic acid content in the sunflower mutant CAS-14. Crop Sci 46: 22–29.

Pfenning M, Palfay G, Guillet T (2008) The CLEARFIELD technology—A new broad-spectrum post-emergence weed control system for European sunflower growers. J Plant Dis Protec 21(Spl Iss): 647–652.

Plocik A, Layden J, Kessli R (2004) Comparative analysis of NBS domain sequences of NBS-LRR disease resistance genes from sunflower, lettuce and chicory. Mol Phylogenet Evol 31: 153–163.

Porfirova S, Bergmüller E, Tropf S, Lemke R, Dormann P (2002) Isolation of an *Arabidopsis* mutant lacking vitamin E and identification of a cyclase essential for all tocopherol biosynthesis. Proc Natl Acad Sci USA 99: 12495–12500.

Prieto-Dapena P, Almoguera C, Rojas A, Jordano J (1999) Seed-specific expression patterns and regulation by ABI3 of an unusual late embryogenesis-abundant gene in sunflower. Plant Mol Biol 39(3): 615–627.

Prieto-Dapena P, Castano R, Almoguera C, Jordano J (2006) Improved resistance to controlled deterioration in transgenic seeds. Plant Physiol 142: 1102–1112.

Prieto-Dapena P, Castano R, Almoguera C, Jordano J (2008) The ectopic overexpression of a seed-specific transcription factor, HaHSFA9, confers tolerance to severe dehydration in vegetative organs. Plant J 54: 1004–1014.

Pugliesi C, Ceconi F, Mandolfo A, Baroncelli S (1991) Plant regeneration and genetic variability from tissue cultures of sunflower (*Helianthus annuus* L.). Plant Breed 106: 114–121.

Radwan O, Bouzidi MF, Vear F, Philippon J (2003) Identification of non-TIR-NBS-LRR markers linked to the *Pl5/Pl8* locus for resistance to downy mildew in sunflower. Theor Appl Genet 106: 1438–1446.

Radwan O, Bouzidi MF, Nicolas P, Mouzeyar S (2004) Development of PCR markers for the *Pl5/Pl8* locus for resistance to *Plasmopara halstedii* in sunflower, *Helianthus annuus* L. from complete CC-NBS-LRR sequences. Theor Appl Genet 109: 176–185.

Radwan O, Mouzeyar S, Nicolas P, Boudizi MF (2005) Induction of a sunflower CC-NBS-LRR resistance gene analogue during incompatible interaction with *Plasmopara halstedii*. J Exp Bot 56: 567–575.

Radwan O, Gandhi S, Heesacker A, Whitaker B, Taylor C, Plocik A, Kesseli R, Kozik A, Michelmore RW, Knapp SJ (2008) Genetic diversity and genomic distribution of homologs encoding NBS-LRR disease resistance proteins in sunflower. Mol Genet Genom 280: 11–115.

Rahim M, Jan CC, Gulya TJ (2002) Inheritance of resistance to sunflower downy mildew races 1, 2, and 3 in cultivated sunflower. Plant Breed 121: 57–60.

Roche J, Hewezi T, Bouniols A, Gentzbittel L (2007) Transcriptional profiles of primary metabolism and signal transduction-related genes in response to water stress in field-grown sunflower genotypes using thematic cDNA microarrays. Planta 226: 601–617.

Roeckel-Drevet P, Gagne G, Mouzeyar S, Gentzbittel L, Philippon J, Nicolas P, Tourvieille de Labrouhe D, Vear F (1996) Collocation of downy mildew (*Plasmopara halstedii*) resistance genes in sunflower (*Helianthus annuus* L.). Euphytica 91: 225–228.

Rodriguez RH, Salaberry MT, Echeverria MM (1998) Inheritance of chlorophyll deficiency in sunflower. Helia 21: 109–114.

Rodriguez-Concepcion M, Ahumada I, Diez-Juez E, Sauret-Gueto S, Lois LM, Gallego F, Carretero-Paulet L, Campos N, Boronat A (2001) 1-Deoxy-D-xylulose 5-phosphate reductoisomerase and plastid isoprenoid biosynthesis during tomato fruit ripening. Plant J 27: 213–222.

Rogers CE, Thompson TE, Seiler GJ (1982) Sunflower species of United States. Natl Sunflower Assoc, Bismark, ND, USA, pp 1–75.

Rojas A, Almoguera C, Jordano J (1999) Transcriptional activation of a heat shock gene promoter in sunflower embryos: synergism between ABI3 and heat shock factors. Plant J 20: 601–610.

Rojas A, Almoguera C, Carranco R, Scharf KD, Jordano J (2002) Selective plant gene expression studies. Plant Physiol 129: 1207–1215.

Rubinelli P, Hu Y, Ma H (1998) Identification, sequence analysis and expression studies of novel anther-specific genes of *Arabidopsis thaliana*. Plant Mol Biol 37: 607–619.

Rousselin P, Molinier J, Himber C, Schontz D, Prieto-Dapena P, Jordano J, Martini N, Weber S, Horn R, Ganssmann M, Grison R, Pagniez M, Toppan A, Friedt W, Hahne G (2002) Modification of sunflower oil quality by seed-specific expression of a heterologous Δ9-stearoyl-[acyl carrier protein] desaturase gene. Plant Breed 121: 108–116.

Rueda EC, Dezar CA, Gonzalez DH, Chan RL (2005) *Hahb-10*, a sunflower homeobox-leucine zipper gene, is regulated by light quality and quantity, and promotes early flowering when expressed in *Arabidopsis*. Plant Cell Physiol 46 (12): 1954–1963.

Sackston WE (1992) On a TREADMILL: Breeding sunflower for resistance to disease. Annu Rev Phytopathol 30: 529–551.

Salas JJ, Youssar L, Martinez-Force E, Garces R (2008) The biochemical characterization of a high-stearic acid sunflower mutant reveals the coordinated regulation of stearoyl-acyl-carrier protein desaturases. Plant Physiol Biochem 46: 109–116.

Salvini M, Bernini A, Fambrini M, Pugliesi C (2005) cDNA cloning and expression of the phytoene synthase gene in sunflower. J Plant Physiol 162: 479–484.

Sandmann G (2001) Carotenoid biosynthesis and biotechnological application. Arch Biochem Biophys 385: 4–12.

Santini S, Cavallini A, Natali L, Minelli S, Maggini F, Cionini PG (2002) Ty1-*copia* and Ty3-*gypsy*-like DNA sequences in *Helianthus* species. Chromosma 111: 192–200.

Sarda X, Tousch D, Ferrare K, Cellier F, Alcon C, Dupuis JM, Casse F, Lamaze T (1999) Characterization of closely related delta-TIP genes encoding aquaporins which are differentially expressed in sunflower roots upon water deprivation through exposure to air. Plant Mol Biol 40: 179–191.

Sattler SE, Cahoon EB, Coughlan SJ, DellaPanne D (2003) Characterization of tocopherol cyclases from higher plants and cyanobacteria: Evolutionary implications for tocopherol synthesis and function. Plant Physiol 132: 2184–2195.

Sattler SE, Cheng Z, DellaPenna D (2004) From *Arabidopsis* to agriculture: engineering improved vitamin E content in soybean. Trends Plant Sci 9 (8): 365–367.

Sawahel W, Hagran A (2006) Generation of white mold disease-resistant sunflower plants expressing human lysozyme gene. Biol Plant 50(4): 683–687.

Scharf KD, Siddique M, Vierling E (2001) The expanding family of *Arabidopsis thaliana* small heat stress proteins and a new family of proteins containing alpha-crystallin domains (Acd proteins). Cell Stress Chaperones 6: 225–237.

Schnabel U, Engelmann U, Horn R (2008) Development of markers for the use of the PEF1-cytoplasm in sunflower hybrid breeding. Plant Breed 127: 587–591.

Schuppert GF, Tang X, Slabaugh MB, Knapp SJ (2006) The sunflower high-oleic mutant *Ol* carries variable tandem repeats of *FAD2-1*, a seed-specific oleoyl-phosphatidyl choline desaturase. Mol Breed 17: 241–256.

Serrano-Vega MJ, Venegas-Caleron M, Garces R, Martinez-Force E (2003) Cloning and expression of fatty acids biosynthesis key enzymes from sunflower (*Helianthus annuus* L.) in *Escherichia coli*. J Chromatogr B 786: 221–228.

Serrano-Vega MJ, Garces R, Martinez-Force E (2005) Cloning, characterization and structural model of a FatA-type thioesterase from sunflower seeds (*Helianthus annuus* L.). Planta 221: 868–880.

Serieys H (1999) Identification, study and utilization in breeding programs of new CMS sources. FAO progress report. Helia 22 (Spl Iss): 71–84.

Shintani DK, DellaPenna D (1998) Elevating the vitamin E content of plants through metabolic engineering. Science 282: 2098–2100.

Siculella L, Palmer JD (1988) Physical and gene organization of mitochondrial DNA in fertile and male sterile sunflower. CMS associated alterations in structure and transcription of the *atpA* gene. Nucl Acids Res 16(9): 3787–3799.

Slabaugh MB, Yu JK, Tang X, Heesacker A, Hu X, Lu G, Bidney D, Han F, Knapp SJ (2003) Haplotyping and mapping large cluster of downy mildew resistance gene candidates in sunflower using multilocus intron fragment length polymorphisms. Plant Biotechnol J 1: 167–185.

Soldatov KI (1976) Chemical mutagenesis in sunflower breeding. In: Proc 7th Int Sunflower Conf, 27 June–3 July 1976, Krasnodar, USSR, pp 352–357.

Stintzi A, Browse J (2000) The *Arabidopsis* male-sterile mutant, *opr3*, lacks the 12-oxophytodienoic acid reductase required for jasmonate synthesis. Proc Natl Acad Sci USA 97: 10625–10630.

Taira M, Evrard JL, Steinmetz A, Dawid IB (1995) Classification of LIM proteins. Trends Genet 11: 431–432.

Tang S, Yu J-K, Slabaugh MB, Shintani DK (2002) Simple sequence repeat map of the sunflower genome. Theor Appl Genet 105: 1124–1136.

Tang S, Hass CG, Knapp SJ (2006) *Ty3/gypsy*-like retrotransposon knockout of a 2-metyl-6-phytyl-1,4-benzoquinone methyltransferase is non-lethal, uncovers a cryptic paralogous mutation, and produces novel tocopherol (vitamin E) profiles in sunflower. Theor Appl Genet 113: 783–799.

Tanksley SD, Ganal MW, Martin GB (1995) Chromosome landing: a paradigm for map-based cloning in plants with large genomes. Trends Genet 11: 63–68.

Tautz D (1989) Hypervariability of simple sequences as a general source for polymorphic DNA markers. Nucl Acids Res 17: 6463–6471.

Tioni MF, Gonzalez DH, Chan RL (2003) *Knotted1*-like genes are strongly expressed in differentiated cell types in sunflower. J Exp Bot 54(383): 681–690.

Tioni MF, Viola IL, Chan RL, Gonzalez DH (2005) Site-directed mutagenesis and footprinting analysis of the interaction of the sunflower KNOX protein HAKN1 with DNA. FEBS J 272: 190–202.

Thompson GA, Scherer DE, Aken SFV, Kenny JW, Young HL, Shintani DK, Kridl JC, Knauf VC (1991) Primary structure of the precursor and mature forms of stearoyl-acyl carrier protein desaturase from safflower embryos and requirement of ferredoxin for enzyme activity. Proc Natl Acad Sci USA 88: 2578–2582.

Thompson JD, Gibson TJ, Plewniak F, Jeanmougin F, Higgins DG (1997) The ClustalX windos interface: Flexible strategies for multiple sequence alignment aided by quality analysis tools. Nucl Acids Res 24: 4876–4882.

Tranel PJ, Wright TR (2002) Resistance of weeds to AHAS-inhibiting herbicides: What have we learned? Weed Sci 50: 700–712.

Tron AE, Welchen E, Gonzalez DH (2004) Engineering the loop region of a homeodomain-leucine zipper protein promotes efficient binding to a monomeric DNA binding site. Biochemistry 43: 15845–15851.

Valle EM, Gonzalez DH, Gago G, Chan RL (1997) Isolation and expression pattern of *hahr1*, a homeobox-containing cDNA from *Helianthus annuus*. Gene 196: 61–68.

Vear F, Gentzbittel L, Philippon J, Mouzeyar S, Mestries E, Roeckel-Drevet P, Tourvielle de Labrouhe D, Nicolas P (1997) The genetic of resistance to five races of downy mildew (*Plasmopara halstedii*) in sunflower (*Helianthus annuus* L.). Theor Appl Genet 95: 584–589.

Venegas-Caleron M, Martinez-Force E, Graces R (2008) Lipid characterization of a wrinkled sunflower mutant. Phytochemistry 69: 684–691.

Vera-Ruiz EM, Velasco L, Leon AJ, Fernandez-Martinez JM, Pèrez-Vich B (2006) Genetic mapping of the *Tph*1 gene controlling beta-tocopherol accumulation in sunflower seeds. Mol Breed 17: 291–296.

Vranceanu VA (1970) Advances in sunflower breeding in Romania. In: Proc 4th Int Sunflower Conf, 23–25 June 1970, Memphis, TN, USA, pp 136–148.

Vranceanu VA, Pirvu N, Stoenescu FM (1981) New sunflower downy mildew resistance genes and their management. Helia 4: 23–27.

Wang Z, Zou Y, Li X, Zhang Q, Chen L, Wu H, Su D, Chen Y, Guo J, Luo D, Long Y, Zhong Y, Liu YG (2006) Cytoplasmic male sterility of rice with Boro II cytoplasm is caused by a cytotoxic peptide and is restored by two related PPR motif genes via distinct modes of mRNA silencing. Plant Cell 18: 676–687.

Way JC, Chalfie M (1988) *mec-3*, a homeobox-containing gene that specifies differentiation of the touch receptor neurons in *C. elegans*. Cell 54: 5–16.

Weber S, Horn R, Friedt W (2000) High regeneration potential in vitro of sunflower (*Helianthus annuus* L.) lines derived from interspecific hybridization. Euphytica 116: 271–280.

Weber S, Friedt W, Landes N, Molinier J, Himber C, Rousselin P, Hahne G, Horn R (2003) Improved *Agrobacterium*-mediated transformation of sunflower (*Helianthus annuus* L.): assessment of macerating enzymes and sonication. Plant Cell Rep 21: 475–482.

West MA, Harada JJ (1993) Embryogenesis in higher plants: an overview. Plant Cell 5: 1361–1369.

White AD, Graham MA, Owen MDK (2003) Isolation of acetolactate synthase homologs in common sunflower. Weed Sci 50: 432–437.

Whitham S, Dinesh-Kumar SP, Choi D, Hehl R, Corr C, Baker B (1994) The product of tobacco mosaic virus resistance gene N: similarity to Toll and interleukin-1 receptor. Cell 78: 1101–1115.

Yu JK, Tang SX, Slabaugh MB, Heesacker A, Cole G, Herring M, Soper J, Han F, Chu WC, Webb DM, Thompson L, Edwards KJ, Berry S, Leon AJ, Grondona M, Olungo C, Maes N, Knapp SJ (2003) Towards a saturated molecular genetic linkage map for cultivated sunflower. Crop Sci 43: 367–387.

Zimmer DE, Kinman ML (1972) Downy mildew resistance in cultivated sunflower and its inheritance. Crop Sci 12: 749–751.

Molecular Breeding

Begoña Pérez-Vich[1*] and *Simon T. Berry*[2]

ABSTRACT

In the last few years, there has been a tremendous advance in molecular genetics and genomics of sunflower but the practical applications in breeding programs remain limited. The purpose of this chapter is to show how this molecular information is evolving into a valuable tool for sunflower breeding. The requirements and objectives for sunflower molecular breeding are discussed, as well as applications such as germplasm characterization, enhancement of genetic variability, and marker-assisted selection are overviewed. Finally, molecular breeding in private sector programs is summarized.

Keywords: molecular breeding; germplasm characterization; marker-assisted selection; gene introgression; gene pyramiding

7.1 Introduction

Plant breeding during the 20th century has played an essential role in transforming sunflower (*Helianthus annuus* L.) into a major oilseed crop. The two most important developments were the dramatic increase in seed oil percentage achieved by breeders in the former Soviet Union (FSU), and the development of a cytoplasmic male sterility system, combined with fertility restoration by nuclear genes, which enabled the commercial production of hybrid seed (see for review, Fick and Miller 1997). The subsequent development of short-stemmed, high yielding hybrid cultivars with high oil

[1]Instituto de Agricultura Sostenible (CSIC), Apartado 4084, E-14080 Córdoba, Spain.
[2]Limagrain UK Limited, Woolpit Business Park, Windmill Avenue, Woolpit, Bury St. Edmunds, Suffolk, IP30 9UP, UK.
*Corresponding author: *bperez@ias.csic.es*

content, which were well adapted to mechanical cropping, have turned sunflower into a cash crop and its oil into a major commodity on the world trade markets. In addition to these improvements achieved through conventional breeding, recent years have witnessed an impressive number of advances in molecular genetics and genomics in sunflower that have enhanced the breeding potential of the crop and greatly increased our understanding of its genetic make up. The purpose of this chapter is to show how this molecular information has evolved into a valuable tool for sunflower breeding, and how it has been applied for sunflower improvement.

7.2 Overview of Components and Objectives for Sunflower Molecular Breeding

The key components required for an efficient system for molecular breeding are the identification and characterization of suitable genetic markers, and the establishment of marker-trait associations for characters of agronomic importance. In sunflower, as in other plant species, genetic markers evolved from morphological markers through isozymes to DNA markers, which in turn evolved from hybridization-based detection to polymerase chain reaction (PCR) amplification and more recently to new sequence-based systems. DNA marker development, characterization, and genetic map construction in sunflower has been reviewed in other chapters of this volume, and only a brief outline will be given here. Three types of DNA markers have been developed in sunflower. The first type are anonymous DNA markers (also known as random or neutral markers), of unknown function but with the ability to detect polymorphisms across the genome. These include restriction fragment length polymorphism (RFLP), random amplified polymorphic DNA (RAPD), amplified fragment length polymorphism (AFLP), and simple sequence repeats (SSRs) or microsatellite markers. Genetic maps have been constructed with all these types of markers, independently or in combination (see for review, Knapp et al. 2001 and Paniego et al. 2007). The second type of DNA markers developed in sunflower are gene-targeted markers. These markers are derived from polymorphism within genes, and include InDel (insertion-deletion) markers based on sequenced sunflower RFLP-cDNA probes (Yu et al. 2003; Kolkman et al. 2007), single nucleotide polymorphisms (SNPs) and SSR markers developed from sunflower expressed sequence tags (ESTs) (Lai et al. 2005; Pashley et al. 2006; Heesacker et al. 2008), and target region amplification polymorphism (TRAP) markers (Hu et al. 2004; Hu 2006). All of these different types of gene-based assays have been integrated into the existing SSR maps (Yu et al. 2003; Lai et al. 2005; Hu 2006). Finally, the third type of DNA markers are functional markers or so-called "perfect markers", which detect the causal DNA sequence differences between alleles of a gene underlying a given phenotype. Examples of this type of markers in sunflower are currently rare, but they have been developed for traits determining oil quality (Tang et al. 2006b) and herbicide resistance (Kolkman et al. 2004).

The ideal DNA marker system for molecular breeding should meet the following requirements: the ability to detect high level of polymorphism, codominance, abundance (i.e., whole genome coverage), high reproducibility, suitability for high-throughput analysis and multiplexing, technical simplicity, cost effectiveness, require small amounts of DNA, and be user-friendly (such as their suitability for use in different genotyping systems and facilities). The second and third generation of sunflower markers including SSRs, gene-targeted and functional markers generally satisfy these requirements, thus providing an excellent platform for molecular breeding programs. In addition, the development of a reference map using both an internationally accepted linkage group nomenclature system and publicly available markers, for cross-referencing maps and mapped gene locations, is also an essential prerequisite. In sunflower, the reference genetic map is the one reported by Tang et al. (2002) based on a recombinant inbred line (RIL) population derived from RHA280 x RHA801, which comprises a critical mass of public SSR markers. Additionally, the "immortal" RIL mapping resource enables the on-going placement of new markers, such as InDels derived from RFLP cDNA probes (Yu et al. 2003), SNPs based on ESTs (Lai et al. 2005) or telomere sequence repeat-derived markers (Hu 2006). It should be noted that the linkage group nomenclature of this reference map will be used throughout this chapter.

Determination of significant marker-trait associations is essential for any molecular breeding program. A number of major genes underlying simply inherited traits have been mapped in sunflower, and molecular markers linked to them or based on candidate genes have been described. For oil quality, the *Es1*, *Es3*, and *Ol* genes controlling modified fatty acid profiles, and the *Tph1* and the *Tph2* genes, associated with altered tocopherol profiles, were located on linkage groups (LGs) 1, 8, 14, and 1, and 8, respectively. Markers based on stearoyl-acyl carrier protein desaturase (for *Es1*), oleoyl phosphatidyl-choline desaturase (for *Ol*), 2-methyl-6-phytyl-1,4-benzoquinone/2-methyl-6-solanyl-1,4-benzoquinone methyltransferase (for *Thp1*) and gamma-tocopherol methyltransferase (for *Tph2*) genes were subsequently developed (Pérez-Vich et al. 2002, 2006; Hass et al. 2006; Schuppert et al. 2006; Tang et al. 2006b). For disease resistance, the major *Pl* genes conferring resistance to downy mildew (*Plasmopara halstedii* [Farl.] Berl. and de Toni), and the *Or5* gene associated with resistance to race E of broomrape (*Orobanche cumana* Wallr.) have been mapped. The *Pl* genes were mapped to different clusters, one on LG 8 that included *Pl1*, *Pl2*, *Pl6*, and *Pl7*, another one on LG 13 which included *Pl5* and *Pl8*, and a third one on LG 1 for Pl_{arg} (Vear et al. 1997; Bert et al. 2001; Bouzidi et al. 2002; Dussle et al. 2004). The *Or5* gene was located on a telomeric region of LG 3 (Lu et al. 1999, 2000; Tang et al. 2003; Pérez-Vich et al. 2004). Other genes such as *Ms6*, *Ms7*, *Ms9*, *Ms10*, and *Ms11*, determining nuclear male sterility, and *S*, controlling self-incompatibility, were located on LG 16, 6, 10, 11, 8, and 17, respectively (Gandhi et al. 2005; Pérez-Vich et al. 2005; Chen et al. 2006; Capatana et al. 2008; Li et al. 2008).

In addition, molecular mapping studies in sunflower have identified genes or chromosomal regions that control quantitative traits, so-called quantitative trait loci (QTL). Three to eight QTLs determining oil content and its components (i.e., percentage of kernel weight in relation to whole achene weight and kernel oil concentration) have been mapped (Leon et al. 1995a, 2003; Mestries et al. 1998; Mokrani et al. 2002; Bert et al. 2003; Tang et al. 2006a). The QTLs on LG 10, 16, and 17 were reported to be centered on the phenotypic loci *B* (apical branching), *hyp* (pigmentation of the achene hypodermis layer), and *P* (pigmentation of the achene phytomelanin layer), respectively (Tang et al. 2006a). With respect to disease resistance, lists of QTLs for resistance to white rot (*Sclerotinia sclerotiorum* [Lib.] de Bary), Phomopsis (*Diaporthe helianthi* Munt-Cvet. et al.) and black stem (*Phoma macdonaldii* Boerma) are becoming available (Bert et al. 2002, 2004; Al-Chaarani et al. 2002; Micic et al. 2004, 2005a, b; Rönicke et al. 2005; Abou Alfadil et al. 2007; Darvishzadeh et al. 2007; Yue et al. 2008). QTLs for other complex traits such as somatic embryogenesis and *in vitro* organogenesis (Flores-Berrios et al. 2000) or photosynthesis and water status (Hervé et al. 2001; Poormohammad Kiani et al. 2007) have also been reported.

It is obvious that the targets for molecular breeding in sunflower do not (and should not) differ markedly from those of conventional breeding. These objectives may vary with specific programs, but the main emphasis is high seed yield, high oil content, and improved oil quality for special markets. Seed yield, and to a lesser extent, oil content depend on many factors including a suitable agronomic type, tolerance to abiotic stresses, and resistance to diseases and insects, and these, consequently, have also become important marker assisted objectives.

7.3 Germplasm Characterization

Traditionally, germplasm characterization has been assessed by measuring the variation in phenotypic traits. This approach has certain limitations, as the number of traits that can be routinely measured is limited and they are subject to environmental influences. Molecular markers circumvent these issues and provide accurate and detailed information which complements phenotypic evaluation. Molecular characterization of cultivated sunflower was initially carried out using biochemical methods. Studies by Torres (1983) resolved enzymatic systems such as alcohol dehydrogenase and acid phosphatase and concluded that they could be used for evaluation of genetic similarity of sunflower cultivars. Quillet et al. (1992) described the discrimination of 52 distantly related inbred lines from different origins using eight polymorphic isozyme loci, whereas Tersac et al. (1994) were able to group 39 cultivated sunflower populations according to their geographical origin using nine polymorphic isozymes. Finally, Carrera et al. (2002) reported that eight polymorphic isozyme loci were capable of separating the majority of maintainer lines (B) from restorer lines (R) in a survey of 25 elite inbred lines.

Enzymatic systems have also been used to assess genetic variation in both domesticated and wild sunflower populations. Rieseberg and Seiler (1990) used this technique, in addition to RFLP analysis of chloroplast DNA, for an evolutionary study of cultivated sunflower in relation to wild *Helianthus annuus*. They found extensive isozyme and chloroplast DNA polymorphism in the wild accessions and a virtual monomorphism in cultivated lines, concluding that domesticated sunflower must have evolved from a very limited gene pool. Cronn et al. (1997) undertook a larger survey of 146 germplasm accessions from the US National Plant Germplasm System (NPGS) sunflower collection using 20 allozyme loci and demonstrated that domesticated sunflowers (ranging from Native American landraces to elite cultivars of both confectionary and oilseed types) and wild sunflowers formed two nearly independent groups. Again, the wild *H. annuus* group was genetically more diverse than the domesticated group and this diversity was geographically structured.

Despite their value in these initial studies, biochemical markers have not been used extensively to characterize sunflower germplasm due to their poor genome coverage and low levels of polymorphism. For example, the previously mentioned study of Cronn et al. (1997) revealed only 53 alleles at 20 allozyme loci, with an average of 1.39 alleles per locus, in a large survey of over 700 achenes, representing 114 domesticated accessions. Pizarro et al. (2000) and Popov et al. (2002) compared isozyme and RAPD markers to assess genetic diversity in 21 and 30 sunflower inbred lines, respectively, and demonstrated the greater discriminatory power of the DNA-based marker systems.

The characterization of genetic structures in cultivated sunflower was one of the first aims of genetic fingerprinting using DNA-based markers. Initial studies using RFLPs consistently separated sunflower inbred lines into sterility maintainer (B-line) and fertility restorer (R-line) groups (Berry et al. 1994; Gentzbittel et al. 1994; Zhang et al. 1995), reflecting the breeding strategies that maximize heterosis. These results were later confirmed using AFLPs and SSRs (Hongtrakul et al. 1997; Paniego et al. 2002; Yu et al. 2002). Tang and Knapp (2003) performed phylogenetic analyses on 19 elite inbred lines and 28 domesticated (Native American landraces and open-pollinated cultivars, including oilseed, confectionary and ornamental types) and wild (*H. annuus*) germplasm accessions using 122 SSRs distributed throughout the sunflower genome. The authors found an extraordinary allelic diversity in the Native American landraces and wild populations, and progressively less allelic diversity in germplasm produced by successive cycles of domestication and breeding. These findings suggest that the contemporary oilseed sunflower pool could profit from an infusion of novel alleles from the reservoir of latent genetic diversity present in both wild populations and Native American landraces. DNA-based markers have also been used to characterize specific sunflower collections. For example, Dong et al. (2007) evaluated a confectionary sunflower collection of 70 accessions from China with AFLP markers. They found an absence of duplicated entries and a lack of a consistent classification of the material according to their geographical origin.

Zhang et al. (1995) used RFLPs to screen four inbred lines for intra-line polymorphisms, and found RFLPs within them. However, these four lines presented a good uniformity for morphological characters in the field, which led the authors to conclude that the polymorphisms stemmed from residual heterozygosity or outcrossing. Zhang (1995) proposed using RFLPs for distinctness, uniformity, and stability (DUS) testing in sunflower, taking into account inter-line variability for distinctness, and intraline variability for uniformity and stability. Later on, Zhang et al. (2005) also demonstrated the usefulness of SSRs for diversity assessment of sunflower inbred lines, and suggested that a set of mapped SSR markers with a high PIC (polymorphism information content) value could be very useful for varietal description, purity testing, hybrid formula verification, and varietal identification in the process of seed certification. These molecular markers could also play an important role in protecting plant breeders' rights in sunflower and in assessing whether or not an inbred line was essentially derived [UPOV (International Union for the Protection of New Varieties of Plants) Convention revised in 1991].

Molecular markers have also been employed to characterize different sources of cytoplasmic male sterility (CMS). The development of new CMS systems and their respective fertility restoration genes is an important breeding objective in sunflower. Commercial hybrid production is currently based on a single source of CMS, cmsPET1 from *H. petiolaris*, which narrows the genetic base of the crop and increases its vulnerability to pests and environmental stresses. Understanding the molecular biology of the various CMS systems helps to classify different types and to identify novel sources. Köhler et al. (1991) and Laver et al. (1991) suggested that a new open reading frame, *orfH522*, in the 3'-flanking region of the mitochondrial (mt) *atpA* gene was associated with the CMS phenotype. It was later demonstrated that the gene product of *orfH522*, a 16-kDa protein, differed between the male fertile and male sterile lines (Horn et al. 1991; Monéger et al. 1994). Further diversity studies of 28 male sterility-inducing cytoplasms and one male fertile cytoplasm of *Helianthus* using nine mtDNA genes and three probes for the open reading frame clearly distinguished CMS sources by their mtDNA organization and CMS mechanism (Horn 2002).

As mentioned above, one of the most important milestones in sunflower breeding has been the development of hybrid cultivars, which benefit from hybrid vigor or heterosis. Genetic distance has been widely studied as predictor of heterosis and hybrid performance in crop plants. In sunflower, Tersac et al. (1994) estimated genetic distance from isozyme data and found that it was not correlated to heterosis. Cheres et al. (2000), using AFLPs and co-ancestries, found that genetic distance alone was a weak predictor of hybrid performance in sunflower. As reported in other crops, such as soybean or maize, genetic distance and F_1 heterozygosity estimated from randomly selected markers are typically not correlated with heterosis (Melchinger et al. 1990; Gizlice et al. 1993). Conversely, approaches based on screening for heterosis-related markers (Melchinger et al. 1990), using specific heterozygosity (Zhang et al. 1994), and identifying favorable combinations of allele and heterotic patterns (Liu and Wu 1998) could be exploited to improve this prediction.

7.4 Molecular Strategies to Create and Identify Novel Variation

Natural biodiversity is an under-exploited sustainable source that can enrich the genetic pool of cultivated sunflower with novel alleles and molecular approaches for exploiting this natural diversity are currently being explored in sunflower. The use of association-based mapping techniques for the discovery of genes and novel alleles controlling complex traits across a broad spectrum of germplasm is being considered. Association mapping, also known as linkage disequilibrium (LD) mapping, is a method that relies on LD to study the relationship between phenotypic variation and genetic polymorphism (Flint-Garcia et al. 2003). Basically it utilizes ancestral recombination events in natural populations, evaluating whether certain alleles are found with specific phenotypes more frequently than expected and so does not require any special population development for trait mapping. The extent of LD is the main factor determining resolution of association mapping, and it is affected by natural selection, domestication and breeding history, mating systems, mutation, migration, genomic rearrangements, and other factors (Buckler and Thornsberry 2002). Until recently, the extent of LD and the application of association mapping in sunflower have not been studied in-depth. Recent reports by Liu and Burke (2006), Kolkman et al. (2007) and Fusari et al. (2008) looking into the patterns of nucleotide diversity in genic loci from wild and cultivated sunflower, demonstrate that SNP frequencies and LD decay are sufficient in wild populations (1 SNP/19.9 bp and LD decay within ~200 bp), exotic germplasm accessions (1 SNP/38.8 bp and LD decay within ~1,100 bp), and modern sunflower cultivars [1 SNP/45.7 bp and LD decay within ~5,500 bp according to Kolkman et al. (2007); 1 SNP/69 bp according to Fusari et al. (2008)] for high-resolution association mapping in sunflower.

Naturally occurring allelic variation at specific candidate genes underlying agronomically important traits is a new resource for the functional analysis of plant genes, and facilitates the identification and selection of new alleles and genes based directly on their DNA sequence. In sunflower, numerous resistance gene candidates (RGCs) have been described as underlying the *Pl* clusters conferring resistance to downy mildew located on LG 8 (Gentzbittel et al. 1998; Gedil et al. 2001; Bouzidi et al. 2002; Slabaugh et al. 2003), LG 13 (Radwan et al. 2003, 2004), and LG 1 (Pl_{arg}; Radwan et al. 2008). These RGCs belong to different sub-classes of the resistance gene products characterized by the presence of leucine-rich repeat (LRR) motifs and a nucleotide binding site (NBS) N-terminal to the LRR domain. Slabaugh et al. (2003) assessed genetic diversity at the LG 8 *Pl* cluster with specific intron fragment length polymorphism (IFLP) designed from a RGC in this cluster, and found an extraordinary level of diversity in this region in wild sunflowers in comparison to domesticated germplasm. They concluded that wild sunflowers represented a rich source of untapped resistance genes. Radwan et al. (2008) have isolated and sequenced a collection of RGCs (NBS-LRR homologs) from *H. annuus*, *H. paradoxus*, *H. deserticola*, *H. tuberosus* and *H. argophyllus*. From these, 167 NBS-LRR loci were mapped in the sunflower genetic map in 44 clusters or singletons, many of them co-localizing to previously mapped downy mildew, rust, and broomrape resistance genes. This process of sequencing and mapping RGC from diverse germplasm will facilitate the discovery of novel resistance genes.

In addition to exploiting natural variation, breeders have also induced novel genetic variation using both chemical and physical mutagens. A relatively new method for identifying mutations in known genes called TILLING (Targeting Induced Local Lesions In Genomes) has renewed the interest in mutation breeding. TILLING is a powerful reverse genetics approach that employs a mismatch-specific endonuclease to detect single base pair (bp) allelic variation in a target gene using a high-throughput assay (Till et al. 2003). This allows the identification of induced point mutation in target genes within a mutagenized population, as well as the identification of natural allelic variation in populations or breeding material, which is known as EcoTILLING (Comai et al. 2004). Its advantages over other reverse genetics techniques include its applicability to virtually any organism, its facility for high-throughput analysis, and its independence from genome size, reproductive system, or generation time. TILLING has been used successfully for the detection of both induced and natural variation in several crops, including wheat (Slade et al. 2005), rice (McNally et al. 2006) and maize (Till et al. 2004). Bauer et al. (2006) and Montemurro et al. (2007) have proposed establishment of TILLING platforms in sunflower. These authors describe the development of TILLING populations using different ethylmethane sulfonate (EMS) treatments, establishment of heteroduplex analysis for mutation detection, and a pilot screen on three genes for fatty acid biosynthesis (Sabetta et al. 2008). Although no results are currently available, the development of TILLING platforms will be a valuable resource for future sunflower improvement.

7.5 Marker-assisted Selection

In a general sense, marker-assisted selection (MAS) involves the selection of plants carrying genomic regions that are involved in the expression of traits of interest using molecular markers. In sunflower, it is clear from other chapters of this volume that there have been major advances in recent years in the development of DNA markers, construction of genetic maps, and mapping of important traits controlled by major genes and/or QTL. Whilst the number of reports on mapped genes continues to grow rapidly, the literature on the practical application of those markers in breeding programs remains very limited. One reason for this is that there are several scientific and logistical issues that must be resolved before a practical MAS strategy can flow from a mapping study, and at each step there will be a certain level of failure. Moreover, MAS programs in sunflower are mainly carried out by private breeding companies, which may have restrictions on publishing their findings. Therefore, in this section we will deal with factors that can enhance the efficiency of MAS, and describe the few examples of practical use of molecular markers in breeding programs that have been published in this crop.

7.5.1 Marker Validation and Refinement

The efficiency of MAS depends on many factors associated with how the underlying marker-trait associations were identified, including the size of the mapping population, the nature of phenotyping, the design and analysis of the experiment, the number and quality of markers used, the proximity between markers and the trait of interest, and the proportion of the phenotypic variance contributed by each QTL, among others. Many of these variables can be optimized through marker validation and refinement.

For markers associated with simply inherited traits, marker validation and reducing the distance between the marker and the gene of interest is fairly straightforward. In these cases, the effect of the genetic background is usually minimal, and the ease of phenotyping makes fine-mapping of the gene simpler (Dwivedi et al. 2007). In mapping studies, genes for simply inherited traits can be mapped with an adequate accuracy in a mapping population of 100–150 individuals. This can then be followed by fine-mapping using larger populations of targeted recombinants, and eventually, by map-based cloning of the target gene. For example, fine-mapping of the *Rf1* gene responsible for restoring pollen fertility in sunflower PET1-based material was accomplished using enlarged mapping populations (Kusterer et al. 2005), followed by a map-based cloning approach (Kusterer et al. 2004; Hamrit et al. 2006a). In addition, validation in different genetic backgrounds for markers associated with major resistance genes has been accomplished for those associated with the Pl_2 gene determining resistance to different downy mildew races (Brahm et al. 2000), with the R_1 and the R_{adv} genes conferring rust resistance (Lawson et al. 1998), and with the *Or5* gene conferring resistance to race E of broomrape (Tang et al. 2003; Pérez-Vich et al. 2004).

Unfortunately, in many instances the marker identified through the process of fine-mapping may not be polymorphic in all the populations tested, thus requiring the identification of alternative markers for those populations. The "perfect" marker for selection would be one that provides 100% accurate prediction of the phenotype in all genetic backgrounds. These markers, described as functional markers in previous sections, are based on the gene mutations that have been demonstrated to be responsible for the trait of interest. In sunflower, functional markers have been developed for oil quality traits and other simple traits. For oil quality, Tang et al. (2006b) determined that a non-lethal knockout mutation in a 2-methyl-6-phytyl-1,4-benzoquinone/2-methyl-6-solanyl-1,4-benzoquinone methyltransferase locus (MT-1) caused by the insertion of a 5.2 kb *Ty3/gypsy*-like retrotransposon was underlying β-tocopherol accumulation in sunflower seeds, and robust STS markers diagnostic for wild type and mutant MT-1 alleles were developed. For herbicide resistance, Kolkman et al. (2004) identified mutations in codons 197 and 205 in the acetohydroxyacid synthase gene *AHAS-1* that confers resistance to sulfonylurea (SU) and imidiazolonone (IMI) herbicides, and developed a SNP genotyping assay diagnostic for the codon 205 mutation.

In the case of QTL markers for complex traits, the situation is far more complicated. Factors such as population structure and size, parental selection and genetic background effects, epistasis, inaccurate phenotyping, and QTL x environment (QTL x E) interactions can all contribute to biasing the estimation of QTL effects and position, thus reducing the likelihood of a successful outcome in a MAS program. QTL validation in independent samples and in different genetic backgrounds and environments is, therefore, necessary before using marker-QTL association in MAS programs. In sunflower, QTL for *Sclerotinia* resistance have been validated across environments (Bert et al. 2002), generations (Micic et al. 2005a; Vear et al. 2008), and genetic backgrounds (Rönicke et al. 2005; Micic et al. 2005b). Micic et al. (2004, 2005a) and Vear et al. (2008) consistently identified QTLs for *Sclerotinia* resistance on LG 8 and LG 16 (stem lesion), and on LG 1 and LG 10 (capitulum lesion), respectively, across early F_3 and later F_6 (RIL population) generations obtained from a cross involving the resistant lines NDBLOS$_{sel}$ (Micic et al. 2005a) or PSC8 (Vear et al. 2008). The QTL on LG 8 was also detected in another mapping population developed from a different resistant source (TUB-5-3234) (Micic et al. 2005b). In addition, Micic et al. (2005b) cross-referenced previous studies of Mestries et al. (1998), Bert et al. (2002), and Rönicke et al. (2005), and found that the same six linkage groups (LG 1, 4, 8, 9, 10, and 13) carried QTLs for resistance against *Sclerotinia sclerotiorum* in more than one of the five mapping populations considered. However, the most important resistance QTLs appear to be on LG 1, LG 9, and LG 10, which had a significant effect in at least three of the five populations. For oil content, QTLs have also been validated across generations, environments, and mapping populations. Oil content in sunflower depends on both the percentage of kernel weight in relation to whole achene weight and on the concentration of oil in the kernel. Mestries et al. (1998) identified at least two QTLs for seed (achene) oil content consistently in F_2 though F_4 generations from the cross GH x PAC2, whereas Bert et al. (2003) detected four QTLs consistently in both F_2 and F_3 generations from the cross between the lines PSC8 and XRQ. Both studies also reported that one of the QTLs associated with oil content co-located with the *B* gene determining apical branching. Leon et al. (2003) validated five QTLs for oil content in the population ZENB8 x HA89, detected previously by Leon et al. (1995a), through the evaluation of replicated progenies in four locations within a target environment. Four of the QTLs were detected in at least two locations. The QTL with the largest effect, which co-located with the *hyp* gene that determines achene hypodermis color, was detected in all four locations. QTLs for seed oil content centered on the *B* and *hyp* loci were also identified by Tang et al. (2006a) in LG 10 and 16, respectively, in a different mapping population developed from the cross RHA280 x RHA801.

QTL effects may also be environmentally sensitive. Variation in expression due to QTL x E interaction remains a major constraint to the discovery of QTL that will confer a consistent advantage across a wide range of environments; however, if a QTL shows QTL x E interaction then the selection of genotypes adapted to specific environments may be possible. In sunflower, Leon et al. (2001) identified QTLs for days from seedling emergence to flowering (DTF) associated with photoperiod (PP) response in a population evaluated in six environments (locations, years, and sowing dates) and they explicitly included a QTL x E interaction in the interval-mapping model. These authors demonstrated that the two QTLs with the strongest effect on DTF, identified across environments, also showed a highly significant QTL x E interaction and were responsive to PP. Identifying these QTLs could potentially allow the conversion of photoperiod-sensitive germplasm to photoperiod insensitive.

Epistasis is another factor that must be considered during validation studies. It causes the allelic effects of one QTL to be dependant on the genotype at a different locus (Holland 2001). Tang et al. (2006a) provided an example of complex interactions among QTLs associated with seed oil concentration in sunflower, which could result from complex networks of the underlying genes. However, these epistatically interacting QTLs for seed oil content were not previously identified in other genetic analyses in sunflower (Leon et al. 1995a, 2003; Mestries et al. 1998; Mokrani et al. 2002; Bert et al. 2003). Studies in *Arabidopsis* suggest that epistatic QTLs are more important than additive QTLs for fitness traits (Malmberg et al. 2005). In contrast, studies designed to explicitly model epistatic interactions in maize revealed that epistasis was of little or only moderate importance for quantitative traits (Mihaljevic et al. 2005; Blanc et al. 2006). Although the relative importance of epistasis is still under debate, it has sometimes reduced the predicted gains from MAS (Holland 2001) and so its effect should clearly be evaluated for MAS programs.

Where possible, mapping of QTL with a high level of resolution will enable the identification of more tightly linked markers and the manipulation, via MAS, of a smaller chromosomal segment carrying the QTL of interest. Development of the high-density sunflower genetic map (0.8 cM/locus) through the mapping of 2,495 high-throughput DNA marker loci (Knapp et al. 2007) will contribute towards this objective. Availability of specific genetic resources such as near-isogenic lines (NILs) differing in a genomic segment containing a target QTL (QTL-NILs), advanced segregating populations such as RILs, or RIL testcross populations (TC-RILs) will also help to estimate with a greater accuracy QTL positions and their effects. In sunflower, Micic et al. (2005a) re-estimated the position and effect of a number of QTLs for *Sclerotinia* resistance in an RIL population developed from F_3 families in which the QTLs were originally identified. However, not all the QTLs identified in the F_3 were found in the RIL population. Pizarro et al. (2006) developed QTL-NILs varying for target QTLs for seed oil concentration by backcrossing a donor parent (a wild *H. annuus*) to a recurrent parent (elite high oil cultivar) combined with MAS. Comparing the phenotypes of the

QTL-NILs with those of the recurrent parent allowed the authors to obtain an accurate evaluation of the effects of the target QTL in an adapted background. On the other hand, TC-RIL populations may contribute to identifying QTL without the confounding effect of diverse segregating backgrounds. Draeger et al. (2006) developed a TC-RIL population fixed for the branching (controlled by the *B* gene) character (i.e., an unbranched population), and were able to identify a QTL affecting seed oil concentration on LG 10 (*soc10.2**) linked in repulsion to another *B*-linked QTL affecting the same trait (*soc10.1*). The *soc10.2** QTL had not been identified previously in the RIL analysis *per se* (Tang et al. 2006a) because its effect was apparently masked by *soc10.1*, or by pleiotropic effects of the *B* gene. Finally, Huang et al. (2007) increased the resolution of QTL mapping by performing the analyses on F_2 segregating populations from crosses between RILs selected for having favorable alleles for somatic embryogenesis at three QTLs on LGs 5, 10, and 13, and another RIL having alleles with negative effects at these QTLs. The number of cross-over events and bins in the QTL regions was higher in the new F_2 maps, making it possible to localize the QTLs to smaller intervals.

The ultimate goal, as for traits controlled by major genes, is to develop molecular markers based on functional genes underlying the QTLs controlling complex traits. Reverse genetic approaches appear to be promising for the identification of markers based on genes underlying QTLs and for enabling the dissection of the genetic basis of complex traits. As a first approach, functional genomic analyses of candidate genes can be performed. For example, Poormohammad Kiani et al. (2007) identified QTLs for water status traits as well as net photosynthesis in a sunflower RIL population. Four RILs were selected for contrasting response to water stress and QTL complement and their differential gene expression was studied for four water-stress associated candidate genes (aquaporin, dehydrin, leafy cotyledon1-like protein, and fructose-1,6 biphosphatase) under well-watered and water-stressed conditions. RILs carrying different genomic regions for some QTLs also had different gene expression patterns for some of the four genes. This candidate gene approach has also been used to identify genes underlying QTLs for salt tolerance in wild sunflower hybrids (*H. annuus* x *H. petiolaris*). Salt tolerance candidate genes were identified in expressed sequence tag (EST) libraries of sunflower and 11 were mapped to an existing QTL framework map from an interspecific BC_2 population. One EST that coded for a Ca-dependent protein kinase mapped to a salt tolerance QTL (Lexer et al. 2003a).

For some traits of interest, *a priori* candidate-gene approaches are not feasible because a biological model does not exist or the number of candidates is too numerous that individual follow-ups are prohibitively expensive. Quantitative expression studies, such as microarray analyses, can reveal regulatory variation in genes determining complex traits. These studies, when combined with QTL mapping, permit the identification of positional candidate genes for a phenotype of interest whose expression varies between the parental lines (Wayne and McIntyre 2002). A number of microarray studies have been carried out in sunflower, which are unraveling genes involved in complex traits such as quantitative resistance to *Phoma macdonaldii* (Alignan et al. 2006), early sunflower seed development (Hewezi et al. 2006a), chilling sensitivity (Hewezi et al. 2006b), chilling and salt stresses (Fernández et al. 2008), and drought tolerance (Roche et al. 2007). In addition, microarray analyses combined with QTL mapping is being used to determine the potential adaptive value of differentially expressed genes in the sunflower species *Helianthus deserticola* (Lai et al. 2006). The use of other profiling techniques such as proteomic approaches that are more suitable for determining changes in protein expression, rather than predicting them from transcript expression data, are also being used in sunflower to identify gene products underlying complex traits. For example, Hajduch et al. (2007) reported 77 protein spots differentially expressed in the high oil line RHA801 versus the low oil line RHA280. Identification of 44 of these proteins indicated that the two main processes affecting low or high oil concentration in these lines were glycolysis and amino acid metabolism. However, the use of such profiling technologies for applied aspects in plant breeding is still rare due to a limited correlation with QTL studies. Jansen and Nap (2001) have proposed the use of gene expression data in QTL analyses by evaluating the expression levels of genes within a segregating population, identification of the so-called expression QTL (eQTL), and co-localization of eQTL and trait QTL in the same population. This approach has demonstrated its utility in understanding complex traits in an increasing number of crops (Dwivedi et al. 2007), and it is also a potential source for the development of "perfect markers".

Finally, if QTLs are mapped accurately to a relatively small region of a chromosome, another approach to determining their underlying genes consists of cloning the QTLs using a map-based cloning strategy. To our knowledge, there are no examples of QTL cloning in sunflower. However, there have been major advances in recent years, such as the development of both bacterial artificial chromosome (BAC) and binary-bacterial artificial chromosome (BIBAC) libraries (Gentzbittel et al. 2002; Özdemir et al. 2004; Feng et al. 2006; Tang et al. 2007), which provide the necessary tools for making such an approach feasible. Additionally, association mapping could be used for higher-throughput QTL cloning by identifying correlations between candidate gene sequence variation and phenotypic variation in breeding lines, without requiring the development of special mapping populations. As mentioned in previous sections, association mapping approaches are currently being explored in sunflower.

7.5.2 *Assay Optimization*

After the development of molecular markers and the validation of their selection efficiency, it is often necessary to optimize the marker assay. This is driven by the need to reduce laboratory costs and turn around times, whilst increasing throughput and minimizing errors. Technologies, which speed up the implementation process, reduce laboratory requirements or errors, and lower the costs associated with scaling-up, are therefore crucial to the success of MAS. In fact, one of the main priorities [included in the "White paper: Priorities for research, education and extension in genomics, genetics and breeding of the Compositae" (The Compositae Genome project, *http://compgenomics.ucdavis.edu/* 2007)] for translating sunflower genomics into practical breeding programs was the reduction in total marker costs. Several advances in sunflower marker technologies have been made in recent years. For SSR markers, PCR multiplexes for a genome-wide framework of SSR marker loci developed by Tang et al. (2003) increased genotyping throughput and reduced reagent costs, which is ideal for genotyping applications that always use the same set of markers. In addition, multi-color assays, SSR primer design to facilitate "pooled amplicon multiplexing" by length in SSR development, and SSR analyses in semi-automated, high-throughput genotyping systems (Tang et al. 2002; Yu et al. 2002) have all resulted in time-saving and reduced costs in routine microsatellite analysis. Currently, different techniques for SNP detection are being used in sunflower to type SNPs in a high-throughput, time-saving and cost-effective fashion. These include denaturing high-performance liquid chromatography (dHPLC) (Lai et al. 2005), single-base extension (SBE) and allele-specific primer extension assays (ASPE) for flow cytometric platforms (Knapp et al. 2007), and high-resolution melting curve analyses (MCA) for real-time PCR platforms (Tang and Knapp 2008).

Improved QTL detection methods that reduce genotyping have also been proposed. Micic et al. (2005b) used selective genotyping (Lebowitz et al. 1987) for detecting QTL for *Sclerotinia* resistance in sunflower. This method exploits the fact that most of the information for QTL effects is in the "tails" of the quantitative trait distribution, allowing the reduction of population sizes to those individuals found in these "tails". It was concluded that selective genotyping could be efficiently used for major QTL detection and analysis of congruency for resistance genes across populations, but the limited sample size and the non-random sampling may lead to biased estimations of the QTL effects.

7.5.3 MAS in Sunflower Breeding Programs

7.5.3.1. Marker-assisted Gene Introgression

One of the most common applications of MAS is marker-assisted/accelerated backcross breeding for gene introgression. Optimally, this is based on positive foreground selection for a donor trait, positive background selection for the recurrent parent genome, and negative background selection against undesirable donor parent alleles (Frisch et al. 1999). In general, marker assistance is expected to provide greater selection efficiency, and/or to shorten the time taken for the backcross breeding scheme compared to conventional methods. Different factors such as the nature of the target locus (specific locus vs. QTL, dominant vs. recessive) and the number, nature, and distance of the markers associated to it should be taken into account in order to optimize the design of the marker-assisted backcross program (Frisch 2005). However, economics is usually the key determinant for the application of molecular genetics in gene introgression programs. Morris et al. (2003) compared the costs of marker-assisted and traditional backcrossing of a single major gene into an elite line and found, as expected, that MAS was faster but more costly than traditional selection. These authors concluded that the cost-effectiveness of DNA markers is dependant on four critical parameters: (1) the relative cost of phenotyping versus genotyping screening, (2) the time savings achieved using MAS, (3) the size and temporal distribution of benefits associated with accelerated release of improved germplasm, and (4) the availability of operating capital to the breeding program. The latter point explains to a large extent why private industry has adopted MAS gene introgression more rapidly than public sector programs.

In fact, marker-assisted backcross breeding in sunflower is mainly being carried out in private companies to accelerate the introgression of target genes into elite germplasm. As stated in their websites, traits such as downy mildew resistance, high oleic acid content in the seed oil, and herbicide resistance are currently the main targets, although complex traits such as resistance to *Sclerotinia*, *Phoma* and Phomopsis are also mentioned. Despite the absence of publications from these programs, it seems that, in general, markers are being routinely used to select for alleles with large effects on traits of a relatively simple inheritance. However, dissection of complex traits into their components is contributing to the implementation of marker-assisted backcross programs for such characters. For oil content, different QTLs underlying its components (percentage of kernel and kernel oil concentration) have been identified, and some of them co-located with the phenotypic loci *B*, *hyp* and *P* (Leon et al. 1996, 2003; Tang et al. 2006a). This fact was explained by Tang et al. (2006a) as a pleiotropic effect of such phenotypic loci on oil content, which were presumably targeted by selection in the transition from large-seeded, low-oil to small-seeded, high oil cultivars, and allowed Leon et al. (1995b) to establish combined marker and phenotypic (based on the *hyp* locus underlying a QTL for percentage of kernel) assisted selection for high oil content in the backcross process, in addition to marker-assisted background selection.

Codominant markers are the most useful for marker-assisted backcrossing because selection among backcross progeny involves identification of heterozygous individuals. Conversely, if a dominant marker is used, it will only be informative if the dominant allele (conferring the presence of a band) is linked to the donor parent allele. For selection of the *Pl6* gene conferring sunflower resistance to downy mildew race 730, Pankovi et al. (2007) proposed the use of *Pl6* tightly linked codominant cleaved amplified polymorphic sequence (CAPS) markers in combination with dominant markers developed from resistance candidate genes for increasing the MAS efficiency in backcross programs.

Wild relatives of sunflower possess a high level of resistance to many biotic and abiotic stresses (Jan and Seiler 2007). Marker-assisted backcross breeding is also a very effective way of transferring genes or QTLs determining these traits from wild donor genotypes into elite breeding lines by reducing both the time needed to produce a commercial cultivar and the risk of undesirable linkage drag. However, current breeding efforts in sunflower are directed more towards the introgression of specific genes from wild species through phenotypic evaluations, for example to diseases (Jan and Seiler 2007), rather than through the use of molecular markers directly associated with the desired trait. To facilitate and accelerate the introgression process and to distinguish new pathogen resistance gene specificities, Slabaugh et al. (2003) have proposed the identification of allelic variation in wild species in specific candidate genes known to be associated with resistance to sunflower pathogens such as RGCs. In addition, QTL and candidate gene analyses in wild sunflower species are contributing towards identifying genes and QTL for adaptation to salt or drought tolerance (Lexer et al. 2003a, b; Kane and Rieseberg 2007) that could be exploited as a source of new genes to be introgressed into cultivated sunflower. DNA-based markers have also been useful for the characterization and verification of interspecific sunflower hybrids (Natali et al. 1998; Binsfeld et al. 2001) and to identify introgressed DNA fragments from wild species in interspecific progenies. Some of these fragments have been related to increased levels of resistance to diseases such as *Sclerotinia sclerotiorum*, Phomopsis, or downy mildew (Besnard et al. 1997; Rönicke et al. 2004; Wieckhorst et al. 2008).

So-called "advanced backcross QTL analysis" (AB-QTL) has been proposed for transferring QTLs for agronomically important traits from unadapted donor lines (wild species, landraces) into established elite inbred lines (Tanksley and Nelson 1996). In this approach, an unadapted donor line is backcrossed to a superior cultivar and the QTL analysis is delayed until the BC_2 or BC_3 generation, whilst a negative selection is carried out during the development of these populations to reduce the frequency of deleterious donor alleles. QTL-NILs can be derived from advanced backcross populations to verify QTL effects, and these same QTL-NILs may also become commercial inbreds improved for one or more quantitative traits. A modified AB-QTL approach has been used in sunflower to improve populations derived from interspecific crosses for resistance to *Sclerotinia sclerotiorum* or Phomopsis (Serieys et al. 2000). In a first step, interspecific BC_1 and BC_2 populations are subjected to a number of cycles of recombination associated with low selection pressure to favor intergenomic rearrangements and to decrease frequency of undesirable traits. In a second step, the selection of S_0 plants is performed based on phenotypic traits and the presence of RAPD bands specific for the parental wild species. After intercrossing, the S_1 polycross progenies are produced and evaluated for *Sclerotinia sclerotiorum* and Phomopsis resistance. This approach allowed the identification of advanced progenies with increased resistance levels transferred from the wild parental species (Serieys et al. 2000). Additionally, advanced backcross populations and QTL-NILs are being used in sunflower to determine whether genomic regions (QTL) derived for wild *H. annuus* germplasm have the potential to further increase oil content in high-oil genetic backgrounds. Preliminary results reported by Pizarro et al. (2006) indicated that the introgressed wild alleles decreased seed oil concentration in the elite genetic background.

MAS is not only being used for gene introgression, but also as an alternative to testcross and progeny testing. Hybrid sunflower production in sunflower is based on a CMS system combined with fertility restoration by nuclear genes. The development of new restorer and maintainer pools for the sterile cytoplasm involves a large amount of test-crossing and progeny testing, which can be replaced by MAS using markers closely linked to fertility restoration genes. Molecular markers linked to different fertility restoration genes have been identified, and some of these genes have been mapped to LG 13 (*Rf1*, for PET1 cytoplasm from *H. petiolaris*) and to LG 7 (*Rf3*) (Gentzbittel et al. 1995; Horn et al. 2003; Yu et al. 2003, Hamrit et al. 2006b; Abratti et al. 2008). In addition, Yue et al. (2007) have demonstrated the utility of a TRAP marker developed from a sunflower EST that showed homology to a *Petunia* fertility restorer gene to recover *rf1/rf1* genotypes with a greater efficiency in progeny tests for confection of B lines. These markers will be useful for efficiently building up new restorer and maintainer pools for new CMS sources and for the identification of fertility restored plants in breeding programs.

7.5.3.2 Gene Pyramiding

Gene pyramiding is a genotype building strategy for simultaneous manipulation of several genes or favorable alleles at QTL when they are originally present in multiple parents (Hospital 2003). The final objective would be to construct an ideal genotype by accumulating all the favorable alleles at these loci in a single line. This strategy involves several initial crosses between the parents. For example, Hospital (2003) proposed a two-step procedure for the combination of four genes (*G1* to *G4*) into a single line. In the first step, two lines that are homozygous for two target genes (*G1/G2* versus *G3/G4*) are developed by crossing pairs of lines (L1 x L2 versus L3 x L4), followed by selection on the basis of linked markers of homozygotes among F_2 or RIL progeny. In the second step, such individuals are crossed to produce individuals that are homozygous at all target genes.

Gene pyramiding has been proposed as a useful approach to increase the durability of resistance to pest and diseases, or to increase the level of abiotic stress tolerance. Phenotyping assays have been used to pyramid downy mildew and *Sclerotinia* stem rot resistance genes in sunflower (Tourvieille de Labrouhe et al. 2004; Feng et al. 2007). For downy mildew control, the preliminary results reported by Tourvieille de Labrouhe et al. (2004) comparing different methods for obtaining durable resistance indicated that gene pyramids were less effective in reducing the appearance of new races compared to other control methods such as the use of combinations of resistance *Pl* genes by alternation or in "multi-hybrids". Vear (2004) suggested that in order to increase the durability of these major genes, they needed to be backed up by quantitative, non-race specific resistance QTL. Markers will be essential for this strategy because it would be impossible to select for these QTLs phenotypically when they are combined with major resistance genes. Additionally, molecular markers will be very useful for pyramiding tightly-linked resistance genes within the same RGC cluster (Slabaugh et al. 2003). For partial resistances such as *Sclerotinia* and Phomopsis, a very important step towards the improvement of the levels of resistance is the use of MAS to combine different resistance QTL. Studies reporting the identification, validation, and fine-mapping of resistance QTL, as described in previous sections, will contribute to the use of MAS for resistance QTL pyramiding.

7.6 Molecular Breeding in Private Sector Breeding Programs

Until relatively recently, sunflower had lagged behind other economically important crops both in terms of the available molecular tools and its genomic resources. The first linkage maps were based on RFLP markers and these were generated by several private companies and/or consortia, each using a different LG nomenclature system (Berry et al. 1995; Gentzbittel et al. 1995; Jan et al. 1998). Fortunately, for both the sake of public research and ease of use, these early maps have been superseded by one based on SSR markers. The vast majority of these microsatellites are in the public domain thanks to the efforts of Prof. Steven Knapp (The University of Georgia, Athens, USA) and co-workers, but this may not be the situation for the next generation of SNP-based markers. However, in the short to medium term SSRs will remain the marker system of choice in commercial MAS programs.

In most breeding companies, sunflower is regarded as relatively unimportant when compared to crops such as maize, but unlike maize there are many important traits in sunflower that are controlled by single, major genes. Most breeding companies are able to deploy any published marker-trait linkages extremely rapidly in their own breeding programs. As discussed earlier in this chapter, many disease resistance genes have been mapped and since sunflower is very prone to fungal attack, the manipulation of these resistances via MAS is seen as essential in providing hybrid yield stability. For example, the marker-assisted introgression of *Verticillium dahliae* resistance allowed the yield testing of the iso-hybrids with and without the resistance QTL. The resistant iso-hybrid out-yielded the susceptible control by 54% for seed yield and 65% for oil yield under *Verticillium* attack (Alberto Leon pers. comm.). In some environments certain resistance genes, such as those controlling resistance to the parasitic weed *Orobanche cumana* in Spain and Turkey or resistance to sunflower rust (*Puccinia helianthi*) in Australia, are a prerequisite for commercialization. Unfortunately, pathogen populations can rapidly evolve to overcome new sources of single gene resistance and so gene pyramiding using markers is essential to provide some durability. We are beginning to understand the linkage arrangement of resistance gene clusters in the sunflower genome; however, they are often located in regions of high recombination (e.g., near the telomeres or in gaps in the genetic map), which makes the identification of tightly-linked flanking markers more problematic. A further difficulty is that with the current marker resolution it is impossible to select for recombinants within these resistance gene clusters.

Wild, diploid annual species have mainly been used as a source of novel resistance genes for cultivated sunflower, but often little is known about their genome organization and so gene transfers may be unstable due to differences in chromosome structure (i.e., translocations). A key consideration when screening accessions for novel resistance genes is whether the resistance is dominant, additive or recessive. Recessive or additive resistances will require the marker-assisted introgression of the gene or genes into both the male and female parent of a given hybrid, thus doubling the amount of marker work. That said it is now possible, with the use of flanking markers, to manipulate recessive genes very efficiently in comparison to running conventional progeny tests after each backcross. Marker-assisted backcrossing plays an important part in many commercial programs with markers being used for both the selection of plants carrying the gene(s) of interest and for recurrent parent background. Hospital et al. (1992) demonstrated that it is not possible to optimize selection for the gene of interest and the recurrent parent background simultaneously. They suggested that the process should be subdivided into: a) the reduction in length of the donor segment around the gene and b) the recovery of the recurrent parent genotype. Depending on population sizes, a single gene can be introgressed in three backcrosses and a self with, for example, the emphasis in the first generation on a favorable crossover on one side of the target gene, then in the next cycle on the other side and then the final generation and self to "clean up" the background and to fix the introgressed region.

Marker-assisted backcross programs are dependent on the characterization of proprietary germplasm with a set of high quality SSR markers that give a good coverage of the sunflower genome. Each year, new lines and hybrids are genotyped to identify DNA polymorphisms and to assess the genetic diversity of the material available to the breeders for crossing. As discussed earlier, it is easy to distinguish elite male and female lines using DNA markers, but hybrid sunflower breeding only dates backs to the late 1960s and so, as yet, no heterotic groups exist within these two major groups of elite commercial oilseed germplasm. However, some breeders use the genetic distance between lines as a basis for making diverse B and R line breeding populations. They will also make some crosses between B and R lines and in these cases markers for the branching and CMS restorer genes are useful tools for screening the progenies.

Routine marker application in segregating breeding populations requires: the accurate sampling of leaf discs from the field for DNA extraction in the laboratory and high throughput, cheap diagnostic tests for the genes of agronomic importance. Knowledge of the major gene/QTL complement of the parental lines enables the identification of populations that will segregate for specific loci and the breeder then prioritizes which crosses are to be screened with markers. Marker deployment in the F_2 or F_3 generation depends on the economic importance of the trait, the robustness of the phenotypic screen, the codominance/dominance of the linked markers and whether there is an opportunity within crosses to pyramid genes/QTL for different traits. Obviously the more traits that segregate in a given cross, the bigger the population size has to be. Rapid methodologies have been developed for extracting DNA from thousands of plants, but PCR on a locus by locus basis, fragment separation and data capture are still relatively time-consuming, depending on the lab set up. The advantages of SNPs are that it is possible to assay from 10's to 1,000's of loci in a single reaction, there is no need for any electrophoresis and data capture is almost completely automated. SNPs are also far more abundant than SSRs and so there is also a potential for running genome-wide association or linkage disequilibrium (LD) studies rather than conventional QTL detection.

Classical QTL analyses rely on the development of experimental biparental mapping populations that generally have little or no commercial value. The population is then phenotyped, mapped with genetic markers and the QTL identified over several environments, but always in the same genetic background. Some common, major QTLs have been identified in different sunflower populations, as described in previous sections, but in general it is not known whether a QTL will function when introgressed into different elite lines. This whole process is rather slow and cumbersome and so a lot of effort is now being invested within private industry on LD mapping in order to identify QTL that function in a wider range of genetic backgrounds. The success of this strategy is dependent on the extent of LD within the genome and the current estimate for cultivated sunflower is, on average, less than six kilobases (Kolkman et al. 2007), which equates to a fraction of centiMorgan. However, due to selection pressure, LD may extend further around loci of economic interest such as the recessive branching gene on LG 10 in elite male inbred lines (Fig. 7-1). Molecular technology is evolving extremely rapidly and very high density maps have or will become available for most economically important crop species, and eventually even the complete

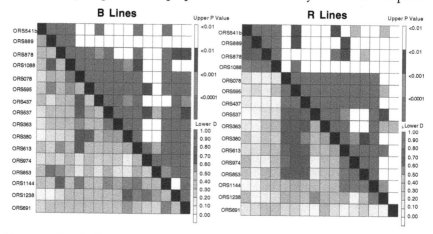

Figure 7-1 Two TASSEL (Bradbury et al. 2007) outputs, which were produced by analyzing the male and female lines separately from a total panel of 192 elite inbreds, genotyped with 16 public SSR markers on LG 10. Several regions of high LD, between tightly-linked SSR markers (as measured by the D'statistic—*shown below the black diagonal line*), can clearly be seen in the R lines (*shown in the right-hand panel*), but these regions are completely absent from the B lines (*shown in the left-hand panel*). The recessive branching gene (*b1*) is located in between ORS1088 and ORS437 and selection for this trait in elite R lines has resulted in a reduction of the allelic diversity at linked SSR loci (Combes and Berry, unpub data).

genomic DNA sequence, as in the case of rice, but in the future more emphasis will have to be placed on accurate phenotyping in order to identify robust marker-trait associations. In fact, Monsanto recently reported the combined use of both phenotypic and genotypic data in a marker-assisted recurrent selection (MARS) program, which significantly enhanced the genetic gain for quantitative traits such as grain yield, grain moisture and kernel oil in one of their European sunflower breeding programs (Eathington et al. 2007).

7.7 Limitations and Prospects of MAS

In the last decade, there have been major advances in the development of molecular tools and genomic resources in sunflower. Marker-trait linkages have been reported for a diverse array of traits, a critical mass of reports of validation in different genetic backgrounds and environments has emerged, and DNA marker analyses are being used to dissect complex traits and effects, such as epistasis and genotype x environment interactions. However, the application of this knowledge to practical sunflower breeding programs remains limited. There is a general lack of reports on MAS, and only a very small proportion of marker studies are being used in MAS programs for quantitative traits. This situation may change in the near future. A shift from linked markers to diagnostic gene-based assays, the development of high-throughput and cost effective assays and polymorphism detection methods, the use of robust methods to accurately define QTL positions and effects and the advent of novel approaches to create and identify genetic variation, such as TILLING or association mapping, are some issues that are currently being addressed in sunflower. This work is contributing to build implementation systems to understand, predict and manipulate genes, their interaction, and gene networks, and to integrate them into the breeding pipeline. For this reason, the future of molecular breeding in sunflower looks promising, and there is no doubt that MAS will become fully integrated into classical sunflower breeding programs over the next decade.

References

Abou Alfadil T, Poormohammad Kiani S, Dechamp-Guillaume G, Gentzbittel L, Sarrafi A (2007) QTL mapping of partial resistance to Phoma basal stem and root necrosis in sunflower (*Helianthus annuus* L.). Plant Sci 172: 815–823.

Abratti G, Bazzalo ME, León A (2008) Mapping a novel fertility restoration gene in sunflower. In: Proc 17th Int Sunflower Conf, June 8–12, 2008, Córdoba, Spain, pp 617–621.

Alignan M, Hewezi T, Petitprez M, Dechamp-Guillaume G, Gentzbittel L (2006) A cDNA microarray approach to decipher sunflower (*Helianthus annuus*) responses to the necrotrophic fungus *Phoma macdonaldii*. New Phytol 170: 523–536.

Al-Chaarani GR, Roustaee A, Gentzbittel L, Mokrani L, Barrault G, Dechamp-Guillaume G, Sarrafi A (2002) A QTL analysis of sunflower partial resistance to downy mildew

(*Plasmopara halstedii*) and black stem (*Phoma macdonaldii*) by the use of recombinant inbred lines (RILs). Theor Appl Genet 104: 490–496.

Bauer E, Renz B, Schön C-C, Hahn V (2006) Establishing a TILLING resource for sunflower. In: Proc 7th Eur Conf Sunflower Biotechnol, 3–6 Sept 2006, Gengenbach, Germany, p 23.

Berry ST, Allen RJ, Barnes SR, Caligari PDS (1994) Molecular-marker analysis of *Helianthus annuus* L. 1. Restriction fragment length polymorphism between inbred lines of cultivated sunflower. Theor Appl Genet 89: 435–441.

Berry ST, Leon AJ, Hanfrey CC, Challis P, Burkholz A, Barnes SR, Rufener GK, Lee M, Caligari PDS (1995) Molecular-marker analysis of *Helianthus annuus* L. 2. Construction of an RFLP linkage map for cultivated sunflower. Theor Appl Genet 91: 195–199.

Bert PF, Tourvieille De Labrouhe D, Philippon J, Mouzeyar S, Jouan I, Nicolas P, Vear F (2001) Identification of a second linkage group carrying genes controlling resistance to downy mildew (*Plasmopara halstedii*) in sunflower (*Helianthus annuus* L.). Theor Appl Genet 103: 992–997.

Bert PF, Jouan I, Tourvieille de Labrouhe D, Serre F, Nicolas P, Vear F (2002) Comparative genetic analysis of quantitative traits in sunflower (*Helianthus annuus* L.). 1. Characterisation of QTL involved in resistance to *Sclerotinia sclerotiorum* and *Diaporthe helianthi*. Theor Appl Genet 105: 985–993.

Bert PF, Jouan I, Tourvieille de Labrouhe D, Serre F, Nicolas P, Vear F (2003) Comparative genetic analysis of quantitative traits in sunflower (*Helianthus annuus* L.). 2. Characterisation of QTL involved in developmental and agronomic traits. Theor Appl Genet 107: 181–189.

Bert PF, Dechamp-Guillaume G, Serre F, Jouan I, De Labrouhe DT, Nicolas P, Vear F (2004) Comparative genetic analysis of quantitative traits in sunflower (*Helianthus annuus* L.). 3. Characterisation of QTL involved in resistance to *Sclerotinia sclerotiorum* and *Phoma macdonaldii*. Theor Appl Genet 109: 865–874.

Besnard G, Griveau Y, Quillet MC, Serieys H, Lambert P, Vares D, Bervillé A (1997) Specifying the introgressed regions from *H. argophyllus* in cultivated sunflower (*Helianthus annuus* L.) to mark Phomopsis resistance genes. Theor Appl Genet 94: 131–138.

Binsfeld PC, Wingender R, Schnabl H (2001) Cytogenetic analysis of interspecific sunflower hybrids and molecular evaluation of their progeny. Theor Appl Genet 102: 1280–1285.

Blanc G, Charcosset A, Mangin B, Gallais A, Moreau L (2006) Connected populations for detecting quantitative trait loci and testing for epistasis: an application in maize. Theor Appl Genet 113: 206–224.

Bouzidi MF, Badaoui S, Cambon F, Vear F, Tourvieille de Labrouhe D, Nicolas P, Mouzeyar S (2002) Molecular analysis of a major locus for resistance to downy mildew in sunflower with specific PCR-based markers. Theor Appl Genet 104: 592–600.

Bradbury PJ, Zhang Z, Kroon DE, Casstevens TM, Ramdoss Y, Buckler ES (2007) TASSEL: Software for association mapping of complex traits in diverse samples. Bioinformatics 23: 2633–2635.

Brahm L, Röcher T, Friedt W (2000) PCR-based markers facilitating marker assisted selection in sunflower for resistance to downy mildew. Crop Sci 40: 676–682.

Buckler ES, Thornsberry JM (2002) Plant molecular diversity and applications to genomics. Curr Opin Plant Biol 5: 107–111.

Capatana A, Feng J, Vick BA, Duca M, Jan CC (2008) Molecular mapping of a new induced gene for nuclear male sterility in sunflower (*Helianthus annuus* L.). In: Proc 17th Int Sunflower Conf, June 8–12, 2008, Córdoba, Spain, pp 641–644.

Carrera A, Pizarro G, Poverene M, Feingold SE, León A, Berry ST (2002) The genetic mapping of five isozyme loci on the public sunflower RFLP linkage map. Genet Mol Biol 25: 65–72.

Chen J, Hu J, Vick BA, Jan CC (2006) Molecular mapping of a nuclear male-sterility gene in sunflower (*Helianthus annuus* L.) using TRAP and SSR markers. Theor Appl Genet 113: 122–127.

Cheres MT, Miller JF, Crane JM, Knapp SJ (2000) Genetic distance as a predictor of heterosis and hybrid performance within and between heterotic groups in sunflower. Theor Appl Genet 100: 889–894.

Comai L, Young K, Till BJ, Reynolds SH, Greene EA, Codomo CA, Enns LC, Johnson JE, Burtner C, Odden AR, Henikoff S (2004) Efficient discovery of nucleotide polymorphisms in populations by ecotilling. Plant J 37: 778–786.

Cronn R, Brothers ME, Klier K, Wendel JF, Bretting PK (1997) Allozyme variation in domesticated *Helianthus annuus* and wild relatives. Theor Appl Genet 95: 532–545.

Darvishzadeh R, Poormohammad Kiani S, Dechamp-Guillaume G, Gentzbittel L, Sarrafi A (2007) Quantitative trait loci associated with isolate specific and isolate nonspecific partial resistance to *Phoma macdonaldii* in sunflower. Plant Pathol 56: 855–861.

Dong GJ, Liu GS, Li KF (2007) Studying genetic diversity in the core germplasm of confectionary sunflower (*Helianthus annuus* L.) in China based on AFLP and morphological analysis. Russ J Genet 43: 627–635.

Draeger D, Tang S, Leon A, Hahn V, Knapp SJ (2006) The discovery of quantitative trait loci for seed oil concentration linked in repulsion to the branching locus in an unbranched sunflower hybrid testcross population. In: Proc 7th Eur Conf Sunflower Biotechnol, 3–6 Sept 2006, Gengenbach, Germany, p 51.

Dussle CM, Hahn V, Knapp SJ, Bauer E (2004) $Pl_{(Arg)}$ from *Helianthus argophyllus* is unlinked to other known downy mildew resistance genes in sunflower. Theor Appl Genet 109: 1083–1086.

Dwivedi SL, Crouch JH, Mackill DJ, Xu Y, Blair MW, Argot M, Upadhyaya HD, Ortiz R (2007) The molecularization of public sector crop breeding: Progress, problems and prospects. Adv Agron 95: 163–318.

Eathington SR, Crosbie TM, Edwards MD, Reiter RS, Bull JK (2007) Molecular markers in a commercial breeding program. Crop Sci 47: S154–S163.

Feng J, Vick BA, Lee M, Zhang H, Jan CC (2006) Construction of BAC and BIBAC libraries from sunflower and identification of linkage group-specific clones by overgo hybridization. Theor Appl Genet 113: 23–32.

Feng J, Seiler GJ, Gulya Jr TJ, Jan CC (2007) Advancement of pyramiding new *Sclerotinia* stem rot resistance genes from *H. californicus* and *H. schweinitzii* into cultivated sunflower. In: Proc 29th Sunflower Res Workshop, 10–11 Jan, 2007, Fargo, ND, USA: *http://www.sunflowernsa.com/research/research-workshop/documents/ Feng_etal_Pyramid_2007.pdf*

Fernandez P, Di Rienzo J, Fernandez L, Hopp HE, Paniego N, Heinz RA (2008) Transcriptomic identification of candidate genes involved in sunflower responses to chilling and salt stresses based on cDNA microarray analysis. BMC Plant Biol 8: 11.

Fick GN, Miller JF (1997) Sunflower breeding. In: AA Schneiter (ed) Sunflower Technology and Production. Agronomy Series. ASA, CSSA and SSSA, Madison, WI, USA, 35: 395–439.

Flint-Garcia SA, Thornsberry JM, Buckler ES (2003) Structure of linkage disequilibrium in plants. Annu Rev Plant Biol 54: 357–374.

Flores-Berrios E, Gentzbittel L, Kayyal H, Alibert G, Sarrafi A (2000) AFLP mapping of QTLs for in vitro organogenesis traits using recombinant inbred lines in sunflower (*Helianthus annuus* L.). Theor Appl Genet 101: 1299–1306.

Frisch M (2005) Breeding strategies: Optimum design of marker-assisted backcross programs. In: H Lörz, G Wenzel (eds) Molecular Marker Systems in Plant Breeding and Crop Improvement. Springer, Berlin and Heidelberg, Germany, pp 319–334.

Frisch M, Bohn M, Melchinger AE (1999) Comparison of selection strategies for marker-assisted backcrossing of a gene. Crop Sci 39: 1295–1301.

Fusari C, Lía V, Hopp HE. Reinz RA, Paniego N (2008) Identification of single nucleotide polymorphisms and analysis of linkage disequilibrium in sunflower elite inbred lines using the candidate gene approach. BMC Plant Biol 8: 7.

Gandhi SD, Heesacker AF, Freeman CA, Argyris J, Bradford K, Knapp SJ (2005) The self-incompatibility locus (*S*) and quantitative trait loci for self-pollination and seed dormancy in sunflower. Theor Appl Genet 111: 619–629.

Gedil MA, Slabaugh MB, Berry ST, Johnson R, Michelmore R, Miller J, Gulya T, Knapp SJ (2001) Candidate disease resistance genes in sunflower cloned using conserved nucleotide-binding site motifs: Genetic mapping and linkage to the downy mildew resistance gene *Pl1*. Genome 44: 205–212.

Gentzbittel L, Zhang YX, Vear F, Griveau B, Nicolas P (1994) RFLP studies of genetic relationships among inbred lines of the cultivated sunflower, *Helianthus annuus* L.: Evidence for distinct restorer and maintainer germplasm pools. Theor Appl Genet 89: 419–425.

Gentzbittel L, Vear F, Zhang YX, Berville A (1995) Development of a consensus linkage RFLP map of cultivated sunflower (*Helianthus annuus* L.). Theor Appl Genet 90: 1079–1086.

Gentzbittel L, Mouzeyar S, Badaoui S, Mestries E, Vear F, Tourvieille D, Nicolas P (1998) Cloning of molecular markers for disease resistance in sunflower, *Helianthus annuus* L. Theor Appl Genet 96: 519–525.

Gentzbittel L, Abbott A, Galaud JP, Georgi L, Fabre F, Liboz T, Alibert G (2002) A bacterial artificial chromosome (BAC) library for sunflower, and identification of clones containing genes for putative transmembrane receptors. Mol Genet Genom 266: 979–987.

Gizlice Z, Carter TE Jr, Burton JW (1993) Genetic diversity in North American soybean: I. Multivariate analysis of founding stock and relation to coefficient of parentage. Crop Sci 33: 614–620.

Hajduch M, Casteel JE, Tang S, Hearne LB, Knapp SJ, Thelen JJ (2007) Proteomic analysis of near isogenic sunflower varieties differing in seed oil traits. J Proteom Res 6: 3232–3241.

Hamrit S, Schnabel U, Warber D, Kusterer B, Lazarescu E, Özdemir N, Friedt W, Horn R (2006a) Identification and DNA fingerprinting of positive BAC-clones at the restorer locus *Rf1* in sunflower. In: Proc 7th Eur Conf Sunflower Biotechnol, 3–6 Sept 2006, Gengenbach, Germany, p 34.

Hamrit S, Engelmann U, Schnabel U, Warber D, Kurutz S, Kusterer B, Lazarescu E, Özdemir N, Friedt W, Abratti G, Leon A, Horn R (2006b) Comparative mapping of restorer genes restoring pollen fertility in the presence of different CMS cytoplasms in the genus *Helianthus*. In: Proc 7th Eur Conf Sunflower Biotechnol, 3–6 Sept 2006, Gengenbach, Germany, p 10.

Hass CG, Tang S, Leonard S, Traber MG, Miller JF, Knapp SJ (2006) Three non-allelic epistatically interacting methyltransferase mutations produce novel tocopherol (vitamin E) profiles in sunflower. Theor Appl Genet 113: 767–782.

Heesacker A, Kishore VK, Gao W, Tang S, Kolkman JM, Gingle A, Matvienko M, Kozik A, Michelmore RM, Lai Z, Rieseberg LH, Knapp SJ (2008) SSRs and INDELs mined from the sunflower EST database: abundance, polymorphisms, and crosstaxa utility. Theor Appl Genet 117: 1021–1029.

Hervé D, Fabre F, Flores-Berrios E, Leroux N, Al-Chaarani G, Planchon C, Sarrafi A, Gentzbittel L (2001) QTL analysis of photosynthesis and water status traits in sunflower (*Helianthus annuus* L.) under greenhouse conditions. J Exp Bot 362: 1857–1864.

Hewezi T, Petitprez M, Gentzbittel L (2006a) Primary metabolic pathways and signal transduction in sunflower (*Helianthus annuus* L.): Comparison of transcriptional profiling in leaves and immature embryos using cDNA microarrays. Planta 223: 948–964.

Hewezi T, Léger M, El Kayal W, Gentzbittel L (2006b) Transcriptional profiling of sunflower plants growing under low temperatures reveals an extensive down-regulation of gene expression associated with chilling sensitivity. J Exp Bot 57: 3109–3122.

Holland JB (2001) Epistasis and plant breeding. Plant Breed Rev 21: 27–92.

Hongtrakul V, Huestis GM, Knapp SJ (1997) Amplified fragment length polymorphism as a tool for DNA fingerprinting sunflower germplasm: Genetic diversity among oilseed inbred lines. Theor Appl Genet 95: 400–407.

Horn R (2002) Molecular diversity of male sterility inducing and male-fertile cytoplasms in the genus *Helianthus*. Theor Appl Genet 104: 562–570.

Horn R, Köhler RH, Zetsche K (1991) A mitochondrial 16 kDa protein is associated with cytoplasmic male sterility in sunflower. Plant Mol Biol 17: 29–36.

Horn R, Kusterer B, Lazarescu E, Prüfe M, Friedt W (2003) Molecular mapping of the *Rf1* gene restoring pollen fertility in PET1-based F$_1$ hybrids in sunflower (*Helianthus annuus* L.). Theor Appl Genet 106: 599–606.

Hospital F (2003) Marker-assisted breeding. In: HJ Newbury (ed) Plant Molecular Breeding. Blackwell Scientific Publ, London, UK, pp 30–56.

Hospital F, Chevalet C, Mulsant P (1992) Using markers in gene introgression breeding programs. Genetics 132: 1199–1210.

Hu J (2006) Defining the sunflower (*Helianthus annuus* L.) linkage group ends with the Arabidopsis-type telomere sequence repeat-derived markers. Chrom Res 14: 535–548.

Hu J, Chen J, Bervillé A, Vick BA (2004) High potential of TRAP markers in sunflower genome mapping. In: Proc 16th Int Sunflower Conf, 29 Aug–2 Sept 2004, Fargo, ND, USA, pp 665–671.

Huang X, Nabipoura A, Gentzbittel L, Sarrafi A (2007) Somatic embryogenesis from thin epidermal layers in sunflower and chromosomal regions controlling the response. Plant Sci 173: 247–252.

Jan CC, Seiler G (2007) Sunflower. In: RJ Singh (ed) Oilseed Crops. Genetic Resources, Chromosome Engineering, and Crop Improvement. CRC Press, Boca Raton, FL, USA, pp 103–165.

Jan CC, Vick BA, Miller JF, Kahler AL, Butler ET (1998) Construction of an RFLP linkage map for cultivated sunflower. Theor Appl Genet 96: 15–22.

Jansen RC, Nap JP (2001) Genetical genomics: The added value from segregation. Trends Genet 17: 388–391.

Kane NC, Rieseberg LH (2007) Selective sweeps reveal candidate genes for adaptation to drought and salt tolerance in common sunflower, *Helianthus annuus*. Genetics 175: 1803–1812.

Knapp SJ, Berry ST, Rieseberg LH (2001) Genetic mapping in sunflowers. In: RL Phillips, IK Vasil (eds) DNA Markers in Plants. Kluwer, Dordrecht, The Netherlands, pp 379–403.

Knapp SJ, Heesacker A, Tang S, Gandhi S, Kishore VK, Kolkman JM, Freeman-Barrios C, Bushman BS, Schuppert GF, Leon A, Berry ST, Lai Z, Rieseberg LH (2007) Genetic mapping of 2,495 high-throughput DNA marker loci in sunflower. In: Plant, Animal & Microbe Genome XV Conf, San Diego, CA, USA: *http://www.intl-pag.org/15/abstracts/PAG15_P05r_649.html*

Köhler RH, Horn R, Lössl A, Zetsche K (1991) Cytoplasmic male sterility in sunflower is correlated with the co-transcription of a new open reading frame with the *atpA* gene. Mol Gen Genet 227: 369–376.

Kolkman JM, Slabaugh MB, Bruniard JM, Berry ST, Shaun Bushman B, Olungu C, Maes N, Abratti G, Zambelli A, Miller JF, Leon A, Knapp SJ (2004) Acetohydroxyacid synthase mutations conferring resistance to imidazolinone or sulfonylurea herbicides in sunflower. Theor Appl Genet 109: 1147–1159.

Kolkman JM, Berry ST, Leon AJ, Slabaugh MB, Tang S, Gao W, Shintani DK, Burke JM, Knapp SJ (2007) Single nucleotide polymorphisms and linkage disequilibrium in sunflower. Genetics 177: 457–468.

Kusterer B, Friedt W, Lazarescu E, Prüfe M, Özdemir N, Tzigos S, Horn R (2004) Map-based cloning strategy for isolating the restorer gene *Rf1* of the PET1 cytoplasm in sunflower (*Helianthus annuus* L.). Helia 27: 1–14.

Kusterer B, Horn R, Friedt W (2005) Molecular mapping of the fertility restoration locus *Rf1* in sunflower and development of diagnostic markers for the restorer gene. Euphytica 143: 35–42.

Lai Z, Livingstone K, Zou Y, Church SA, Knapp SJ, Andrews J, Rieseberg LH (2005) Identification and mapping of SNPs from ESTs in sunflower. Theor Appl Genet 111: 1532–1544.

Lai Z, Gross B, Zou Y, Andrews J, Rieseberg LH (2006) Microarray analysis reveals differential gene expression in hybrid sunflower species. Mol Ecol 15: 1213–1227.

Laver HK, Reynolds SJ, Moneger F, Leaver CJ (1991) Mitochondrial genome organization and expression associated with cytoplasmic male sterility in sunflower (*Helianthus annuus*). Plant J 1: 185–193.

Lawson WR, Goulter KC, Henry RJ, Kong GA, Kochman JK (1998) Marker-assisted selection for two rust resistance genes in sunflower. Mol Breed 4: 227–234.

Lebowitz RJ, Soller M, Beckmann JS (1987) Trait-based analyses for the detection of linkage between marker loci and quantitative trait loci in crosses between inbred lines. Theor Appl Genet 73: 556–562.

Leon AJ, Lee M, Rufener GK, Berry ST, Mowers RP (1995a) Use of RFLP markers for genetic linkage analysis of oil percentage in sunflower seed. Crop Sci 35: 558–564.

Leon AJ, Berry ST, Rufener GK, Mowers RP (1995b) Oil producing sunflowers and production thereof. US Patent US005476524.

Leon AJ, Lee M, Rufener GK, Berry ST, Mowers RP (1996) Genetic mapping of a locus (*Hyp*) affecting seed hypodermis color in sunflower. Crop Sci 36: 1666–1668.

Leon AJ, Lee M, Andrade FH (2001) Quantitative trait loci for growing days to flowering and photoperiod response in sunflower (*Helianthus annuus* L.). Theor Appl Genet 102: 497–503.

Leon AJ, Andrade FH, Lee M (2003) Genetic analysis of seed oil concentrations across generations and environments in sunflower (*Helianthus annuus* L.). Crop Sci 40: 404–407.

Lexer C, Lai Z, Rieseberg LH (2003a) Candidate gene polymorphisms associated with salt tolerance in wild sunflower hybrids: Implications for the origin of *Helianthus paradoxus*, a diploid hybrid species. New Phytol 161: 225–233.

Lexer C, Welch ME, Durphy JL, Rieseberg LH (2003b) Natural selection for salt tolerance quantitative trait loci (QTLs) in wild sunflower hybrids: implications for the origin of *Helianthus paradoxus*, a diploid hybrid species. Mol Ecol 12: 1225–1235.

Li C, Feng J, Ma F, Vick BA, Jan CC (2008) Identification of molecular markers linked to a new nuclear male-sterility gene *ms7* in sunflower (*Helianthus annuus* L.). In: Proc 17th Int Sunflower Conf, June 8–12 2008, Córdoba, Spain, pp 651–654.

Liu A, Burke JM (2006) Patterns of nucleotide diversity in wild and cultivated sunflower. Genetics 173: 321–330.

Liu XC, Wu JL (1998) SSR heterotic patterns of parents for making and predicting heterosis in rice breeding. Mol Breed 4: 263–268.

Lu YH, Gagne G, Grezes-Besset B, Blanchard P (1999) Integration of a molecular linkage group containing the broomrape resistance gene *Or5* into an RFLP map in sunflower. Genome 42: 453–456.

Lu YH, Melero-Vara JM, García-Tejada JA, Blanchard P (2000) Development of SCAR markers linked to the gene *Or5* conferring resistance to broomrape (*Orobanche cumana* Wallr.) in sunflower. Theor Appl Genet 100: 625–632.

Malmberg RL, Held S, Waits A, Mauricio R (2005) Epistasis for fitness related quantitative traits in *Arabidopsis thaliana* grown in the field and in the greenhouse. Genetics 171: 2013–2027.

McNally KL, Wang H, Naredo EB, Raghavan C, Atienza G, Leung H (2006) EcoTILLING in rice. In: Plant, Anim & Microb Genom XIV Conf, San Diego, CA, USA: *http://www.intl-pag.org/pag/14/abstracts/PAG14_W258.html*

Melchinger AE, Lee M, Lamkey KR, Hallauer AR, Woodman WL (1990) Genetic diversity for restriction fragment length polymorphisms and heterosis for two diallel sets of maize inbreds. Theor Appl Genet 80: 488–496.

Mestries E, Gentzbittel L, Tourvieille de Labrouhe D, Nicolas P, Vear F (1998) Analyses of quantitative trait loci associated with resistance to *Sclerotinia sclerotiorum* in sunflower (*Helianthus annuus* L.) using molecular markers. Mol Breed 4: 215–226.

Micic Z, Hahn V, Bauer E, Schön CC, Knapp SJ, Tang S, Melchinger AE (2004) QTL mapping of *Sclerotinia* midstalk rot resistance in sunflower. Theor Appl Genet 109: 1474–1484.

Micic Z, Hahn V, Bauer E, Schön CC, Melchinger AE (2005a) QTL mapping of resistance to *Sclerotinia* midstalk rot in RIL of sunflower population NDBLOSsel × CM625. Theor Appl Genet 110: 1490–1498.

Micic Z, Hahn V, Bauer E, Melchinger AE, Knapp SJ, Tang S, Schön CC (2005b) Identification and validation of QTL for *Sclerotinia* midstalk rot resistance in sunflower by selective genotyping. Theor Appl Genet 111: 233–242.

Mihaljevic R, Utz HF, Melchinger AE (2005) No evidence for epistasis in hybrid and per se performance of elite European flint maize inbreds from generation means and QTL analyses. Crop Sci 45: 2605–2613.

Mokrani L, Gentzbittel L, Azanza F, Fitamant L, Al-Chaarani G, Sarrafi A (2002) Mapping and analysis of quantitative trait loci for grain oil content and agronomic traits using AFLP and SSR in sunflower (*Helianthus annuus* L.). Theor Appl Genet 106: 149–156.

Monéger F, Smart CJ, Leaver CJ (1994) Nuclear restoration of cytoplasmic male sterility in sunflower is associated with the tissue-specific regulation of a novel mitochondrial gene. EMBO J 13: 8–17.

Montemurro C, Sabetta W, Alba V, Sunseri F, Alba E, Simeone R, Blanco A (2007) Development of a tilling sunflower population. In: Proc 51st Italian Soc Agri Genet Annu Congr, 23–26 Sept 2007, Riva del Garda, Italy, Poster Abstr A 58.

Morris M, Dreher K, Ribaut JM, Khairallah M (2003) Money matters (II): Costs of maize inbred line conversion schemes at CIMMYT using conventional and marker-assisted selection. Mol Breed 11: 235–247.

Natali L, Giordani T, Polizzi E, Pugliesi C, Fambrini M, Cavallini A (1998) Genomic alterations in the interspecific hybrid *Helianthus annuus*×*Helianthus tuberosus*. Theor Appl Genet 97: 1240–1247.

Özdemir N, Horn R, Friedt W (2004) Construction and characterization of a BAC library for sunflower (*Helianthus annuus* L.). Euphytica 138: 177–183.

Paniego N, Echaide M, Muñoz M, Fernández L, Torales S, Faccio P, Fuxan I, Carrera M, Zandomeni R, Suárez EY, Hopp HE (2002) Microsatellite isolation and characterization in sunflower (*Helianthus annuus* L.). Genome 45: 34–43.

Paniego N, Heinz R, Fernández P, Talia P, Nishinakamasu V, Hopp HE (2007) Sunflower. In: C Kole (ed) Genome Mapping and Molecular Breeding in Plants. vol 2: Oilseeds. Springer, Berlin, Heidelberg, New York, pp 153–177.

Pankovi D, Radovanovi N, Joci S, Škori D (2007) Development of co-dominant amplified polymorphic sequence markers for resistance of sunflower to downy mildew race 730. Plant Breed 126: 440–444.

Pashley CH, Ellis JR, McCauley DE, Burke JM (2006) EST databases as a source for molecular markers: lessons from *Helianthus*. J Hered 97: 381–388.

Pérez-Vich B, Fernández-Martínez JM, Grondona M, Knapp SJ, Berry ST (2002) Stearoyl-ACP and oleoyl-PC desaturase genes cosegregate with quantitative trait loci underlying stearic and oleic acid mutant phenotypes in sunflower. Theor Appl Genet 104: 338–349.

Pérez-Vich B, Akhtouch B, Knapp SJ, Leon AJ, Velasco L, Fernández-Martínez JM, Berry ST (2004) Quantitative trait loci for broomrape (*Orobanche cumana* Wallr.) resistance in sunflower. Theor Appl Genet 109: 92–102.

Pérez-Vich B, Berry ST, Velasco L, Fernández-Martínez JM, Gandhi S, Freeman C, Heesacker A, Knapp SJ, Leon AJ (2005) Molecular mapping of nuclear male sterility genes in sunflower. Crop Sci 45: 1851–1857.

Pérez-Vich B, Velasco L, Grondona M, Leon AJ, Fernández-Martínez JM (2006) Molecular analysis of the high stearic acid content in sunflower mutant CAS-14. Theor Appl Genet 112: 867–875.

Pizarro G, Carrera AD, Poverene M (2000) Comparative analysis of genetic relationships in sunflower inbred lines based on isozymic, RAPD, and pedigree data. In: Proc 15th Int Sunflower Conf, 12–15 June, Toulouse, France, vol 2, pp E111–E116.

Pizarro G, Gandhi S, Freeman-Barrios C, Knapp SJ (2006) Quantitative trait loci for seed oil concentration and other seed traits are linked to the self-incompatibility locus in sunflower. In: Proc 7th Eur Conf Sunflower Biotechnol, 3–6 Sept 2006, Gengenbach, Germany, p 50.

Poormohammad Kiani S, Grieu P, Maury P, Hewezi T, Gentzbittel L, Sarrafi A (2007) Genetic variability for physiological traits under drought conditions and differential expression of water stress-associated genes in sunflower (*Helianthus annuus* L.). Theor Appl Genet 114: 193–207.

Popov VN, Yu Urbanovich O, Kirichenko VV (2002) Studying genetic diversity in inbred sunflower lines by RAPD and isozyme analyses. Russ J Genet 38: 785–790.

Quillet MC, Vear F, Branlard G (1992) The use of isozyme polymorphism for identification of sunflower (*Helianthus annuus*) inbred lines. J Genet Breed 46: 795–804.

Radwan O, Bouzidi MF, Vear F, Philippon J, Tourvieille de Labrouhe D, Nicolas P, Mouzeyar S (2003) Identification of non-TIR-NBS-LRR markers linked to *PL5/PL8* locus for resistance to downy mildew in sunflower. Theor Appl Genet 106: 1438–1446.

Radwan O, Bouzidi MF, Nicolas P, Mouzeyar S (2004) Development of PCR markers for the Pl5/Pl8 locus for resistance to *Plasmopara halstedii* in sunflower, *Helianthus annuus* L. from complete CC-NBS-LRR sequences. Theor Appl Genet 109: 176–185.

Radwan O, Gandhi S, Heesacker A, Whitaker B, Taylor C, Plocik A, Kesseli R, Kozik A, Michelmore RW, Knapp SJ (2008) Genetic diversity and genomic distribution of homologs encoding NBS-LRR disease resistance proteins in sunflower. Mol Genet Genom 280: 111–125.

Rieseberg L, Seiler G (1990) Molecular evidence and the origin and development of the domesticated sunflower (*Helianthus annuus*, Asteraceae). Econ Bot 44: 79–91.

Roche J, Hewezi T, Bouniols A, Gentzbittel L (2007) Transcriptional profiles of primary metabolism and signal transduction-related genes in response to water stress in field-grown sunflower genotypes using a thematic cDNA microarray. Planta 226: 601–617.

Rönicke S, Hahn V, Horn R, Gröne I, Brahm L, Schnabl H, Friedt W (2004) Interspecific hybrids of sunflower as a source of *Sclerotinia* resistance. Plant Breed 123: 152–157.

Rönicke S, Hahn V, Vogler A, Friedt W (2005) Quantitative trait loci analysis of resistance to *Sclerotinia sclerotiorum* in sunflower. Phytopathology 95: 834–839.

Sabetta W, Montemurro C, Alba E, Stein N, Gottwald S, Sunseri F, Blanco A, Alba V (2008) Early steps into sunflower tilling. In: Proc 52nd Italian Soc Agri Genet Annu Congr, 14–17 Sept 2008, Padova, Italy, Poster Abstr A 22.

Schuppert G, Tang S, Slabaugh M, Knapp SJ (2006) The sunflower high-oleic mutant oil carries variable tandem repeats of FAD2-1, a seed-specific oleoyl-phosphatidyl choline desaturase. Mol Breed 17: 241–256.

Serieys H, Tagmount A, Kaan F, Griveau Y, Tersac M, André T, Grave H, Pinochet X, Berville A (2000) Evaluation of an interspecific sunflower population issued from the perennial species *H. occidentalis* ssp. *plantagineus* for resistance to *Diaporthe helianthi* and *Sclerotinia sclerotiorum* in relation with phenotypic and molecular traits. In: Proc 15th Int Sunflower Conf, June 12–15, 2000, Toulouse, France, pp 25–30.

Slabaugh MB, Yu J-K., Tang S, Heesacker A, Hu X, Lu G, Bidney D, Han F, Knapp SJ (2003) Haplotyping and mapping a large cluster of downy mildew resistance gene candidates in sunflower using multilocus intron fragment length polymorphisms. Plant Biotechnol J 1: 167–185.

Slade AJ, Fuerstenberg SI, Loeffler D, Steine MN, Facciotti D (2005) A reverse genetic, nontransgenic approach to wheat crop improvement by TILLING. Nat Biotechnol 23: 75–81.

Tang S, Knapp SJ (2003) Microsatellites uncover extraordinary diversity in Native American land races and wild populations of cultivated sunflower. Theor Appl Genet 106: 990–1003.

Tang S, Knapp SJ (2008) SNP marker development for sunflower using Luminex System and Lightcycler480. In: Plant, Anim & Microb Genom XVI Conf, San Diego, CA, USA: *http://www.intl-pag.org/16/abstracts/PAG16_P03e_168.html*

Tang S, Yu JK, Slabaugh MB, Shintani DK, Knapp SJ (2002) Simple sequence repeat map of the sunflower genome. Theor Appl Genet 105: 1124–1136.

Tang S, Heesacher A, Kishore VK, Fernández A, Sadik ES, Cole G, Knapp SJ (2003) Genetic mapping of the *Or5* gene for resistance to *Orobanche* race E in sunflower. Crop Sci 43: 1021–1028.

Tang S, Leon AJ, Bridges WC, Knapp SJ (2006a) Quantitative trait loci for genetically correlated seed traits are tightly linked to branching and pericarp pigment loci in sunflower. Crop Sci 46: 721–734.

Tang S, Hass CG, Knapp SJ (2006b) *Ty3/gypsy*-like retrotransposon knockout of a 2-methyl-6-phytyl-1,4- benzoquinone methyltransferase is non-lethal, uncovers a cryptic paralogous mutation, and produces novel tocopherol (vitamin E) profiles in sunflower. Theor Appl Genet 113: 783–799.

Tang S, Saski C, Muñoz-Torres M, Atkins M, Tomkins J, Kuehl J, Boore J, Knapp SJ (2007) A large-insert bacterial artificial chromosome library for sunflower. In: Plant, Anim & Microb Genom XV Conf, San Diego, CA, USA: *http://www.intl-pag.org/15/abstracts/PAG15_P02a_49.html*

Tanksley SD, Nelson JC (1996). Advanced backcross QTL analysis: A method for the simultaneous discovery and transfer of valuable QTLs from unadapted germplasm into elite breeding lines. Theor Appl Genet 92: 191–203.

Tersac M, Blanchard P, Brunel D, Vincourt P (1994) Relationships between heterosis and enzymatic polymorphism in populations of cultivated sunflowers (*Helianthus annuus* L.). Theor Appl Genet 88: 49–55.

Till BJ, Reynolds SH, Greene EA, Codomo CA, Enns LC, Johnson JE, Burtner C, Odden AR, Young K, Taylor NE, Henikoff JG, Comai L, Henikoff S (2003) Large-scale discovery of induced point mutations with high throughput TILLING. Genom Res 13: 524–530.

Till BJ, Reynolds SH, Weil C, Springer N, Burtner C, Young K, Bowers E, Codomo CA, Enns LC, Odden AR, Greene EA, Comai L, Henikoff S (2004) Discovery of induced point mutations in maize by TILLING. BMC Plant Biol 4: 12.

Torres A (1983) Sunflower (*Helianthus annuus* L.). In: SD Tanksley, TJ Orton (eds) Isozymes in Plant Genetics and Breeding, part B. Elsevier Science Publ, Amsterdam, The Netherlands, pp 329–338.

Tourvieille de Labrouhe D, Walser P, Mestries E, Gillot L, Penaud A, Tardin MC, Pauchet I (2004) Sunflower downy mildew resistance gene pyramiding alternation and mixture: First results comparing the effects of different varietal structures on changes in the pathogen. In: Proc 16th Int Sunflower Conf, 29 Aug–2 Sept 2004, Fargo, ND, USA, pp 111–116.

Vear F (2004) Breeding for durable resistance to the main diseases of sunflower. In: Proc 16th Int Sunflower Conf, 29 Aug–2 Sept 2004, Fargo, ND, USA, pp 15–28.

Vear F, Gentzbittel L, Philippon J, Mouzeyar S, Mestries E, Roeckel-Drevet P, Tourvieille de Labrouhe D, Nicolas P (1997) The genetics of resistance to five races of downy mildew (*Plasmopara halstedii*) in sunflower (*Helianthus annuus* L.). Theor Appl Genet 95: 584–589.

Vear F, Jouan-Dufournel I, Bert PF, Serre F, Cambon F, Pont C, Walser P, Roche S, Tourvieille de Labrouhe D, Vincourt P (2008) QTL for capitulum resistance to *Sclerotinia sclerotiorum* in sunflower. In: Proc 17th Int Sunflower Conf, June 8–12, 2008, Córdoba, Spain, pp 605–610.

Wayne ML, McIntyre LM (2002) Combining mapping and arraying: An approach to candidate gene identification. Proc Natl Acad Sci USA 99: 14903–14906.

Wieckhorst S, Hahn V, Dußle CM, Knapp SJ, Schön CC, Bauer E (2008) Fine mapping of the downy mildew resistance locus Pl_{ARG} in sunflower. In: Proc 17th Int Sunflower Conf, June 8–12, 2008, Córdoba, Spain, pp 645–649.

Yu JK, Mangor J, Thompson L, Edwards KJ, Slabaugh MB, Knapp SJ (2002) Allelic diversity of simple sequence repeat markers among elite inbred lines in cultivated sunflower. Genome 45: 652–660.

Yu JK, Tang S, Slabaugh MB, Heesacker A, Cole G, Herring M, Soper J, Han F, Chu WC, Webb DM, Thompson L, Edwards KJ, Berry ST, Leon A, Olungu C, Maes N, Knapp SJ (2003) Towards a saturated molecular genetic linkage map for cultivated sunflower. Crop Sci 43: 367–387.

Yue B, Miller JF, Hu J (2007) Experimenting with marker-assisted selection in confection sunflower germplasm enhancement. In: Proc 29th Sunflower Res Workshop, 10–11 Jan, 2007, Fargo, ND, USA: *http://www.sunflowernsa.com/research/research-workshop/documents/Yue_Experiment_Marker_07.pdf*

Yue B, Radi SA, Vick BA, Cai X, Tang S, Knapp SJ, Gulya TJ, Miller JF, Hu J (2008) Identifying quantitative trait loci for resistance to Sclerotinia head rot in two USA sunflower germplasms. Phytopathology 98: 926–931.

Zhang LS, Le Clerc V, Li S, Zhang D (2005) Establishment of an effective set of simple sequence repeat markers for sunflower variety identification and diversity assessment. Can J Bot 83: 66–72.

Zhang Q, Gao YJ, Yang SH, Ragab RA, Saghai Maroof MA, Li ZB (1994) A diallel analysis of heterosis in elite hybrid rice based on RFLPs and microsatellites. Theor Appl Genet 89: 185–192.

Zhang YX (1995) Evaluation of the potential of RFLPs for the study of distinctness, uniformity, and stability sunflower. In: The Working Group on Biochemical and Molecular Techniques and DNA-profiling in Particular. International Union for the Protection of New Varieties of Plants, Wageningen, The Netherlands, pp 2–9.

Zhang YX, Gentzbittel L, Vear F, Nicolas P (1995) Assessment of inter- and intra-inbred line variability in sunflower (*Helinathus annuus*) by RFLPs. Genome 38: 1040–1048.

Oil Composition Variations

André Bervillé

ABSTRACT

This chapter deals with the range of fatty acid composition and variation, the nature of mutations, and environmental factors that may modify oil composition. Sunflower oil contains fatty acids and minor compounds such as tocopherols and phytosterols. Current knowledge of fatty acid, tocopherol and phytosterol synthesis and signalling pathways in model plants is discussed. Each major fatty acid variation, eventual mutants, and their roles are unravelled as to how they function in the pathway. The role that eventual abiotic stresses play is developed and discussed. Specific attention is given to oil modifications brought about by mutants with economic importance to the industry (oil rich in palmitic or stearic acid) or to prevent heart diseases (oil rich in oleic acid or well balanced in polyunsaturated fatty acids). Variation in minor compounds observed in sunflower oil are also discussed and the possibility to breed sunflower for these compounds.

Keywords: fatty acids; mutants; oil composition; oleic acid; palmitic acid; stearic acid

8.1 Introduction

This chapter deals with the range of variation in composition observed in sunflower oils including the oil phenotypes and the nature of the mutation which are revealed as well as the environmental factors that may modify oil composition in this crop.

INRA, Department of Genetics and Plant Breeding, UMR DIAPC, 1097, 2 Place Viala, 34060 Montpellier cedex 1, France; e-mail: *Andre.Berville@supagro.inra.fr*

The standard reference for the oil composition in sunflower is the FAO's Codex Standards for Fats and Oils Derived from Edible Fats and Oils. Table 8-1 shows a variable range due to differences in cultivars and environmental influences in comparison to some other crop oils.

Traditional sunflower oil composition has a wide range of variation for plant seed oils. It is the main oil (48%) consumed in Europe. However, the trend is now to favor oil with high oleic (HO) acid content (HOAC) to prevent heart diseases. This HO oil broadly mimics olive oil, which is a fruit oil, although the cost is much lower. Other types of oil also exist for sunflower, and this range of variation will be examined and the potential to grow this crop with many oil compositions will be described.

Table 8-1 Average percent composition of sunflower oil in comparison to some other crude seed and fruit oils.

Fatty acid C: /others	Traditional Sunflower	Rapeseed/ Canola	Soybean	Safflower	Olive virgin	High oleic Sunflower
Palmitic acid 16:0	5–8	2.5–7.0	8.0–13.5	5.3–8.0	7.5–20	2.6–5.0
Stearic acid 18:0	2–7	0.8–3.0	2.0–5.4	1.9–2.9	0.5–5.0	2.9–6.2
Oleic acid 18:1	14–40	51.0–70.0	17.7–28.0	8.4–21.3	55.0–83.0	75–90.7
Linoleic acid 18:2	48–74	15.0–30.0	49.8–59.0	67.8–83.2	3.5–21.0	2.1–17
Linolenic 18:3	0.1–0.8	5.0–14.0	5.0–11.0	<0.1	<0.9	< 0.3
20:0	< 0.5	0.1–4.5	0.3–0.6	0.2–0.4	<0.6	< 0.5
22:0	0.3–1.5	<2.0	<0.7	<1.0	<0.2	0.5–1.6
Tocopherols mg/kg	440–1520	430–2680	600–3370	240–670	<4.5	450–1120
Phytosterols mg/kg	2400–5000	4500–11300	1800–4500	2100–4600	<1000	1700–5200

8.1.1 Main Features of Lipids

Oil is composed of several lipids, the major component being triacylglycerols (TAG), but tocopherols, sterols, phosphatides, and waxes are present in small amounts and all are stored in the seeds. In the plant, their roles are to protect the embryo from oxidative stress and to release available energy to facilitate germination.

Oil composition has consequences on the uses of the oil, effects on health, and prevention of heart diseases. For industrial purposes, the main source of fat is animals. To compete with animal fats that are inexpensive, plant fats should have technological advantages and specificities in composition

(Bessoule and Moreau 2004). For end products such as coffee and chocolate, the lipid composition affects the quality (taste and flavor) of the final product after cooking.

Historically, comprehensive examinations of oil diversity have been performed to enhance knowledge about the diversity of fatty acids and other compounds, but also to provide new compounds to industries. Biotechnology has modified oil composition since the mid-1970s, first in *Brassica*. Thereafter, applications spread rapidly to other oil crops such as soybean, sunflower, and cotton.

8.1.2 Fatty Acids

The structure of fatty acid synthase (FAS) in animals and fungi has been unravelled as a 262 kDa multipeptide functional complex. It releases saturated fatty acids, either C16:0 or C18:0, that are further processed in metabolism. However, animal metabolism is unable to synthesize C18:2 (linoleic acid) and C18:3 (linolenic acid), and as a consequence these fatty acids need to be found in the diet.

The FAS complex in plants is composed of eight polypeptides and is localized exclusively in plastids. Its complex structure is probably responsible for the lack of mutants that interrupt only one step in fatty acid synthesis. Mutations affecting FAS usually have complex effects.

Brown et al. (2006) have purified the fatty acid biosynthetic enzymes in *Arabidopsis thaliana* type-II FAS in which separate enzymes catalyze sequential reactions. Genes encoding all of the plant FAS components have been identified and the structure of a number of the individual proteins determined. Kachroo et al. (2003) have identified a plastidial fatty acid signalling pathway in the *Arabidopsis ssi2/fab2* mutant involved in defense mechanisms against the necrotrophic pathogen *Botrytis cinerea*. The mutation encodes stearoyl-acyl carrier protein desaturase (S-ACP-DES), and results in the reduction of oleic acid (18:1) levels in the mutant plants and also leads to the constitutive activation of NPR1-dependent and -independent defense responses. The alteration of the prokaryotic fatty acid signalling pathway in plastids required several mutations in fatty acid biosynthetic pathways that cause an increase in the levels of 18:1 in specific compartments of the cell. A loss-of-function mutation in the soluble chloroplastic enzyme glycerol-3-phosphate acyltransferase (*ACT1*) completely reverses salicylic acid and jasmonate-mediated phenotypes in *ssi2*. In conclusion, fatty acid signalling plays an essential role in the regulation of *ssi2*-mediated defense.

Most saturated and unsaturated fatty acids are synthesized for building membranes and to allow cellular functions. There are two main categories of plants according to composition of membrane lipids. One is due to a majority of "palmitic" acid, whereas the other is due to a majority of "linolenic" acid (Mongrand et al. 1998). Membrane lipid composition is modulated by the environmental temperature, where "palmitic" membranes are typical of tropical species and "linolenic" membranes are more northern. This means that deep constraints limit the variation range for lipid membranes to enable biological functions.

Desaturation of fatty acids occurs sequentially from C18:0, C18:1, C18:2, to C18:3 or from C16:0 to C16:1 and so on. For seed oils these steps occur in the developing embryo. Elongases add the acetate motif to fatty acids to lengthen them to C:20, C:22, and higher (Table 8-1). All these steps that are detailed in many reviews enhance the diversity of storage lipids (Roscoe et al. 2001).

Storage lipids are much more variable in composition and in quantity than structural lipids. Humans have consumed many plant oils that are not in use now, but some are still utilized for health or body care (e.g., *Camellina, Oenothera*, flax). Fruit oils (olive) and seed oils (oil crops) are not stored in the same structure. In the seed, oil droplets are surrounded by a membrane made up of a phospholipid monolayer embedded with oleosin proteins (Cummins et al. 1993). In fruit oil, this membrane is absent and the oil droplet floats in the tissue.

8.1.3 *The Role of Mutations to Unravel Fatty Acid Biosynthesis Pathways*

A mutation interrupts one biochemical pathway and causes the accumulation of the compound before the block. Thus, mutations have been found very useful to enhance knowledge about biosynthetic pathways. Practically, a mutant releases a specific oil composition that could be of value in the market. However, biochemical pathways are complex networks and one single mutation may not have the expected effect on one compound only. Consequently, to unravel a mutation mechanism and to determine which enzyme is involved may be more or less difficult. Regulation of a biochemical pathway is generally due to a feedback effect by the end product on one of the initial enzymes. Thus, a mutation may lead to deregulation of the pathway with complex effects on the accumulated products. From a genetic point of view, this is called as epistasis, but it does not suggest any mechanism. The biochemical mechanism of several mutations have now been unravelled in model and cultivated plants. Sobrino et al. (2003) reviewed several experiments performed in Cordoba to modify sunflower oil composition that will be detailed further.

8.1.4 Types of Mutation

Most frequently, a mutation is a change in the coding DNA sequence (exon) of an enzyme that induces one amino acid deletion, insertion, or substitution that makes the enzyme no longer functional. If the mutation is in a non-coding sequence (regulation sites, intron) and affects the transcription ability, the transcript of the enzyme will be absent, and therefore the enzyme will not be synthesized. The mutation may also delete a part of the entire gene and this will have the same consequence. All these mutations are recessive since this change does not affect the function of any normal gene in a diploid heterozygous plant, and is called a *cis* effect.

In some cases, the mutation may affect a regulatory peptide encoded by another gene. The regulatory peptide is modified, and the mutation may affect several functions (syndrome effect), as it has been reported for transcription factors (Schmidt et al. 1992), and will behave as dominant. Such a mutation has *trans* effect. A heterozygous plant for the mutation may be more or less affected depending on the effect of the normal allele, which competes with the mutated one.

Due to insufficient natural variation for oil composition in crop plants, breeders have used artificial mutagenesis, either physical (γ- or X-rays) or chemical (a series of compounds: ethylmethane sulfonate, EMS; diethyl sulphate, DES; N-nitroso-N-methylurea, NMU) to induce mutations. Mutations are random and the treatment enhances their rates of appearance. However, a mutagen may favor some base changes or DNA rearrangements such as deletions or insertions. They may also activate silenced transposon or retrotransposon sequences. As a result, the exact type of mutation cannot be predicted according to the agent and a rigorous study is required to determine the exact changes for a given induced mutation.

In the past few years, fatty acid pathways have been modified using transgenic strategies. We can distinguish when the insertion of an extra copy of a desaturase is in the same direction, because in most cases this causes overexpression of the gene. The mutation is, therefore, partially dominant but also its effect largely depends on the cultivar's genetic background. This, in most cases, will enhance the level of the next product. In another situation, a transgene can knock out a gene by insertion of another (extra) copy in the reverse (or same) direction elsewhere in the genome. This causes post-transcriptional gene silencing (PTGS) and the absence of the transcript. The level of the product before the blocked step will be enhanced. The insertion has a *trans* effect on all copies of the wild gene, and the mutation will be dominant.

8.1.5 Differences between Seed and Fruit Oils

When made by the embryo, oil composition is due to both the paternal and maternal genomes. The genetics of the trait has to be studied on the seed oil (phenotype) from the seed before planting and not from the seed harvested in the next generation. Thus, it is required to genotype each seed before sowing to correlate the phenotype with the molecular markers. This is true for oil in the embryo, including both seed oil and the fleshy fruit (pericarp) oil. Pericarp and seed oil of one fruit do not have the same composition, as has been demonstrated for olive (H. Sommerlatte, pers. comm.). Fleshy pericarp oil is due to the maternal genome only, whereas seed oil includes the effect of the genes from the pollen. Consequently, on a sunflower plant, each seed harvested on an F_1 hybrid plant may have a different oil composition, whereas on an olive tree, all the fleshy pericarps will have the same oil composition, but not the shell and embryo oil (Breton et al. 2009).

In the next section, we will examine which genetic events may have led to the diversity of sunflower oils.

8.2 Mutations Leading to One Main Fatty Acid Accumulated in Oil

8.2.1 Palmitic Acid

Enhancement of palmitic acid is required for some industrial applications whereas other applications require reducing both saturated fatty acids, palmitic and stearic. Thus breeding has followed both these paths.

8.2.2.1 Palmitic Acid in Other Species

High palmitic acid content has been obtained in leaf tissues in *Arabidopsis* (Browse et al. 1989). An increased level of palmitic acid in the oil has also been obtained in soybean. Two mutant soybean lines with palmitic acid contents of >18% were developed by treatment of 'A1937' seeds with NMU and "Elgin" seeds with EMS. The mutant lines, A1937NMU-85 and ElginEMS-421, were crossed to determine their genetic relationship for elevated palmitic acid content (Fehr et al. 1991). The inheritance of the trait was controlled by two loci.

In *Brassica napus*, an induced mutant from European winter oilseed rape with increased palmitic acid content was phenotypically characterized and genetically analyzed (Schnurbusch et al. 2000). The mutant showed a palmitic acid content of 9.2% compared to 4.5% in the parental cultivar. In contrast, the oleic acid content decreased from 61.6% to 44.2%, whereas the linoleic and linolenic acid contents increased. The mutant plants grew poorly and their seed oil content was only 31.2% compared to 42.8% in the parental cultivar. The inheritance of the mutant was oligogenic and was determined by at least four genes. In the F_2 generation, palmitic acid content was negatively correlated with oil content.

Wilson et al. (2001) have produced soybean cultivars with oil that varied from less than 4% to about 35% palmitic acid, compared to about 11% palmitic acid in typical cultivars. A number of recessive alleles associated with these phenotypes have been described that represent different mutations at the *Fap* loci. These metabolic studies narrowed the identification of fap_1, fap_2, and fap_{nc} alleles to the genes that encode or regulate 3-keto-acyl-ACP synthetase II, 16:0-ACP thioesterase, 18:0-ACP desaturase, or 18:1-ACP thioesterase enzymes. This hypothesis was strongly supported by Northern blot assays that revealed a significant reduction in the accumulation of transcripts corresponding to the 16:0-ACP thioesterase in germplasm homozygous for the fap_{nc} allele.

8.2.2.2 Palmitic Acid in Sunflower

Pérez-Vich et al. (2000) have screened and studied sunflower genotypes with increased levels of palmitic acid (C16:0) in the seed oil. Several mutants were studied: CAS-5 displays more than 25% of the total oil fatty acids as C16:0, whereas the parental line BSD-2-691 displays 5.4% C16:0. The segregation fit a model of two alleles at one locus with partial dominance for the low content. To determine the inheritance of the high C16:0 content in the sunflower mutant line, the mutant was reciprocally crossed with standard sunflower line HA-89 (5.7% C16:0) and with its parental line, and F_1, F_2 and BC_1F_1 seeds were obtained from the crosses of CAS-5 with those lines. The cross with HA-89 revealed a segregation that fit a ratio of 19:38:7 for low (<7.5%), middle (7.5–15%), and high (>25%) C16:0 content, respectively. This segregation was explained on the basis of three loci (P1, P2, and P3) each having two alleles showing partial dominance for low content. Two of the loci revealed diversity present in the sunflower lines.

In contrast, low palmitic acid content has also been targeted. Miller and Vick (1999) have determined the mode of inheritance of low stearic and low palmitic acid content found in three sunflower mutant lines treated with two mutagens, NMU and EMS. Two lines, HA 821 LS-1 and RHA 274 LS-2, displayed lower stearic acid content (4.1% and 2.0%), compared to 4.7% for their respective parental lines. Segregation ratios of F_2 and testcross progenies indicated that the low stearic acid content in HA 821 LS-1 was controlled by one gene, designated *fas1*, with additive gene action. The low stearic acid content in RHA 274 LS-2 was controlled by two genes with additive gene action. The first gene was designated *fas2*, and the second gene was temporarily designated *fasx*. The allele *fap1* was identified in RHA 274 LP-1 to control low palmitic acid content with additive gene action. In practice, palmitic acid content and stearic acid content are frequently inversely correlated.

8.2.2.3 Palmitoleic Acid

Salas et al. (2007) have induced a mutation (a high-palmitoleic acid sunflower mutant) accumulating up to 20% of *n*-7 fatty acids. This line produces oil with a complex TAG composition, containing species that have not been previously identified in sunflower. In this regard, palmitoleic acid was esterified in an unexpected way in the three positions of the TAG molecules. The polar glycerolipid composition of the mutant was also studied, in order to identify and quantify the changes in membrane lipids imposed by the sunflower enzymatic machinery during the accumulation of the unusual *n*-7 fatty acids. The high-palmitoleic mutant accumulated important quantities of *n*-7 fatty acids in the polar lipid fraction, especially in the phosphatidylcholine lipid class. However, the total polar lipid content of these lines was not affected. On the other hand, the mutations responsible for the *n*-7 lipid accumulation induced an important decrease in the oil yield of the new mutant.

8.2.2 Stearic Acid

8.2.2.1 Stearic Acid in Other Species

Lightner et al. (1994) obtained a high stearic level in *Arabidopsis* leaves, and reduced growth of the plants. The fatty acid composition of corn oil has also been modified for stearic acid content (Jellum and Widstrom 1983). The inheritance of stearic acid was studied in crosses between standard inbred lines with approximately 2% stearic acid and three strains of an introduced genotype (PI 175334) with unusually high stearic acid of about 10%. Results from single kernel oil analyses of the parents, F_1, F_2, BC_1, and BC_2 generations strongly suggested the involvement of a major single gene recessive for high stearic acid in these crosses. Transgressive segregation for high stearic acid indicated the presence of one or more modifying genes of minor influence on stearic acid.

8.2.2.2 Stearic Acid in Sunflower

Pérez-Vich et al. (2004) have mapped quantitative trait loci (QTL) conferring increased C18:0 content in CAS-20 in an F_2 mapping population developed from a cross between HA-89 (wild type low C18:0) and CAS-20. A genetic linkage map of 17 linkage groups (LGs) comprising 80 RFLP and 19 SSR marker loci from this population was used to identify QTLs controlling fatty acid composition. Three QTLs affecting C18:0 content were identified on LG 3, LG 11, and LG 13, with all alleles for increased C18:0 content inherited from CAS-20. These QTLs jointly explained 43.6% of the C18:0 phenotypic variation. On the basis of positional information, the QTL on LG 11 was suggested to be a *SAD6* locus. The results presented show that increased C18:0 content in sunflower seed oil is not a simple trait, and the markers flanking these QTLs constitute a powerful tool for plant breeding programs.

Pérez-Vich et al. (2006) have also studied the inheritance of high stearic acid content in the sunflower mutant CAS-3 and CAS-14. In contrast to CAS-3, high stearic acid expression in CAS-14 seeds is temperature-dependent and not uniformly distributed in the seed. The trait in CAS-3 has been found to be governed by two genes, *Es*1 and *Es*2. To study the inheritance of high stearic acid content in CAS-14 and CAS-3, crosses were made with P21, a nuclear male sterile (NMS) line with a wild type fatty acid profile. The genetic analysis included the evaluation of the F_1, F_2, F_3, BC_1F_1, and BC_1F_2 seed generations. Crosses between P21 and CAS-14 revealed that the high stearic acid trait was controlled by a single recessive gene designated *Es*3. The analysis of the F_3 and BC_1F_2 (to P21) generations demonstrated a repulsion-phase linkage between the *Es*3 and the *Ms* loci, the latter conferring the NMS trait. The frequency of recombination between *Es*3 and *Ms* was estimated to be 0.09. Crosses between CAS-3 and CAS-14 demonstrated that both lines possess alleles for high stearic acid content at different loci, as transgressive segregants with low stearic acid content were observed in all generations. Genetic recombination of *es*1 and *es*3 alleles did not result in an increment of the maximum stearic acid content in the seeds compared with the maximum levels produced by the *es*3 alleles alone. Further studies of CAS-14 have been done with the *Ol* mutation (high oleic).

Competition assays carried out with CAS-5, a mutant with a higher content of palmitic acid in the seed oil, indicated that a modified FatA-type thioesterase is involved in the mutant phenotype (Martínez-Force et al. 2000).

8.2.3 Oleic Acid

8.2.3.1 Oleic Acid in Other Species

Del Río-Celestino and De Haro-Bailón (2007) have studied the inheritance of high oleic acid content in the seed oil of mutant Ethiopian mustard lines, obtained by mutagenesis. Oleic acid segregation indicated control of accumulation by two segregating genetic systems, one acting on chain elongation from C18:1 to C22:1 and a *fad*2 gene involving desaturation from C18:1 to linoleic acid (C18:2). In addition, C18:1 was influenced by one additional locus (tentatively named *OL*) involved in control of desaturation of C18:1 to form C18:2. Transgressive recombinants were obtained from the cross L-1630×L-25X-1, with about a three-fold increase of the C18:1 content over that of the parents (> 64%) and free of C22:1 content, which represents a high potential for commercial exploitation. Other studies by Schierholt and Becker (2000) with winter oilseed rape mutant lines have also concluded that the high oleic trait is dependent on the *fad2* locus.

8.2.3.1.1 Cotton

Liu et al. (2002) have obtained high-oleic and high-stearic cotton seed oils using gene silencing. They have applied hpRNA-mediated PTGS in cotton to down-regulate key fatty acid desaturase genes and develop nutritionally improved high-oleic (HO) and high-stearic (HS) cottonseed oils. Silencing of the *ghFAD2-1* Δ12-desaturase gene raised oleic acid content from 13% to 78% and silencing of the *ghSAD-1* Δ9-desaturase gene substantially increased stearic acid from the normal level of 2% to as high as 40%. Additionally, palmitic acid was significantly lowered from 26% to 15% in both HO and HS lines. Intercrossing the HS and HO lines resulted in a wide range of unique intermediate combinations of palmitic, stearic, oleic, and linoleic contents.

8.2.3.1.2 Peanut

Jung et al. (2000a, b) have studied a peanut (*Arachis hypogea*) mutant that displays shrunken seeds with poor seed oil but with a high oleate content. Apparently, the mutation was spontaneous. Detailed studies have revealed that the high oleic trait is due to two changes in two oleate desaturases. One mutation causes lower activity of the oleate desaturase and a second mutation decreases transcript accumulation of the oleate desaturase. Two oleate desaturase genes are explained by the allopolyploid origin of *A. hypogea*. However, for this species the mutations have a deep genetic load and the trait has not been exploited for the crop.

8.2.3.2 Oleic Acid in Sunflower

Two high-oleic mutations have been reported in sunflower (Soldatov et al. 1976; Ivanov and Giorgev 1981). The source studied by Petakov et al. (2000) has not been released and has not been as intensely studied as the Pervenets source of Nikolova et al. (2001).

The detailed story of the mutation has been obtained after long discussions with Soldatov both in France and at Krasnodar in Russia (Lacombe et al. 2002). The main feature of the story is that Soldatov screened the mutation in a heterozygous plant and therefore the mutation should act as dominant. Soldatov has named the population with high oleic acid content (HOAC) as Pervenets, and we will further refer to the mutation as the Pervenets mutation. Other researchers have named the mutation *Ol*, which is a generic term. The *Ol* term is not adopted here because a dominant mutation does not affect a structural gene (as *Ol* for oleate), but a regulatory gene, which remains unknown.

8.2.3.2.1 Biochemical Changes Induced by the Pervenets Mutation

Oil enzyme modifications have been investigated by Garcés et al. (1989) and Garcés and Mancha (1991) by comparing traditional and Pervenets sources (Sarmiento et al. 1994, 1998). Desaturases are very hydrophobic membrane-bound enzymes, and comparisons of *in vitro* to *in vivo* activities may be questionable, such as comparisons of temperature effects. However, the results published by this team appear solid and reliable. They attributed the reduced conversion of oleic to linoleic acid to a deficiency in enzymatic activity corresponding to an oleoyl-phosphatidyl choline (PC) desaturase. Apparently, the modification is manifested only at this stage, but there is evidence that other oleoyl-phosphatidyl choline desaturase enzymes also exist in sunflower (Martínez-Rivas et al. 2001). Fatty acids to be desaturated should be attached to PC or to an acyl carrier protein (ACP). In contrast, the *in vitro* temperature effect detected on the desaturase activity was not as effective *in vivo* (Garcés et al. 1992; Sarmiento et al. 1999).

Other desaturase activities appeared unmodified between the traditional sunflower and the Pervenets sources. This has targeted a further molecular approach by Kabbaj et al. (1995, 1996a) on the desaturase steps before and after the presumed blockage point, involving namely the stearoyl-ACP- and oleoyl-PC-desaturases.

Several biosynthetic pathway routes may lead to oleic acid. Researchers have hypothesized that the block in the pathway by Pervenets was compensated by an alternative route. In this respect, all genes of the alternate pathways become candidates as modifiers of the high oleic acid trait. However, their direct involvement in the trait may be difficult to pinpoint (Perez-Vich et al. 2006).

8.2.3.2.2 Genetic Analysis of the HOAC Trait

Many genetic analyses have been performed on the inheritance of the HOAC trait in many genetic backgrounds. In most cases, we have already seen that one is expecting a single mutation. However, for high stearic and palmitic acid levels, we have discovered that the diversity in sunflower revealed a second locus interfering with the mutation (epistatic effect). Genetic studies on the Pervenets mutation revealed that one to five genes could interfere with the HOAC trait. Therefore, the methodologies used to reveal loci must be examined.

Since the progenies of a cross between a low oleic and a high oleic line segregated as continuous for oleic acid content (as a quantitative trait) depending on the population size and the oleic acid distribution of F_2 plants, and considering the dominance of the HO trait, authors have made two (3HO:1LO), three (1HO:2Mid:1LO), or more classes for oleic acid content. This implies a number of loci. All the studies have been reviewed in Lacombe and Bervillé (2000).

However, by the year 2000, individuals that carry the Pervenets mutation could not be identified. Thus, the studies dealt unambiguously with the inheritance of the HOAC trait. Authors tried to determine the locus that produced the main effect, which should be the locus at which the mutation has occurred, in comparison to the effect of loci with other minor effects, also called modifiers.

Lacombe and Bervillé (2001) have shown for the first time that the HOAC in half-cotyledons and a rearrangement in an oleate desaturase cosegregated (genetically linked) in an F_2 population of a cross between a LO with a HO line. Only the kernels that carried the Pervenets rearrangements displayed HOAC as high as 90%, compared to 83% in the female parent. However, many kernels without the Pervenets mutation displayed a range of variation for oleic acid content from 15 to 50%, although the LO maternal line was fixed at 28% oleic acid.

Only one QTL approach to the inheritance of HOAC has been published (Perez-Vich et al. 2004). It followed both the stearic and the oleic acid content in F_2 progenies. Each trait, high stearic and high oleic acid content displayed QTLs in the regions of a stearoyl- and oleoyl- desaturase, respectively. Other minor QTLs have been supposed corresponding to alternate pathways (thioesterase acyl carrier protein) (Pérez-Vich et al. 2004, 2006).

It is clear from all these studies that all HO lines carry the same Pervenets mutation. In contrast, other types of factors also act on the oleic acid content, such as the genetic backgrounds and modifier genes that have not yet been elucidated.

8.2.3.2.3 Expression Induced Changes by Pervenets Mutation

Since the enzymatic function of the oleate desaturase has been shown to be lacking in HOAC sunflower, Kabbaj et al. (1996b) studied whether the corresponding transcript is present or not. The oleate desaturase was first cloned in *Arabidopsis* by Arondel et al. (1992) and later cloned in other plant species. Unfortunately, the sunflower transcript did not display enough homology with *Arabidopsis* to be cloned directly. A cDNA library was constructed from developing sunflower embryos and probed by stearate desaturase cDNA and oleate desaturase cDNA from diverse species. Only with a stearate desaturase cDNA and a partial oleate desaturase clone from olive, a partial cDNA for the stearate desaturase and complete cDNA for the oleate desaturase were isolated in sunflower.

Further studies in sunflower revealed expression of these genes in the developing embryos between 10 to 20 days after fertilization. Kabbaj et al. (1996a, b) demonstrated with the stearate desaturase transcript as control that the oleate desaturase transcript was lacking in HOAC sunflower, whereas it was present in the control.

Hongtrakul et al. (1998a, b) showed by RT-PCR that the oleate desaturase transcript was less expressed in HOAC sunflower than in traditional sunflower and they also used the stearate desaturase transcript level in both types as control. The absence of the transcript explains the absence of the protein. However, this does not indicate whether the gene is present or not. To look for the presence of the gene researchers have used RFLP, the most available technique at this time.

8.2.3.2.4 DNA Changes Correlated with the Pervenets Mutation

Lacombe et al. (1998) and Hongtrakul et al. (1998b) reported an RFLP between traditional and HOAC lines carrying the Pervenets mutation using oleate desaturase cDNA as a probe. Signals have been interpreted as a duplication of the oleate desaturase gene by Hongtrakul et al. (1998b), whereas Lacombe et al. (2000) did not postulate a complete duplication of the oleate gene.

Southern blots of genomic DNA from traditional sunflower restricted separately with *Eco*RI, *Hind*III, *Bam*HI, or *Sac*I, and hybridized with the oleate desaturase cDNA as a probe displayed RFLP profiles (Fig. 8-1). Only one fragment was clearly hybridized, suggesting that only one gene is present in the sunflower genome. We cannot exclude that several copies could exist. In contrast, all HOAC Pervenets lines displayed an RFLP profile that was different from traditional sunflower (Fig. 8-1b). With *Eco*RI, an extra fragment of 8 kb (Pervenets insertion) appeared in addition to the wild *Eco*RI 5.85 kb fragment and was strongly hybridized with the probe. This means that it should carry a sequence homologous to the oleate desaturase. With *Hind*III, the wild 8.0 kb fragment disappeared and was replaced by a 16 kb fragment. With *Sac*I and *Bam*HI, the wild fragments disappeared and were replaced by another lengthened fragment. The double restriction *Eco*RI and *Hind*III revealed a 2.1 kb fragment and the 8 kb Pervenets insertion. Taken all together, these facts suggest that an insertion of 8 kb occurred on the 8.7 *Hind*III fragment in the region of the *Eco*RI site.

Lacombe et al. (1999) attempted to clone genomic fragments from an HOAC line that hybridized with the oleate desaturase probe corresponding to the Pervenets RFLP. Unfortunately, they did not succeed in identifying hybridizing clones that displayed the oleate desaturase rearrangement. They

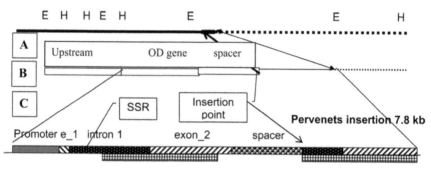

Figure 8-1 Organization of the oleate-desaturase region and details on the Pervenets mutation.

- A: Thick line: region sequenced (17,770 bp); E, H: restriction site for *Eco*RI and *Hind*III, respectively. Thick bar marked by the thick and light arrows pinpoints the Oleate desaturase repeated sequence. Dotted line marks the unknown sequence with deduced restriction sites.
- B: Scheme for the location of the *OD* gene. The line indicates the oleate-desaturase repeat. Dotted line indicates unknown sequence.
- C: Enlargement for the *OD* gene, the spacer region and the oleate-desaturase repeat portion.

The promoter region, exon 1(e_1), intron 1 with the SSR-OD1, exon 2 and spacer are represented by boxes with motives and the bar below indicate the regions with the direct repeat.

also used several library constructions without success. However, they cloned an oleate desaturase gene belonging to a microsomal oleate desaturase (according to the targeted sequence) that displays all the features of a functional gene (Lacombe et al. 2002). The microsomal oleate desaturase (MOD) gene displayed a 1,683 bp intron carrying an SSR motif (ATT) repeated 17 times in RHA 345 (HA-INRA-OD1). The motif is repeated 16 times in 83HR4, allowing the origin of the oleate desaturase allele in each RIL to be determined. Moreover, with the hypothesis that Pervenets insertion is very close to the oleate desaturase sequence, recombination between the SSR locus and Pervenets was not expected.

Lacombe et al. (2002) turned to a PCR approach hypothesizing that the distance between the wild oleate desaturase sequence and the oleate desaturase sequence on the Pervenets insertion could be determined by PCR. Forward and reverse primers designed on the oleate desaturase cDNA were used in pairwise combinations to anchor the wild gene (F) and the Pervenets sequence R. The reverse primers were designed assuming the insert was in the same direction or in the opposite direction of the wild oleate desaturase gene. Long PCRs were run and the amplification products were hybridized (Southern blots) with the full oleate desaturase cDNA. A 4.3 kb fragment was revealed and cloned into a plasmid. It was sequenced and the following features were revealed: 1) the end of the oleate desaturase coding sequence; 2) a spacer sequence with no hit in the databases; 3) the end of intron 1 of the *MOD* gene sequence; and 4) the exon2 *MOD* sequence was interrupted at the end of the R7 primer designed on the cDNA sequence. These sequences are included in the 18,990 bp fragment entirely sequenced and registered as EF469194 in GENBANK (Lacombe et al. 2009).

Further primer pairs were designed in the spacer region and the repeated *MOD* region to obtain a specific fragment of the Pervenets insertion (Bervillé et al. 2004, 2005, patent applied). These specific Pervenets fragments were verified to be specific to HOAC lines derived from the Pervenets population. The final organization of the Pervenets mutation is displayed in Fig. 8.1.

Further studies have been developed using a set of recombinant inbred lines (RILs) constructed from 83HR4 and RHA 345, which are two restorer lines used to avoid the Pet1 cytoplasm and the cytoplasmic male sterility in the progenies. Two hundred and fifty F_2 plants were produced and Lacombe et al. (2002) recovered 174 F_6 RILs. Five seeds per F_6 family were phenotyped on half a cotyledon (GC) to determine the oleic acid content and then sown in the field. Only, one plant per family was retained to study the segregation of the oleic acid content. All plants were genotyped for the following fragments: RFLP *Eco*RI LO:5.85 kb, Pervenets 5.85 kb + 8 kb and *Hind*III traditional (LO) 8.7 kb and Pervenets (HO) 16 kb; SSR-INRA-OD1 (LO 237 bp / Pervenets 240 bp); and one Pervenets-specific fragment, F: Fc-a_7 (caaaccaccacccactaac) and R: R-(ggttctgggtctgggtctggtt) of 902 bp. The control was the fragment F2-(tcgctaacccgttcgttctc) R2- (caaagcccacagtgtcgtc) of 173 bp designed on the *MOD* cDNA.

With these markers we attributed clearly which RIL carried the Pervenets mutation as judged by the Pervenets specific insertion fragment. Moreover, linkage disequilibrium between the SSR-INRA-MOD1 locus and the Pervenets insertion prevents any recombination in this region. As previously, the oleic acid content was determined on half a cotyledon. The results from crossing phenotypes and marker analyses, not yet published, can be summarized as follows:

1) 125 RILs were obtained with an oleic acid content of < 50% and 35 with oleic acid content of > 50%

2) The SSR analysis showed that the ratio of the lines carrying the 237 bp (82 lines) versus the lines carrying the 240 bp allele (78) was in agreement with 1:1, as expected ($\chi^2 P$ <0.6),

   All the lines carrying the 237 bp allele were LO with an oleic acid content < 50%.

   Out of the 78 lines carrying the 240 bp allele, 35 were with oleic acid content > 50% whereas 43 were with oleic acid content < 50%. The ratio 43/35 fits an 1:1 ratio, suggesting that one independent locus controls the HOAC in the RIL carrying Pervenets. The Pervenets mutation is not sufficient for a high oleic acid content.

3) RFLP 5.85 *Eco*RI and RFLP 8.0 *Hind*III: 82 RILs displayed both the 5.85 kb *Eco*RI and the 8.0 *Hind*III fragments. However, 14 RILs displayed unusual RFLPs, 10 RILs displayed an extra 5.0 kb *Eco*RI instead of 8 kb, and 4 RILs gave a 13 kb *Hind*III fragment instead of a 16 kb fragment.

4) PCR Pervenets specific fragment: 78 RILs displayed a 902 bp fragment as expected. All lines displayed the 173 bp MOD fragment.

Skoric et al. (1996) reported unstable expression of the *Ol* gene for HOAC in one progeny, but they recovered the HOAC trait in subsequent progenies. Considered all together, the data suggested another independent locus that enables the Pervenets mutation for HOAC.

8.2.3.2.5 A Model as the Pervenets Mutation Functions

The three main features of the Pervenets mutation are: 1) its expression is mainly dominant over the traditional allele; 2) it corresponds to the duplication of part of intron and exon 2 of the *MOD* gene; and 3) the absence of the *MOD* transcript suggests a silencing mechanism as post-transcriptional gene silencing.

Up to now, all mutations that modify fatty acid levels have been found to be recessive against the wild type gene. In a few cases, it may behave as codominant. No cases of regulation by silencing have been found for fatty acid metabolism in plants. However, in cotton using transgene methods, Liu et al. (2002) caused a total knockout of oleate desaturase leading to HOAC oil in this species. This points out that a silencing model is conceivable and functioning in plants. Moreover, a dominant suppressor of silencing has been found in rice (Kusaka et al. 1993).

Sunflower breeders have frequently experienced the unexpected behavior of the Pervenets mutation. According to breeders, the Pervenets mutation seems to disappear or to be unstable. In the RIL family, a genetic factor independent of the *MOD* locus with the Pervenets mutation directs the oleic acid level. Because it behaves as a suppressor of the Pervenets mutation (Lacombe and Bervillé 2001), we have called it "olesup". We have shown that in RILs carrying both the Pervenets mutation and "olesup"— in RILs we cannot say whether it is dominant or recessive—the *MOD* transcript level is restored. This fact means that "olesup" could act directly on the silencing mechanism (Lacombe et al. 2009).

A suppressor explains perfectly that the Pervenets mutation may "disappear" (the plant's oil composition is indistinguishable from traditional sunflower) or be unstable when it segregates. We also have observed some rearrangements in the Pervenets mutation. In this case, each RIL family loses the Pervenets mutation and reverts to a traditional sunflower. However, our molecular tools show that it remains as an insertion of approximately 1 kb instead of the 8 kb, but the fragment is not detectable due to its yet unknown sequence, however, it lengthens the 8.0 kb *Hind*III fragment to 9.0 kb.

Gene silencing in plants is environmentally susceptible and the variation of oleic acid content in commercial hybrids is not surprising. But we can predict that oleic acid variation is not due to the oleate desaturase but probably to a silencing mechanism that affects the final transcript stability.

8.2.3.2.6 Other Molecular and Biochemical Studies on the Pervenets Mutation

Martínez-Rivas et al. (2001) have characterized three oleate desaturase cDNAs. Two genes are constitutively expressed and one is embryo specific. The latter probably corresponded to *MOD* whereas the two others correspond to other loci. Only the *MOD* gene is affected by the Pervenets mutation, and consequently most of the lipid pattern is normal in HO sunflower as shown by Lagravère (1999) and Lagravère et al. (2000). To reconcile the results from Martínez-Rivas et al. (2001) with those of Lacombe et al. (2001) on the number of *MOD* genes in sunflower, we propose that the 5.85 kb *Eco*RI fragment carrying the *MOD* gene could be duplicated at three loci. The *MOD* locus expressed in the embryo is not affected by the Pervenets mutation as we have proposed that it is silenced.

Lagravère et al. (2004) proposed a new hypothesis based on biochemical differences found between traditional and HO sunflowers. They suggest that two different oleate desaturases could function in sunflower seeds to explain the remaining linoleic acid in HO sunflower.

Merrien et al. (2005) have studied environmental factors that modify the oleic acid content in HO sunflower oil. They measured variations in the coldest temperature during the time of oil deposition and desaturation. Temperature may also affect the intensity of gene silencing on fatty acid composition. More studies are, therefore, required to unravel whether breeding HO sunflower has to be done by targeting either the oleate desaturase or the silencing suppressors. However, in this review we show that most data favor the second hypothesis.

Schuppert et al. (2006) have studied in detail the *MOD* region using RT-PCR strategy. They have found most of the features already presented for the Pervenets mutation organization. However, some details seem different and in their map, the 5.85 kb *Eco*RI fragment is by 4 kb. An explanation of the difference is that they did not clone the genomic fragment and they concatenated the sequences of the PCR products. Because of the other loci revealed by Martínez-Rivas et al. (2001), Schuppert et al. (2006) may have concatenated fragments from different loci. There are many sequence repeats in the *MOD* region and they may have shifted a sequence to the wrong position (Bervillé et al. 2004). However, broadly, there is a good fit between the two maps in the *MOD* region. Bervillé et al. (2004, 2005) have found the ATT SSR in the intron of the *MOD*. The SSR polymorphisms (four alleles) are very useful to map the locus and to follow linkage disequilibrium with the Pervenets insertion. Tang et al. (2006) have shown that the ORS1180 is tightly linked to the Pervenets locus, but there are still many recombination events that are worrisome for breeding.

8.2.4 Linoleic and Linolenic Acids

8.2.4.1 Linoleic and Linolenic Acids in Other Species

Linoleic acid content is an alternative to oleic acid content as we have shown for sunflower and other crops. High linoleic acid oil for industry could be of value, but when over 70% in seed oil, plant growth seems affected.

In soybean, Wang (2006) has found one plant containing zero linolenic acid content with the half-seed method in the strain 0358. It was the first report of a zero linolenic acid soybean. The material will be very useful for breeding cultivated varieties with low linolenic acid content and especially for studying gene action for linolenic acid in soybean by the aid of biotechnology.

Poneleit and Alexander (1965) have studied the inheritance of linoleic and oleic acids in maize. Their results suggest that desaturation of stearic acid to oleic acid is under simple Mendelian control. High linoleic acid content is recessive to low linoleic acid.

When linolenic acid content is too high in the oil, it becomes easily oxidized, and it is unstable. Moreover, the ratio C18:3 to C18:2 and C18:1 in the diet may cause health disorders. This is the case in soybean and rapeseed oils. Silva-Gesteira et al. (2003) have studied the inheritance and the nature and magnitude of gene effects on soybean seed linolenic acid level. They crossed an accession BARC-12 (low linolenic acid content) with a commercial Brazilian cultivar CAC-1 (normal linolenic acid content). Means and variances of F_1, F_2, and F_3 generations have been studied and results demonstrate that linolenic acid content in soybean is under the genetic control of a small number of genes.

Scarth and Tang (2006) have modified *Brassica* oil using conventional and transgenic approaches. The conventional approach to fatty acid modification has explored natural or induced mutations occurring in the same plant species or close relatives within the *Brassica* genus. These mutations have been shown to be associated with a few enzymes in the biosynthetic pathway of the fatty acids. Several types of *Brassica* oil with significantly altered levels of the long chain fatty acid, viz., erucic acid (C22:1) and medium chain fatty acids such as oleic acid (C18:1) and linolenic acid

(C18:3) have been developed for different end uses through conventional breeding. When the necessary genetic variation is not available within *Brassica* species, gene transfer by genetic transformation has been applied, as this approach is not restricted by the sexual incompatibility barrier across species. The fatty acids targeted by the transgenic approach included fatty acids with various carbon chain lengths ranging from C8 to C22, with different numbers of double bonds, and with various functional groups such as epoxy and hydroxy fatty acids. A commercial specialty oil with high level of a novel fatty acid, lauric acid (C12:0) was produced as a result of the transfer of a *FatB* thioesterase gene from a distantly related plant species that produces seed oil with high level of this unusual fatty acid. Considerable progress has been achieved in altering the relative levels of the fatty acids found in *Brassica* oils for increased health and economic benefits and in developing *Brassica* oils, which contain other unusual fatty acids, mainly through genetic transformation. Although the use of natural or induced mutations in the fatty acid biosynthesis within *Brassica* remains a valid option for oil modification, the transgenic approach will play an increasingly important role in the development of *Brassica* oils with altered novel fatty acid composition.

8.2.4.2 Linoleic and Linolenic Acids in Sunflower

In sunflower, as in most Asteraceae seed oils (*Carthamus, Vernonia,*) linolenic acid exists as a trace. Since it is a common feature to the whole taxon, this suggests that this function has been lost when the Asteraceae merged. In contrast, the whole plant tissues display common linolenic acid content in membranes and leaf tissues. The FAD3 enzyme is, therefore, not expressed in the seed whereas it is highly expressed in other tissue. Since some programs would seek to enhance linolenic acid content in sunflower, the underlying mechanism has to be unravelled to start these programs.

In high-oleic sunflower some RILs display a very low oleic level (10–12%) but at this moment we do not know if the trait is heritable to breed high-linoleic sunflower (Lacombe et al. 2002). Another interesting feature is the mid-oleic level of RIL families that did not carry the Pervenets mutation. Up to now, we have no explanation for such a high level of oleic acid content in traditional sunflower.

8.2.5 Wild Annual Sunflowers for Improving Oil Content and Quality in Cultivated Sunflower

We have shown through this review that sunflower oil has the potential to be improved for nutritional and industrial purposes through selection and breeding. The diversity in cultivated sunflower has been broadened by mutations. Seiler (2007) focused on genes from wild *Helianthus* species, resulting in a continuous improvement in agronomic traits. Interest in using wild species in breeding programs has increased, but concerns about the introduction of low oil concentration and quality from the wild species persist. Two annual desert species, *Helianthus anomalus* Blake and *H. deserticola* Heiser, are excellent candidates for increasing oil concentration and enhancing quality based on their adaptation to desert environments. Seiler (2005) reported that the only *H. deserticola* population collected had an average oil concentration of 33%, whereas the two populations of *H. anomalus* had an oil concentration of 43 and 46%, the highest concentration recorded in any wild sunflower species. A linoleic acid concentration of 54% in *H. deserticola* was more typical for a desert environment. The linoleic fatty acid concentration in the oil of *H. anomalus* populations was uncharacteristically high for an Asteraceae and a desert environment, approaching 70%. Further research will be needed to determine the inheritance of the fatty acids and oil concentration in annual *Helianthus* species to determine whether the traits are transferable to cultivated sunflower.

8.3 Tocopherols

Tocopherols prevent oil from becoming rancid (Table 8-1). Sunflower oil is the richest oil for α-tocopherol (vitamin E). Several tocopherols are found in the seeds and their antioxidant property is inversely correlated with their vitamin E activity: α-tocopherol, β-tocopherol, γ-tocopherol and δ-tocopherol and the vitamin E activity is decreased 10-fold at each step.

Hunter and Cahoon (2007) have defined strategies to enhance vitamin E in oilseeds. They provided a recent review of tocopherol and tocotrienol biosynthesis, focusing on branch points and metabolic engineering to enhance and alter vitamin E content and composition in oilseed crops.

Moreover, a few elements of their biosynthetic pathway are known. α-tocopherol is largely predominant in sunflower oil (95%). Demurin et al. (1996) have characterized the genetic variability of tocopherol composition in sunflower seeds developed as isogenic lines in different backgrounds. They named the genes *Tph1* and *Tph2*, which undergo natural variation. They have shown that '*tph1*' (mutated allele) reduces α- and enhances β-tocopherol. The '*tph2*' allele reduces α-tocopherol and β-tocopherol, but enhances γ-tocopherol with a trace of δ-tocopherol. Together, *tph1+tph2* release an equilibrated set of the four compounds.

Garcia-Moreno et al. (2006) have mapped the *Tph1* locus by bulked segregant analysis on LG 1, enabling further marker-assisted selection and positional cloning. They performed genetic and molecular analysis of high γ-tocopherol content in sunflower. Four sources of high γ-tocopherol content

(> 85%) have been developed. The first studies have concluded that the trait in both lines was determined by recessive alleles at the *Tph2* locus. Bulked segregant analysis identified two simple sequence repeat (SSR) markers on LG 8 linked to *Tph2*. A large linkage group was constructed by genotyping additional markers. *Tph2* mapped closely to linked PCR-based markers. The location of the *Tph2* gene on the sunflower genetic map will be useful for marker-assisted selection and further characterization of tocopherol biosynthesis in sunflower seeds.

Tang et al. (2006) have shown that the Ty3/gypsy-like retrotransposon knockout of a 2-methyl-6-phytyl-1,4-benzoquinone methyltransferase produces novel tocopherol (vitamin E) profiles in sunflower. The *m* (*Tph1*) mutation partially disrupted the synthesis of α-tocopherol in sunflower seeds and disrupted a methyltransferase activity necessary for the synthesis of α- and γ-tocopherol. It corresponded to a nonlethal knockout mutation of MT-1 caused by the insertion of a 5.2-kb Ty3/gypsy-like retrotransposon in exon 1. MT-1 and *m* cosegregated and mapped to LG 1. MT-1 was not transcribed in mutant homozygotes (*m/m*). They isolated two 2-methyl-6-phytyl-1,4-benzoquinone/2-methyl-6-solanyl-1,4-benzoquinone methyltransferase (MPBQ/MSBQ-MT) paralogs from sunflower (MT-1 and MT-2) and uncovered a cryptic codominant mutation (*d*). The *m* locus was epistatic to the *d* locus, that is, the *d* locus had no effect in *m*[+] *m*[+] and *m*[+] *m* individuals, but significantly increased γ-tocopherol percentages in *m/m* individuals. MT-2 and *d* cosegregated, MT-2 alleles isolated from mutant homozygotes (*d d*) carried a 30-bp insertion at the start of the 5'-UTR, and MT-2 was more strongly transcribed in seeds and leaves of wild type (*d*[+]/ *d*[+]) than mutant (*d/d*) homozygotes (transcripts were 2.2- to 5.0-fold more abundant in the former than the latter). The double mutant (*m/m*//*d/d*) was nonlethal and produced 24-45% α- and 55-74% β-tocopherol.

8.4 Phytosterols

The sterol biosynthesis pathway produces a large set of phytosterols (sitosterol, campesterol), which are structurally similar to cholesterol and act in the intestine to lower cholesterol absorption. Campesterol also serves as a precursor to the brassinosteroid class of phytohormones. In plants, phytosterols have a structural role in membrane fluidity (Table 8-1). All their functions are not clear; however, in *Arabidopsis*, mutants have helped reveal a role for sterols in plant embryogenesis (Schrick et al. 2002).

Crude sunflower seed oil contains sterols up to 300 mg/100 g of oil, in which β-sitosterol is predominant (60%) and is followed by stigmasterol and campesterol (10% each). A natural variation is observed according to seed lots, but it is still not possible to explain if the variation is due to the genetic background or to environmental factors (Philips et al. 2005).

We have shown that many oil types can be produced by plant species and that sunflower may produce a wide range of oils. We did not focus on the commercial consequences of releasing so many cultivars. Regulations and markets may limit their distribution.

Acknowledgements

Thanks are due to Stéphane Dussert, IRD-Montpellier and Brady Vick, USDA-Fargo, for valuable suggestions and improving the English .

References

Arondel V, Lemieux B, Hang I, Gibson S, Goodman HM, Sommerville CR (1992) Map-based cloning of a gene controlling omega-3 fatty acid desaturation in *Arabidopsis*. Science 258: 1353–1355.

Bervillé A, Lacombe S, Veillet S, Granier C, Léger S, Jouve P (2004) Patent deposit 29 avril 2004 patent application number ep04 291 102.4.

Bervillé A, Lacombe S, Veillet S, Granier C, Léger S, Jouve P (2005) Method of selecting sunflower genotypes with high oleic acid content in seed oil. WO2005106022.

Bessoule JJ, Moreau P (2004) Phospholipid synthesis and dynamics in plant cells. In: G Daum (ed) Topics in Current Genetics. vol 6. Springer-Verlag, Berlin and Heidelberg, Germany, pp 89–124.

Breton C, Souyris I, Villemeur P, Bervillé A (2009) Oil accumulation kinetic along ripening in four olive cultivars varying for fruit size. OCL 16 (2): 1–7.

Brown AP, Affleck V, Fawcett T, Slabas AR (2006) Tandem affinity purification tagging of fatty acid biosynthetic enzymes in *Synechocystis* sp. PCC6803 and *Arabidopsis thaliana*. J Exp Bot 57: 1563–1571.

Browse J, Kunst L, Anderson S, Hugly S, Somerville C (1989) A mutant of *Arabidopsis* deficient in the chloroplast 16:1 / 18:1 desaturase. Plant Physiol 90: 522–529.

Cummins I, Hills MJ, Ross JHE, Hobbs DH, Watson MD, Murphy DJ (1993) Differential, temporal and spatial expression of genes involved in storage oil and oleosin accumulation in developing rapeseed embryos: implications for the role of oleosins and the mechanisms of oil-body formation. Plant Mol Biol 23: 1015–1027.

Del Río-Celestino MR, De Haro-Bailón F (2007) Inheritance of high oleic acid content in the seed oil of mutant Ethiopian mustard lines and its relationship with erucic acid content. J Agri Sci 145: 353–365.

Demurin Y, Skoric D, Karlovic D (1996) Genetic variability of tocopherol composition in sunflower seeds as a basis of breeding for improved oil quality. Plant Breed 115: 33–36.

Fehr WR, Welke GA, Hammond EG, Duvick DN, Cianzio SR (1991) Inheritance of elevated palmitic acid content in soybean seed oil. Crop Sci 31: 1522–1524.

Garcés R, Mancha M (1989) Oleate desaturation in seeds of two genotypes of sunflower. Phytochemistry 28: 2593–2595.

Garcés R, Mancha M (1991) *In vitro* oleate desaturase in developing sunflower seeds. Phytochemistry 30: 2127–2130.

Garcés R, Sarmiento C, Mancha M (1992) Temperature regulation of oleate desaturase in sunflower (*Helianthus annuus* L) seeds. Planta 186: 461–465.

García-Moreno MJ, Vera-Ruiz EM, Fernández-Martínez JM, Velasco L, Pérez-Vich B (2006) Genetic and molecular analysis of high gamma-tocopherol content in sunflower. Crop Sci 46: 2015–2021.

Hongtrakul V, Slabaugh MB, Knapp SJ (1998a) DFLP, SSCP, and SSR markers for delta 9-stearoyl-acyl carrier protein desaturases strongly expressed in developing seeds of sunflower: intron lengths are polymorphic among elite inbred lines. Mol Breed 4: 195–203.

Hongtrakul V, Slabaugh MB, Knapp SJ (1998b) A seed specific Δ12 oleate desaturase gene is duplicated, rearranged, and weakly expressed in high oleic acid sunflower lines. Crop Sci 38: 1245–1249.

Hunter SC, Cahoon EB (2007) Enhancing vitamin E in oilseeds: Unraveling tocopherol and tocotrienol biosynthesis. Lipids 42: 97–108.

Ivanov P, Georgiev I (1981) Biochemical characteristics of some sunflower mutant forms. EUCARPIA Meet, 26–30 Oct 1981, Prauge, Czechoslovakia, pp 256–266.

Jellum MD, Widstrom NW (1983) Inheritance of stearic acid in germ oil of the maize kernel. *J Hered* 74: 383–384.

Jung S, Swift D, Sengoku E, Patel M, Teule F, Powell G, Moore K, Abbott A (2000a) The high oleate trait in the cultivated peanut *Arachis hypogaea* L. I. Isolation and characterization of two genes encoding microsomal oleoyl-PC desaturases. Mol Gen Genet 263: 796–805.

Jung S, Powell G, Moore K, Abbott AG (2000b) The high oleate trait in the cultivated peanut *Arachis hypogaea* L. II. Molecular basis and genetics of the trait. Mol Gen Genet 263: 806–811.

Kabbaj A, Abbott AG, Vervoort V, Bervillé A (1995) Expression of stearate, oleate and linoleate desaturase genes in sunflower with normal and high oleic contents. 3rd Eur Sunflower Biotechnol Conf, 30 Oct–2 Nov 1995 Giessen, Germany, p 15.

Kabbaj A, Abbott AG, Vervoort V, Tersac M, Bervillé A (1996a) Characterization of genes for stearate, oleate and linoleate desaturase and expression in sunflower with normal and high oleic contents, cloning of Δ9 and Δ12 genes and variability in *Helianthus*. 14th Int Sunflower Conf, June 12–20, Beijing, vol 2, pp 1035–1038.

Kabbaj A, Vervoort V, Abbott AG, Tersac M, Bervillé A (1996b) Polymorphism in *Helianthus* and expression of stearate, oleate and linoleate desaturase genes in sunflower with normal and high oleic contents. Helia 19(25): 1–17.

Kachroo A, Lapchyk L, Fukushige H, Hildebrand D, Klessig D, Kachroo P (2003) Plastidial fatty acid signaling modulates salicylic acid- and jasmonic acid-mediated defense pathways in the *Arabidopsis ssi2* mutant. Plant Cell 15: 2952–2965.

Kusaba M, Miyahara K, Iida S, Fukuoka H, Takano T, Sassa H, Nishimura M, Nishio T (1993) Low glutelin content 1: A dominant mutation that suppresses the glutelin multigene family via RNA silencing in rice. Plant Cell 5: 931–940.

Lacombe S, Bervillé A (2000) Problems and goals in studying oil composition variation in sunflower. Proc 15th Int Sunflower Conf, June 12–15, Toulouse, France, pp D16–27.

Lacombe S, Bervillé A (2001) A dominant mutation for high oleic acid content of sunflower (*Helianthus annuus* L.) oil is genetically linked to a single oleate-desaturase RFLP locus. Mol Breed 8: 129–137.

Lacombe S, Tersac M, Kaan F, Griveau Y, Serieys H, Abbott A, Bervillé A (1998) RFLP profiles in low oleic sunflower using a stearoyl ACP and an oleoyl PC desaturase cDNA. 4th Eur Sunflower Biotechnol Conf, Oct 20–23, Montpellier, France, p 81.

Lacombe S, Lambert P, Cellier F, Casse F, Bervillé A (1999) RFLP profiles in low oleic sunflower using SDI-, a stearoyl-ACP, and an oleoyl-PC desaturases cDNAs. Helia 22(30): 19–28.

Lacombe S, Kaan F, Léger S, Bervillé A (2001) An OD and a suppressor loci direct high oleic acid content of sunflower (*Helianthus annuus* L.) oil in the Pervenets mutant. CR Acad Sci Paris, Série III 324: 839–845.

Lacombe S, Léger S, Kaan F, Bervillé A (2002) Genetic, molecular and expression features of the Pervenets mutant leading to high oleic acid content of seed oil in sunflower. OCL-Ol Corps Gras Li 9: 17–23.

Lacombe S, Souyris I, Berville A (2009) An insertion of oleate desaturase homologous sequence silences via siRNA the functional gene leading to high oleic acid content in sunflower seed oil. Mol Genet Genom281: 43–54.

Lagravère T (1999) Agro-physiological determinism of the expression of the very high oleic acid level in the sunflower oil (*Helianthus annuus* L.)). Institut National Polytechnique de Toulouse, Toulouse, FRANCE INIST-CNRS, Cote INIST: T 131273.

Lagravère T, Kleiber D, Dayde J (1998) Cultural practices and sunflower oilseed agronomic performance: facts and the future OCL-Ol Corps Gras Li 5: 477–85.

Lagravère T, Champolivier L, Lacombe S, Kleiber D, Bervillé A, Daydé J (2000) Effect of temperature variations on fatty acid composition of the oil in standard and high oleic sunflower hybrids. Proc 15th Int Sunflower Conf, June 12–15, Toulouse, France, pp A73–78.

Lagravère T, Lacombe S, Kleiber D, Surel O, Calmon A, Bervillé A, Dayde J (2004) Comparison of fatty acid metabolism of two oleic and one conventional sunflower hybrids: a new hypothesis. J Agron Crop Sci 190: 223–229.

Lightner J, Wu J, Browse J (1994) A mutant of *Arabidopsis* with increased levels of stearic acid. Plant Physiol 106: 1443–1451.

Liu Q, Singh S, Green A (2002) High-oleic and high-stearic cottonseed oils: nutritionally improved cooking oils developed using gene silencing. J Am Coll Nutr 21: 205S–211S.

Martínez-Force E, Cantisan S, Serrano-vega MJ, Garces R (2000) Acyl-acyl carrier protein thioesterase activity from sunflower (*Helianthus annuus* L.) seeds. Planta 211: 673–678.

Martínez-Rivas JM, Sperling P, Lühs W, Heinz E (2001) Spatial and temporal regulation of three different microsomal oleate desaturase genes (FAD2) from normal-type and high-oleic varieties of sunflower (*Helianthus annuus* L.). Mol Breed 8: 159–168.

Mongrand S, Bessoule JJ, Cabantous F, Cassagne C (1998) The C16: 3\ C18: 3 fatty acid balance in photosynthetic tissues from 468 plant species. Phytochemistry 49: 1049–64.

Merrien A, Pouzet A, Krouti M, Dechambre J, Garnon V (2005) Contribution à l'étude de l'effet des températures basses sur la composition en acide gras de l'huile des akènes de tournesol. OCL-Ol Corps Gras Li 12: 455–458.

Miller JF, Vick BA (1999) Inheritance of reduced stearic and palmitic acid content in sunflower seed oil. Crop Sci 39: 364–367.

Nikolova V, Ivanov P, Petakov D (1991) Inheritance of sunflower oil quality following crossing of lines with different fatty acid composition. Genet Breed 24: 391–397.

Pérez-Vich B, Garcés R, Fernández-Martínez JM (2000) Epistatic interaction among loci controlling the palmitic and the stearic acid levels in the seed oil of sunflower. Theor Appl Genet 100: 105–111.

Pérez-Vich B, Knapp SJ, Leon AJ, Fernández-Martínez JM, Berry ST (2004) Mapping minor QTL for increased stearic acid content in sunflower seed oil. Mol Breed 13: 313–322.

Pérez-Vich B, Velasco L, Muñoz-Ruz J, Fernández-Martínez JM (2006) Inheritance of high stearic acid content in the sunflower mutant CAS-14. Crop Sci 46: 22–29.

Petakov D, Lacombe S, Koubaa M, Ivanov P, Bervillé A (2000) Molecular characterization of a new oleic sunflower mutation obtained from Peredovick using seed mutagenesis. IAEA Report

Phillips KM, Ruggio DM, Ashraf-Khorassani M (2005) Phytosterol composition of nuts and seeds commonly consumed in the United States. J Agri Food Chem 53: 9436–9445.

Poneleit CG, Alexander DE (1965) Inheritance of linoleic and oleic acids in maize. Science 147: 1585–1586.

Roscoe TJ, Lessire R, Puyaubert J, Renard M, Delseny M (2001) Mutations in the fatty acid elongation 1 gene are associated with a loss of 3-ketoacyl-CoA synthase activity in low erucic acid rapeseed. FEBS Lett 429: 107–111.

Salas JJ, Moreno-Pérez AJ, Martínez-Force E, Garcés R (2007) Characterization of the glycerolipid composition of a high-palmitoleic acid sunflower mutant. Eur J Lipid Sci Technol 109: 591–599.

Sarmiento C, Mancha M, Garcés R (1994) Microsomal polypeptides in sunflower (*Helianthus annuus*). Comparison between normal type before and after cold-induction, and a high oleic acid mutant. Physiol Plant 91: 97–103.

Sarmiento C, Garcés R, Mancha M. (1998) Oleate desaturation and acyl turnover in sunflower (*Helianthus annuus* L.) seed lipids during rapid temperature adaptation. Planta 205: 595–600.

Scarth R, Tang J (2006) Modification of *Brassica* oil using conventional and transgenic approaches. *Crop Sci* 46: 1225–1236.

Schierholt A, Becker HC, Ecke W (2000) Mapping a high oleic acid mutation in winter oilseed rape (*Brassica napus* L.). Theor Appl Genet 101: 897–901.

Schmidt RJ, Ketudat M, Aukerman MJ, Hoschek G (1992) Opaque-2 is a transcriptional activator that recognizes a specific target site in 22-kD zein genes. Plant Cell 4: 689–700.

Schrick K, Mayer U, Martin G, Bellini C, Kuhnt C, Schmidt J, Jürgens G (2002) Interactions between sterol biosynthesis genes in embryonic development of *Arabidopsis*. Plant J 31: 61–73.

Schnurbusch T, Möllers C, Becker HC (2000) A mutant of *Brassica napus* with increased palmitic acid content. Plant Breed 119: 141–144.

Schuppert GF, Tang S, Slabaugh MB, Knapp S (2006) The sunflower high-oleic mutant *Ol* carries variable tandem repeats of FAD2-1, a seed-specific oleoyl-phospatidyl choline desaturase. Mol Breed 17: 241–256.

Seiler GJ (2007) Wild annual *Helianthus anomalus* and *H. deserticola* for improving oil content and quality in sunflower. Ind Crop Prod 25: 95–100.

Silva-Gesteira A da, Schuster I, Chamel José I, Deniz Piovesan N; Soriano Viana JM, Gonçalves de Barros E, Alves Moreira M (2003) Biometrical analyses of linolenic acid content of soybean seeds. Genet Mol Biol 26: 65–68.

Škoriæ D, Demurin Y, Jociæ S (1996) Development of hybrids with various oil quality. Proc 14th Int Sunflower Conf, 12–20 June 1996, Beijing, PR China, pp 54–59.

Sobrino E, Tarquis AM, Cruz Díaz M (2003) Modeling the oleic acid content in sunflower oil. *Agron J* 95: 329–334.

Soldatov KI (1976) Chemical mutagenesis in sunflower breeding. Proc 7th Int Sunflower Conf, June 27–July 3, Krasnodar, vol 1, pp 352–357.

Tang S, Hass CG, Knapp SJ (2006) Ty3/gypsy-like retrotransposon knockout of a 2-methyl-6-phytyl-1,4-benzoquinone methyltransferase is non-lethal, uncovers a cryptic paralogous mutation, and produces novel tocopherol (vitamin E) profiles in sunflower. Theor Appl Genet 113: 783–799.

Wang J (2006) Selection of breeding materials with high linoleic acid and/or low linolenic acid content in soybean. J Am Sci 2: 29–32.

Wilson RF, Marquardt TC, Novitzky WP, Burton JW, Wilcox JR, Kinney AJ, Dewey RE (2001) Metabolic mechanisms associated with alleles governing the 16:0 concentration of soybean oil. J Am Oil Chem Soc 78: 335–340.

Transgenic Sunflower

Miguel Angel Cantamutto* and María Mónica Poverene

ABSTRACT

At the beginning of the new millennium, sunflower relied on traditional non-GM varieties, which could be a possible disadvantage in relation to other oilseed crops. Different transformation protocols allowed obtaining desirable transgenic traits: the oxalate-oxidase expressing gene for fungal disease control, glyphosate tolerance by expressing *Agrobacterium* gene *cp4*, and the Bt toxin gene *cry1* to control Lepidoptera. Ammonium absorption, other herbicide tolerance, *Cry1* variants and the *CpT1* gene for pest control are new targets. Quality can be improved through a modified protein and latex biosynthesis. The high cost of the GM crop approval process, increased by the probability of gene flow with wild relatives, is a constraint for the transgenic sunflower release. Strategies designed to minimize environmental risks and changes in consumer perception are needed before sunflower transgenic varieties become acceptable.

Keywords: disease; pest; herbicide tolerance; transformation; environmental risk

9.1 Introduction

Towards the end of the first biotech decade after the emergence of genetically modified (GM) crops, sunflower still relied on traditional non-GM varieties and faced an unprofitable position compared to other extensively grown crops. Sunflower seed contribution amounts only to 1.24% of the world production of cereal and oilseed grains, estimated as 2.621 million tons for

Agronomy Department, Universidad Nacional del Sur, 8000, Bahía Blanca, Argentina.
*Corresponding author: *mcantamutto@yahoo.com*

2008-09 (USDA 2009a, b). On the other hand, soybean provides 53% of the total oilseed production, while wheat, corn, and rice each contributes more than 20% of cereals. Corn hardly competes with sunflower for land usage, and 48% of total corn production comes from only four of the world's big producers, which have accessed GM varieties. Among them, insect resistant (IR) GM corn varieties have been adopted in Argentina and South Africa, being herbicide tolerant (HT) GM corn is preferred by the farmers in US and Canada (Brookes and Barfoot 2006; James 2008).

Soybean competes with sunflower not only for land usage but also for oilseed markets; 84% of the production comes from six countries that have accessed GM technology, adopted for 99% of acreage in Argentina, 93% in the US and Paraguay, 60% in Canada, and 40% in Brazil. Rapeseed (canola) and cotton also compete with sunflower for oil markets and both show a high adoption of GM varieties, representing 95% and 74% of acreage in the US and Australia, respectively (Brookes and Barfoot 2006).

The increment in GM crop technology adoption in 25 countries reached 15% annually during the last 12 years, raising acreage to more than 125 million hectares. This technology has economically benefited soybean, corn, rapeseed (canola), and cotton farmers, allowing them to produce two crops a year (as wheat-soybean, or wheat-corn) and save insecticide applications (James 2008). However, GM technology has not determined the increment in cereal and some oilseed world production. The countries that contribute mostly to the total production of sunflower (three countries, 51%), wheat (four countries, 52%), rapeseed (three countries, 55%), rice (eight countries, 80%), and corn (eight countries, 76%) raised the total yield independently of GM technology access in the latter part of the past century (Fig. 9-1 a, b). However, soybean world production greatly increased only in those countries, which have adopted GM technology (Fig. 9-1 b).

Sunflower oil is light in taste and appearance and supplies more vitamin E than any other vegetable oil (Chapter 1). It is a combination of monounsaturated and polyunsaturated fats with low saturated fat levels (NSA 2009). The versatility of this healthy oil is recognized by cooks internationally. Sunflower oil is valued for its light taste, frying performance, and health benefits. There are three types of sunflower oils which are available, NuSun, linoleic and high oleic sunflower oil (Jan and Seiler 2007).

Even though sunflower does not command the high prices of other edible oils, for which there is a greater demand in most select markets, its price is remarkably higher than that of its competitors (Fig. 9-2). In Argentina —the main world contributor for the 2003–08 lapse providing 31% of total production (NSA 2009)—both edible oils moved up, but the refined sunflower oil quoted 17% over soybean, reaching up to US$1090 per ton as differential price (Obschatko et al. 2006; BCBA 2009). The increment in crop-derived biofuel use (bioethanol and biodiesel) also raised the international prices in

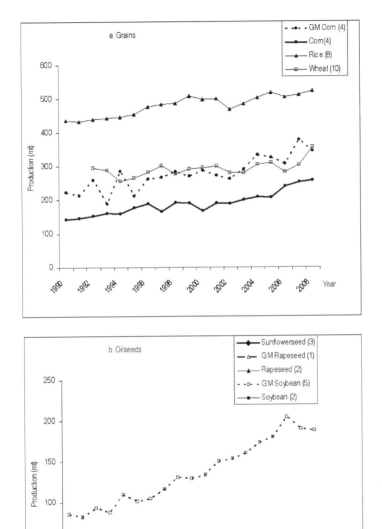

Figure 9-1 Grain (a) and oilseed (b) production of selected countries contributing > 44% of the world production (number of countries between parentheses). Dotted lines correspond to countries where the GM crops are available. Sources: USDA (2009a, b) and James (2008).

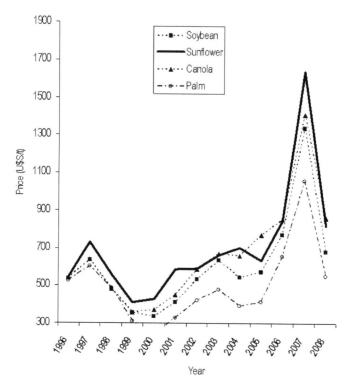

Figure 9-2 Oil prices (US$/t) of main vegetable sources. Palm oil corresponds to Malay market and Rotterdam to the others. Source USDA (2009b).

that period, creating outstanding market conditions for oilseed crops in 2008 (FAO 2008).

Sunflower belongs to the genus *Helianthus* (Asteraceae), which comprise about 50 species native to North America (Heiser et al. 1969). It is a comparatively novel extensive crop, with different uses as oilseed crop, edible confection, birdseed, and ornamental (Jan and Seiler 2007). Cultivated sunflower *H. annuus* L. var. *macrocarpus* derives from the species domestication by Native Americans before the discovery of America, followed by Russian farmers' selection and the breeding done at the former Soviet Union (FSU) experimental stations (Krasnodar, Saratov, and lately Rostov). This intense selection work was successful in the creation of several open-pollinated varieties, rapidly diffused to the entire world in the early 20th century (Vranceanu 1977).

Hybrid cultivars, obtained through crosses among inbred lines, were developed during the second half of the 20th century (Chapter 2). The extensive use of hybrid cultivars becomes possible thanks to cheapening of hybrid seed production through the use of cytoplasmic male sterility (Leclercq

1969). Commercial sunflower hybrids were the outcome of breeding techniques including interspecific crosses, induced mutation, marker-assisted selection, as in other extensive crops. The improvement made the crop a profitable and competitive option for Russia, Ukraine, and Argentina (Chapter 1), comprising 51% of world production estimated in more than 32 million ton for 2008–09 (USDA 2009b).

In the beginning of the new millennium, there are no available GM sunflower varieties for farmers around the world. Along with peanut - a crop in which breeding is handled mostly by private companies (Frey 1996), sunflower is among the five main oilseed crops that lack released transgenic varieties. GM sunflower has been evaluated only under controlled conditions, and its release for commercial use seems unlikely at present. Is this a comparative disadvantage in relation to other oilseed crops? What ecological, biotech, or economic constraints hinder release of GM sunflower? How would GM sunflower modify the productive chain? How should GM sunflower be managed to avoid undesirable environmental impacts? In this chapter, we will try to answer these and other issues related to GM sunflower.

9.2 Released Transgenic Crops

Among the 19 crop species transformed by genetic engineering and released for extensive sowing, soybean, corn, rapeseed, and rice along with cotton stand out because of the wide distribution of GM varieties expressing herbicide tolerance (HT) and/or insect resistance (IR) traits (James 2008). This process is the outcome of an intense biotech-based breeding activity initiated in 1985, which only in the US involves more than 13,000 notifications and permit releases for field experiments (ISB 2007).

By far, the crop that has benefited the most from this novel technology is corn. Among others, the planned IR for obtaining resistance against European corn borer (ECB) *Ostrinia nubilalis,* can be considered one of the most successful cases, where GM technology got rid of yield restraints. GM corn with IR was obtained by transfer of a gene coding for Cry1Ab protein, an insecticidal δ-endotoxin from *Bacillus thuringiensis* (Bt) that has been used in pest control since the beginning of the last century. GM maize varieties expressing *Cry1Ab* gene from Bt were available for the first time in the US after approval of the MON810 event (Monsanto) in 1995. Overcoming the difficulties for ECB control facilitated a fast adoption of this genetically modified organism (GMO) including the European Union (EU) - one of the environments where people largely reject usage of transgenics,- after approval for sowing in Spain since 1998 (AGBIOS 2007).

The second highlight was soybean GM Roundup Ready® variety (Monsanto 2002) tolerant to glyphosate herbicide (GL)—an inhibitor of the

5-enolpyruvylshikimate-3-phosphate (EPSPS) enzyme. RR soybean was obtained through expression of a gene coding for an EPSPS enzyme variant from *Agrobacterium tumefaciens*, insensitive to GL herbicide, inserted by the GTS 40-3-2 event of the same company. It has been available to farmers since 1994–95 in the US, Canada, Japan, and Argentina (Brookes and Barfoot 2006). In the latter case, soybean acreage increased by 10% per year during the first decade and the GM varieties almost reached 100%. This increase was reinforced because it facilitated zero-till practices, enhancing harvest security of the double-crop system (wheat- soybean) and soybean production in less favorable areas (Trigo and Cap 2006). In the US, adoption rate was lower because it did not impact so strongly on the farmers' economic profit, but allowed obtaining an "extra-farm" income due to the simplicity and flexibility of the applied technology (Fernandez-Cornejo et al. 2006).

9.3 Genetic Modification of Sunflower

9.3.1 Early Transformation Methods

Sunflower has been considered as a recalcitrant species to genetic transformation. Despite the lack of an efficient protocol, more than 40 publications have reported success in obtaining transgenic sunflower plants and many field experiments have been performed. The first transgenic sunflower was obtained by Everett et al. (1987) from hypocotyl-derived callus that also produced transgenic shoots through disarmed *Agrobacterium tumefaciens* plasmids. Even though weak and stunted, the transformed plants produced pollen and were identical to the wild type plants of the original inbred line.

Early efforts in sunflower biotechnology included embryo rescue, somatic cell fusion, plant regeneration by direct or indirect organogenesis and embryogenesis from explants, gynogenesis and anther culture (Friedt 1992; Alibert et al. 1994). Chimeric expression of genes was achieved earlier, but recovery of transformed plants from protoplasts remained very difficult until Schrammeijer et al. (1990) succeeded in meristem transformation from split embryonic axes of immature embryos. Since then, sunflower transformation based on microprojectile bombardment of half shoot apices aimed to increase the bacterial infection, in combination with A. *tumefaciens* co-culture and kanamycin as a selective agent (Bidney et al. 1992; Knittel et al. 1994; Malone-Schoneberg et al. 1994; Grayburn and Vick 1995; Laparra et al. 1995; Burrus et al. 1996; Lucas et al. 2000). Biolistic (DNA-coated particle bombardment) efficiency highly relied on the plant material (Hunold et al. 1995). Modifications for improving transgene efficiency, included selectable markers (Escandon and Hahne 1991; Müller et al. 2001), pectinase treatment of explants (Alibert et al. 1999), dehydration and rehydration

(Hewezi et al. 2002), and cotransferring of the *ipt* gene from *Agrobacterium vitis* (Molinier et al. 2002). Two detailed sunflower transformation protocols, one for highly responding genotypes and a second one for recalcitrant genotypes, were published by Hewezi et al. (2004). Gene transfer, earlier recognized as an instrument to overcome sexual incompatibility in interspecific hybridization (Skoric 1992), was used for transformation of sexual (Pugliesi et al. 1993) and somatic interspecific hybrids (Fambrini et al. 1996; Henn et al. 1998; Taski-Ajdukovic 2006); but plant regeneration was genotype-dependent and of low efficiency (Kallerhoff and Alibert 1996; Faure et al. 2002b). The partial genome transfer from perennial *Helianthus* species performed by Binsfeld et al. (2000) emerged as a potential tool for transferring polygenic traits or alien genes. Advances in sunflower genomics made since 1995 have greatly enhanced the development and application of new tools for crop improvement (Paniego et al. 2007).

9.3.2 General Procedures to Transform Sunflower

Detailed techniques for sunflower transformation have been patented (*www.patentstorm.us; www.freepatentsonline.com*). A typical expression vector contains prokaryotic DNA elements and an antibiotic resistance gene, a cloning site for insertion of the exogenous DNA sequence, eukaryotic DNA elements such as a promoter, an optional enhancer, a transcription termination/ polyadenylation sequence, and sequences that are necessary to allow for the eventual integration of the vector into a chromosome. The vector also contains a gene encoding a selection marker, which is functionally linked to promoters that control transcription initiation. The promoter sequences can be constitutive or inducible, environmentally- or developmentally-regulated, or cell- or tissue-specific. A preferred selection marker gene for sunflower transformation is the neomycin phosphotransferase II (*npt*II) gene, which confers resistance to kanamycin when placed under the control of plant regulatory signals. Another useful marker, which allows quantification or visualization of the spatial pattern of expression of the DNA sequence in specific tissues, is the reporter gene β-glucuronidase (*GUS*). Methods for introducing DNA into sunflower cells comprise bacterial infection, artificial chromosome vectors, and direct delivery of DNA. Co-cultivation with *Agrobacterium tumefaciens* or combined with biolistic methods is used for sunflower transformation. Transformed tissues include hypocotyls, apical meristems, and protoplasts (Table 9-1). Cotyledons can often give rise to shoots without an intervening callus stage, producing whole plants more rapidly and efficiently. Regeneration from shoot meristems usually leads to the formation of chimaeras; this can be avoided through transformation protocols based on non-meristematic regeneration (Rao and Rohini 1999; Mohamed et al. 2003). Some examples are hypocotyl regenerants, which

Table 9-1 Explants used for sunflower plant regeneration through organogenesis and somatic embryogenesis.

Tissue	References
Organogenesis	
Cotyledon	Knittel et al. 1991; Chraibi et al. 1991, 1992a, b; Ceriani et al. 1992; Deglene et al. 1997; Baker et al. 1999; Flores Berrios et al. 1999; Dhaka and Kothari 2002; Mayor et al. 2003
Shoot tip or embryonic axis	Bohorova et al. 1986; Lupi et al. 1987;Bidney et al. 1992; Knittel et al. 1994; Molinier et al. 2002
Hypocotyl	Paterson and Everett 1985; Lupi et al. 1987
Immature embryo	Power 1987; Espinasse and Lay 1989; Bronner et al. 1994; Jeannin et al. 1995
Leaves	Greco et al. 1984; Lupi et al. 1987
Protoplasts	Burrus et al. 1991; Krasnyanski and Menczel 1993; Wingender et al. 1996
Somatic embryogenesis	
Hypocotyl	Paterson and Everett 1985; Freyssinet and Freyssinet 1988; Pelissier et al. 1990; Prado and Berville 1990
Immature embryo	Finner 1987; Freyssinet and Freyssinet 1988; Witrzens et al. 1988; Jeannin and Hahne 1991; Bronner et al. 1994; Jeannin et al. 1995; Sujatha and Prabhakaran 2001
Anther	Thengane et al. 1994
Unpollinated ovary	Gelebart and San 1987

can be screened for green fluorescence protein, GFP (Müller et al. 2001; Hess 2006). Rooting proved to be an effective selection method when using kanamycin as a selective agent, with no detected escapes (Radonic et al. 2006). Regeneration largely depends on genotype, showing cytoplasmic and nuclear effects (Nestares et al. 1998, 2002; Vasic et al. 2001; Azadi et al. 2002). Genetic control of organogenesis is polygenic, mainly additive (Sarrafi et al. 1996; Flores-Berrios et al. 2000), and might be improved based on the combining ability of related traits (Bolandi et al. 2000; Mayor et al. 2003, 2006). More efficient protocols have been developed through phenylacetic acid (Dhaka and Kothari 2002), cotransformation with the *ipt* gene of *Agrobacterium vitis*, which is involved in cytokinin synthesis (Molinier et al. 2002), macerating enzymes (Weber et al. 2003), rooting promotion (Koopmann and Kutschera 2005), and biolistics (Mohamed et al. 2006a, b). Post-transcriptional gene silencing, a mechanism that organisms have developed to destroy invasive nucleic acids, can be achieved by RNA hybridized to transgene target sequences (Hewezi et al. 2005). Severe developmental abnormalities following silencing suggested that some genes could have a second role in developmental regulation (Hewezi et al. 2006).

After screening, transgenic plants are established by conventional techniques and represent the T_0 generation. A preferred technique is to establish plantlets by in vitro grafting of regenerated sunflower shoots from split embryonic axis explants. T_0 plants are used as pollen donors to obtain T_1 and T_2 as successive hybrids with sunflower elite lines. The primary transformants are typically small and stunted but T_1 plants are normal. Transgenes are usually dominant in their inheritance and segregate in a regular Mendelian fashion, but several copies of T-DNA are the rule (2 to 10), sometimes as incomplete, truncate segments.

9.3.3 *Intended Transgenic Sunflower Releases*

Since 1991, there was a continuous growth in a number of environmentally-controlled field experiments (Fig. 9-3; Cantamutto and Poverene 2007). However, since 2000 sunflower notifications have declined , both in the US and in Argentina, compared to corn and soybean notifications. Private seed companies have concentrated their efforts in three agronomically important transgenic traits: the oxalate-oxidase expressing gene for fungal disease control, glyphosate tolerance by expressing *Agrobacterium* EPSPS gene *cp4*, and the Bt toxin gene *cry1* for the control of Lepidoptera (Table 9-2).

Oxalic acid is a key component of infection by *Sclerotinia sclerotiorum* and other fungi. The enzyme oxalate oxidase confers disease resistance through the activation of defense genes in transgenic sunflower plants

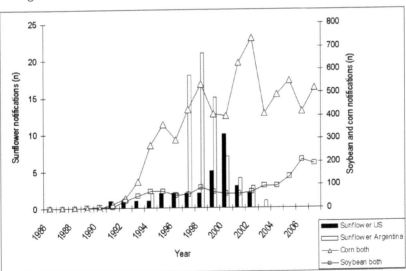

Figure 9-3 Number of GM sunflower environmental release (bars), GM corn and GM soybean in the US and Argentina. Sources: *http: //www.isb.vt.edu/cfdocs/isblists2.cfm?opt=4* and *http: //www.sagpya.mecon.gov.ar/new/0-0/programas/conabia/liberaciones_ogm.php.*

Table 9-2 Notifications of sunflower expressed traits for environmental release in the US, UE and Argentina[1].

Intended effect	Period	Number	Responsible (Country[2])	Brief Description
Fungal resistance	1995–2003	77	Pioneer (FR, AR, US), Syngenta (US), INTA (AR), Zeneca (AR), Advanta (AR), Van der Haven (NL)	Oxalate oxidase (OXO) synthesis by expression of wheat or barley genes conferring resistance to *Sclerotinia sclerotiorum*. Events PHP10335, 10521, TF 22, 23, 24, 28, 34, pHGCA39, 35, pVDH476, PH9755 and others
Resistance to Lepidoptera	1994–2003	55	Mycogen (AR), Pioneer (AR), Van der Have (AR, NL), INTA (AR), Dow (AR)	Bt-derived insect resistance mediated by endotoxin synthesis (Cry1F) from *Bacillus thuringensis*. Events PHP 10092, PHP11486, P 8999
Enhanced protein quality	1989–1993	12	Pioneer (US), Van der Have (US)	Storage protein from Brazil nut (*Bertholletia excelsa*) with high methionine content
Tolerance to glyphosate	1998–2002	6	Monsanto (AR, US), INTA (AR)	5-enolpyruvylshikimate-3-phosphate synthase (EPSPS) synthesis by expression of *Agrobacterium tumefaciens* genes. Events pVCM1,pVNAP1, pMON20999 SFB0193 SFB0216
Tolerance to glufosinate ammonium	1994–1997	4	Zeneca (AR), Van der Have (NL)	Phosphinothricin acetyltransferase (PAT) synthesis by expression of *Streptomyces hygroscopicus* or *S. viridochromogenes* genes. Event TSF43
Resistance to Coleoptera	1994–1997	4	Van der Have (US, NL), Zeneca (AR)	Cowpea (*Vigna unguiculata*) trypsin inhibitor synthesis (CpT1) plus snowdrop lectin (Nptll-SM)
Nitrogen assimilation	1996–1997	4	Zeneca (AR), Van der Have (US, NL)	Ammonium incorporation by asparagine synthetase (AS) or increased N assimilation by nitrate reductase or nitrite reductase synthesis
Rubber yield increased	2002–2005	4	Colorado State University (US)	Enhanced quantity and quality of rubber production by expression of the synthesis complex of Guayule (*Parthenium argentatum*)
Modified stearate content	1997–1999	2	Rustica Prograin Génetique (FR)	High stearate content. Reduction of stearic acid content
Others	1995	1	Van der Have (NL, FR, SP, AR)	Albumin, asparagine, chalcone, chitinase, fructosyltransferase, glucanase or levan sucrase synthesis. Chlorsulphuron tolerance, fungal resistance, male sterility/fertility restoration, drought tolerance, marker system, MAC promoter.

[1]Sources: Cantamutto and Poverene 2007; *http://www.aphis.usda.gov/brs/status/notday.html*; *http://www.sagpya.gov.ar/biotecnologia/conabia* (Accessed May 9, 2009). [2]AR = Argentina, US = United States, FR = France, NL = The Netherlands, SP = Spain.

expressing the wheat germin *gf-2.8 OXO* gene (Bazzalo et al. 2000; Scelonge et al. 2000). The leaves of *OXO* transgenic sunflower constitutively expressed the gene, showing elevated levels of OXO activity, hydrogen peroxide (H_2O_2), salicylic acid, and defense gene expression. OXO releases CO_2 and H_2O_2 from O_2 and oxalic acid (OA). H_2O_2 appears to have roles in signal transduction cascades that coordinate various defense responses, such as induction of hypersensitive response and synthesis of pathogenesis-related proteins and phytoalexins (Hu et al. 2003).

The human lysozyme gene was also found to exert resistance to *Sclerotinia sclerotioum* (Sawahel and Hagran 2006). Zygotic embryos underwent *Agrobacterium*-mediated transformation and white mold resistance was stably inherited. Co-transformed plants expressing the glucanase gene from *Medicago sativa* and the chitinase gene from *Oryza sativa* proved resistant against *Botrytis cinerea* and *S. sclerotiorum* until T_3 generation (Hess 2006).

Some *Bacillus thuringiensis* (Bt) proteins are highly toxic to pests, such as insects, and are specific in their toxic activity, such as Cry1 (Lepidoptera-specific), Cry2 (Lepidoptera- and Diptera-specific), Cry3 (Coleoptera-specific), and Cry4 (Diptera-specific). The Bt *Cry1F* gene was attached to a constitutive plant promoter and T_2 sunflower plants were subjected to enzyme-linked immunosorbent assay (ELISA) and insect bioassay tests, showing resistance to severe attacks of *Rachiplusia nu* larvae (Pozzi et al. 2000).

The transgenic approach has also been used to increase oil quality related changes in the saturate/unsaturated fatty acid profiles in the sunflower fatty acid biosynthesis pathways. A reduced stearic acid sunflower was achieved introducing a castor (*Ricinus communis*) δ-9-stearoyl-(acyl carrier protein) desaturase (Rousselin et al. 2002). *Agrobacterium*-mediated transformation of embryonic axes under the control of seed-specific promoters allowed stable expression of the trait over five generations.

9.4 Overcoming of Crop Constraints through Transgenic Sunflower and New Products

9.4.1 *Low Soil Fertility*

Transgenic traits that are in an experimental stage have focused on overcoming the main constraints of cropping, and have been designed for the control of crop adversities. Sunflower mineral nutrition research has mainly concentrated on macronutrients in general and nitrogen in particular (Table 9-3). Sunflower is a highly nitrogen-dependant crop that unlike soybean does not fix atmospheric nitrogen. Hence, nitrogen accumulation may be affected under low availability conditions, as in zero-till systems in semiarid lands (Lopez-Bellido et al. 2003).

In plants, ammonium absorption, which is an alternative pathway to the nitrogen cycle, is performed through the glutamine synthetase (GS) enzyme. However, in darkness and with a low available C:N ratio, some variants of asparagine synthetase (AS) enzyme, coded by *HAS1* and *HAS1.1* genes provisionally store N as asparagine, thereby preventing ammonium intoxication (Herrera Rodriguez et al. 2004). In GM plants, AS can substitute GS under conditions that limit its activity (such as in *Medicago truncatula*; Carvalho et al. 2000) and act as an alternative N-storing metabolic pathway (as in *Nicotiana tabacum*; Ferrario-Méry et al. 2002). AS expression in GM sunflower may therefore improve N-metabolism and contribute to a more efficient use of this element.

Table 9-3 Research articles about sunflower crop mineral nutrition published during five years (CAB 2002–2007).

Group		Element		Articles	
	#	Detail		#	%
Macronutrientes	4	N-P-K-S		19	46
		only N		16	39
Micronutrients	6	B-Fe-Zn-Ca-Mg-Si		15	37
		only B		9	22
Heavy metals	2	Cd-Se		3	7
Others	3	inorganic amend (lime, gypsum)		3	7
		salt, salinity		2	5
		rhizosphere, mycorrizae		2	5
Total mineral nutrition	15			41	100

9.4.2 Weed Control

Although no-till adoption in sunflower would be limited by the lack of wide range herbicide-tolerant varieties, weed interference seems less attractive to researchers than pests and diseases (Table 9-4). Crop-related weeds comprise at least 86 species, among which the most outstanding is the genus *Orobanche*, especially *O. cumana* (broomrape) a hemiparasitic difficult-to-control weed, widespread in several European regions.

At present, control strategies for broomrape tend to use a specific genetic mechanism obtained from wild species (Fernández Martínez et al. 2000; Labrousse et al. 2004; Velasco et al. 2007). However, the continuous emergence of new races of the weed shows evidence that a process of constant renewal of resistant sources is required to maintain these control strategies.

Some herbicides that are members of the imidazolinone and sulfonylurea families (Group B) inhibiting hydroxyacetic acid synthetase enzyme (AHAS) are useful for controlling broomrape (Alonso et al. 1998). Genes conferring tolerance to these herbicides (Baumgartner et al. 1999; Kolkman et al. 2004) found in wild sunflower populations in the State of Kansas in the US were sexually-transferred to crops to create non-GM sunflowers, under the commercial name of Clearfield® (Tan et al. 2005). Another resistance source, obtained from mutagenesis of domestic sunflower, is promising in order to produce high yielding commercial hybrids, because of the absence of wild background (Sala et al. 2008). Gene expression of tolerance in these new varieties allows herbicide application at the advanced stages of crop development, thus controlling the majority of weeds including broomrape (Gressel et al. 1996). Some other groups of herbicides have also proven effective against this weed, including glufosinate-ammonium (Valkov et al. 1998) and glyphosate (Collin 1999).

There are many cases in the world of weed populations displaying resistance to herbicides that inhibit AHAS (95 cases, including wild *Helianthus, http://www.weedscience.org/summary/MOASummary.asp*) and also to other herbicides. There is a need to keep on searching for new control strategies. Given the absence of glufosinate-ammonium resistance among weeds, a good long-term strategy could involve incorporating this tolerance through GM sunflower. Moreover, two homologous *"bar"* and *"pat"* genes that codify phosphinothricin acetyltransferase (PAT) enzyme have been shown to be safe for this purpose as they do not cause allergy and are rapidly degraded in the intestine (Hérouet et al. 2005). Glufosinate-ammonium tolerance would also be a good strategy for the control of the remaining weeds, mainly cosmopolitan, associated with sunflower crop, with predominance of those of the same family Asteraceae (Table 9-4).

Table 9-4 Research articles about sunflower crop biotic adversities published during five years (CAB 2002–2007).

Group	#	Species Name	Articles #	%
Weeds				
Orobancheceae	6	*Orobanche* spp.	26	35
Chenopodiaceae	4	*Chenopodium album*	12	16
Amaranthaceae	5	*Amaranthus* spp.	11	15
Asteraceae	8	*Ambrosia artemisiifolia*	10	13
Total weeds	86		75	**17**
Pests				
Lepidoptera	27	*Helicoverpa armigera*	22	11
Coleoptera (root)	20	*Agriotes* spp.	9	5
Coleoptera (shoot)	12	*Zygogramma exclamationis*	4	2
Birds (Icteridae)	8	*Agelaius phoeniceus*	3	2
Total pests	103		191	**24**
Diseases				
Fungi	28	*Sclerotinia sclerotiorum*	70	19
		Plasmopara helianthi	51	14
		Diaporthe helianthi	41	11
		Alternaria helianthi	39	10
Viruses	4	Sunflower necrosis virus (SNV)	19	5
Bacteria	3	*Pseudomonas*	5	1
Total diseases	35		260	**59**

9.4.3 Pest Control

More than a hundred insect species live or feed on different structures of the numerous *Helianthus* species. Along with birds, mammals, nematodes and acari, which have caught the attention of recent investigations, insects constitute a highly heterogeneous species group (Table 9.4). At least three of these sunflower enemies would be restricted to the North American *Helianthus* and have specific natural controllers (Cantamutto and Poverene 2007), coleoptera *Zygogramma exclamationis* being the important one. In Europe and South America, most of the insects that affect sunflower are unspecific (Charlet et al. 1997) the main cosmopolitan one being *Helicoverpa armigera*. Sunflower lepidoptera pests constitute a wide group of more than 30 polyphage species that feed mainly on aerial parts of the plants (Rogers 1992) and affect yield by decreasing the photosynthetic area. Another sunflower constraint caused by Arthropoda is failure of stand establishment due to soil larvae, mainly of Coleoptera. These herbivors, which feed on seedling stems and roots at different levels, all correspond to polyphagous species, *Agriotes* being the most widespread genus.

Some animal constraints have only a regional importance, like birds (i.e., blackbird, *http://www.sunflowernsa.com*), acari, and nematodes that are very harmful in some crop regions of America and Asia. Novel pests originate from changes in the production systems. In Argentina with the increase in no-till areas, two previously unnoticed snails of the genus *Deroceras* (Carmona 2001) and crustaceous of the genera *Armadillidium*, *Porcellio* and *Balloniscus* (Ves Losada et al. 2007) have recently become limiting factors for the sunflower crop.

Classical sunflower breeding techniques have succeeded in achieving insect resistance. The European moth (*Homoeosoma nebulella*), which was once the main constraint on the dissemination of this crop in Europe (Chen and Welter 2003) and the tobacco caterpillar (*Spodoptera litura*), an important crop limitation in warm regions (Sujatha and Lakshminarayana 2007) and possibly other pests could be controlled through resistant sources from wild species (Rogers 1992; Charlet et al. 2007).

However, the most frequent methods used for pest control in sunflower involve the use of pesticides. A number of chemical products are recommended to control insects that reduce crop stand. Biotechnology could improve this control by helping in developing insect-resistant GM sunflower. Sustainable management calls for complete knowledge of the biology of the target pest and its relationship with other components of the agro-ecosystem.

The options offered by genetic engineering include GM crops that express gene fragments from insecticide proteins of *Bacillus thuringensis* (Bt endotoxins including Cry1Aa, Cry1Ab, Cry1Ac, Cry1Ca, Cry1Fa, Cry3Aa, and others), the *Vigna unguiculata* trypsin inhibitor (CpT1), lectins, and other metabolic inhibitors. The most widespread Bt proteins show strong activity against Lepidoptera, although some bacterial variants have also proven effective against Diptera (*B. th.* var. *israeliensis*) and Coleoptera (*B. th.* var. *tenebrionis*). There is a strong specificity in the action and expression of Bt endotoxins. Not all the genes that codify Bt proteins are expressed in different plant species. Similarly, not all the Lepidoptera found in a crop are controlled by the same event.

On the other hand, the CpT1 agent is very active against Coleoptera and Orthoptera (Boulter et al. 1989) and is already available in GM crops. Modern biotechnological strategies incorporate the expression of a carrier to improve the toxin penetration and its influx into the insect's haemolymph (Fitches et al. 2004). The ideal GM technology should be environment-friendly, with a wide spectrum of activity with respect to the target insects, but with few, if any, effects on beneficial insects (Hilder and Boulter 1999).

The GM sunflowers released in the environment and authorized for research include two groups of events for insect control. The reported Lepidoptera-resistant varieties express the Bt insecticide protein, which is codified by the *Cry1F* gene. If expressed in the early stages of crop development, this could be a valuable tool for controlling polyphagous moth larvae of the genera *Agrotis* and *Euoxa*, which are present in the main sunflower growing regions (Rogers 1992; Charlet et al. 1997). Control of *Suleima helianthana*, which bores sunflower roots and stems in North America is difficult by the use of GM varieties because the damage caused is seldom significant (Charlet and Brewer 2001).

Bt proteins could also offer excellent possibilities for controlling insect damage to aerial tissues. Lepidoptera that cause important crop damage include *H. armigea*, *Spodoptera* spp., *Diabrotica* spp., *Spilosoma* spp., *Colias lesbia*, *Rachiplusia nu*, and *Vanessa cardui*. These species could be controlled through GM technology based on Cry1 variants of the Bt gene. As these species are highly polyphagous, refuges to prevent the selection pressure for insect resistance would not be indispensable, except in cases where all the crops in a given region were GM varieties with the same expression of Bt proteins.

On the other hand, CpT1 could improve stand establishment in cases where failure is due to the Coleoptera of the genera including *Agriotes* spp., *Melolontha* sp., *Annoxia* sp., and Orthoptera of *Gryllotalpa* spp., *Calolampra* spp. and *Teleogryllus* spp. Larvae of these species exhibit subterranean habits and eat plant roots at different stages of crop development, causing the death of seedlings in early attacks (Rogers 1992; Charlet et al. 1997). To achieve the required impact at crop establishment, the expression of CpT1 toxin should take place early in crop development and involve concentrations that are lethal for the plague. Seedlings are very sensitive to the loss of certain of their parts, so it is therefore important to stop damage as early as possible at the beginning of the attack.

Two beetles cause economically important damage in North America (Charlet and Brewer 2001). *Cylindrocopturus adspersus* causes crop damage mainly by lodging in weakened plants whose stems have been bored; this also facilitates the development of fungi. This pest can be controlled by the application of insecticides, though it would also be interesting to explore the genetic resistance of many wild sunflower species. A similar situation occurs with another Coleoptera, *Smicronyx fluvus*, whose larvae develop inside seeds. This is an oligophagous species, which is adapted to only a few hosts and can be controlled with insecticides, sometimes in combination with crop traps. Some parasitic Hymenoptera and Diptera act as controllers, and genetic resistance could, therefore, be achieved. Females consume head bracts and pollen before oviposition, so the expression of the *CpT1* gene in these tissues would help to reduce adult populations. However, the probability of transgene escape highlights the need for management strategies that limit the induction of insect resistance and the acquisition of transgenes by other wild host plants. This would provide durable resistance without environmental impact.

9.4.4 Disease Control

Most research on sunflower crop production restraints have relied on a few diseases caused by fungi (Table 9-4). Although there are more than 30 reported pathogens, half of the CAB Abstracts indexed publications for 2002–2007 refer to only four. Of these, the polyphagous *Sclerotinia sclerotiorum* has received the main attention because of the severe damage it causes and the difficulty of controlling it. It is responsible for white rot and attacks many plants, including soybean. There have been continuous efforts to develop methods enabling early (Vuong et al. 2004) and advanced selection (Becelaere and Miller 2004a, b) of lines expressing tolerance to *Sclerotinia*. Although there does not seem to be a complete resistance (Pedraza et al. 2004), there is a strong genetic component that allows identification of outstanding tolerant materials (Miller et al. 2006).

Research involving the heads of infected plants has shown that tolerance to white rot is related to the accumulation of phenolic (Prats et al. 2003) and coumarin compounds (Prats et al. 2006) and to the absence of the phytotoxic effect of oxalic acid (Baldini et al. 2002). Oxalic acid concentration increases when tissues are damaged and this can be used as an indirect method for selecting on the basis of tolerance to disease (Vasic et al. 2002).

Biotechnology offers a number of strategies for the control of white rot including defense activation, fungus inhibition, and detoxification (Schnabl et al. 2002; Lu 2003). GM sunflower might present resistance to the damage caused by *Sclerotinia* through overexpression of the oxalate oxidase (OXO) enzyme, which degrades oxalic acid to carbon dioxide and hydrogen peroxide as a hypersensitivity mechanism. The first strategy outlined for sunflower by Lu et al. (1998) was also successful in other host plants. Donaldson et al. (2001) demonstrated that wheat gene expression of the OXO enzyme in soybean cell walls close to the site of pathogen attacks reduced disease progression.

The OXO effect in sunflower seems to be more than a hypersensitivity mechanism. Hu et al. (2003) demonstrated that fungus-related damage promotes defense gene activation that is independent of cell death in GM plants that express the wheat *OXO* gene. The transgenic event TF28 significantly improves white mold resistance in cultivated sunflower (see supporting online material in Burke and Rieseberg 2003). The OXO expression may also reduce the herbivory action of certain insects, as demonstrated in maize under field conditions (Ramputh et al. 2002).

Other diseases that have attracted attention in recent research include more sunflower-specific pathogens, among them downy mildew caused by *Plasmopara halstedii*, an oomycete native to America but has spread all over the world. It is an obligate biotrophic plant pathogen that can be controlled through the development of elite sunflower lines resistant to its different races using molecular marker-assisted selection (Panković et al. 2004). Genetic resistance mechanisms are linked to the lack of expression of pathogenesis-related proteins in resistant lines (Roldan Serrano et al. 2007). PlArg from *Helianthus argophyllus* provides a new source of resistance against *P. halstedii* in sunflower (Dussle et al. 2004). Cultural control methods, including seed treatment with fungicide (i.e., metalaxil), are generally recommended for downy mildew control, and on the same accounts for other two significant diseases, brown-gray stem spot (*Diaporthe helianthi*), anamorph (*Phomopsis helianthi*) and alternaria leaf blight (*Alternaria helianthi*).

9.4.5 Novel Products

Biotechnology offers other potential improvement in the quality of sunflower products and byproducts. The fatty acid composition of some sunflower varieties has been modified through conventional plant breeding and mutagenesis (Lacombe and Berville 2000). Although biotechnology could overcome some of the restrictions in this area and pave the way for further advances (Rousselin et al. 2002), its acceptance by the consumer market must also be carefully considered. The high price of sunflower oil is due to its being perceived as a healthy, high quality product (Chapter 1). Given that consumers in many countries are opposed to GM food, diffusion of GM varieties would probably affect its price and make sunflower products less popular than soybean alternatives. After the oil extraction process, the residual sunflower meal has a low value as feed due to the limited level of methionin, an amino acid that is also scarce in other plant products. The Brazil nut (*Bertholletia excelsa*) is an exception to this general rule, providing high concentration of this amino acid. Its genome has been biotechnologically manipulated in order to improve its amino acid content (Marcellino et al. 1996) and enable it to be transferred to other species. Unfortunately, Brazil nut albumen causes allergy in the natural product and also in GM soybean expressing its traits (Lack 2002). Given that the Codex Alimentarius (*ftp:// ftp.fao.org/es/esn/food/guide_plants*) strongly recommends avoiding the transference of genes that cause allergies, interest in this kind of product for food purposes has declined. The situation for sunflower might, however, be different because the main destination of its meal is animal feed.

Interest in procuring alternative sources of latex has led to a search for increased biosynthesis in sunflower. The goal of this project, which has been exclusively sponsored by a governmental organization, is to commercially produce substitutes for USA imports. The guayule (*Parthenium argentatum*) is a desert shrub that produces a variant of rubber, which does not cause allergy and which therefore has a high economic value. Progress in understanding the regulation of rubber biosynthesis in guayule has made it possible to obtain GM plants that offer profitable yields (Cornish and Scott 2005; Veatch et al. 2005). Such an annual crop should facilitate extensive management, though at present transformation efficiency rates remain extremely low (Cornish et al. 2007).

9.5 Debate about Transgenic Sunflower

Concerning the expansion of area devoted to transgenic crops in 2000, the US National Sunflower Association pointed that sunflower was losing acreage to other row crops, such as soybean and canola. Many farmers found that hybrids of the former crops tolerant to glyphosate were easier to manage, with less chemical usage and better crop quality and yields (*www.sunflowernsa.com*). Some seed companies (e.g, Pioneer, Advanta, Dow Agroscience, and Monsanto) and public institutions (e.g., National Institute of Agricultural Technology of Argentina, INTA) collaborated to accelerate the release of sunflower GM varieties towards a more competitive crop, both in the US and in Argentina. Most traits under field experimentation consisted in *Sclerotinia* resistant, Lepidoptera resistant, and glyphosate tolerant varieties. However, the high cost of the GM crop approval process by the regulation offices (estimated in US$ 0.3–0.5 to 2 million in both countries), plus an additional cost of US$ 5 million to obtain approval by the EU countries—which is the main consumer market—(Fonseca et al. 2004) seemed to hinder the effort. Sunflower would not return the investment as fast as other GM crops. Its seed uses per unit surface and the acreage, that in the US and Argentina represents among one and two million hectares each (*http://faostat.fao.org/site/340/DesktopDefault.aspx?PageID=340*) which are by far lower than in corn or soybean crops.

Sunflower is an insect-pollinated outcrossing species, which hybridizes with several *Helianthus* species (Heiser et al. 1969; OECD 2004). Although first generation hybrids from crosses between sunflower and most annual wild relatives are highly sterile, it is possible to obtain fertile progenies through backcrosses to the cultivated sunflower. This technology is less effective in crosses with perennial species, sexually detached by a number of chromosome rearrangements and/or a different ploidy level (Jan 1997). In spite of that, even without special techniques (such as embryo rescue), cultivated *Helianthus annuus* can easily hybridize with at least 16 wild relatives naturally occurring in the US, and with some introduced in other continents as well (Table 9-5).

Although horizontal gene flow between sunflower and other Asteraceae including *Thitonia* spp., *Vigiera* spp. (Sossey-Alaoui et al. 1998) or *Verbesina* spp. (Encheva and Christov 2005) seems unlikely, the probability and implications of vertical gene flow from *H. annuus*, or diagonal with other sexually compatible *Helianthus* species (Gressel and Al-Ahmad 2005) pose additional constraints to transgenic sunflower release. Genetic exchange with wild relatives set the risk of originating novel weeds, through dispersal of transgenic pollen or seed escapes There are examples in wild radish (*Raphanus raphanistrum*) hybridized with *R. sativus* (Campbell et al. 2006) and in *Agrostis stolonifera* (Reichman 2006).

At the center of origin in North America, introgression and persistence of crop genes in wild *Helianthus* populations have been well known for more than a decade (Heiser 1978; Whitton et al. 1997; Linder et al. 1998; Rieseberg et al. 1999; Burke et al. 2002). Agro-ecological studies on transgenic varieties demonstrated that some transgenes were expected to perform as neutral, for example *Sclerotinia* resistance (Burke and Rieseberg 2003), while others could increase fecundity in wild populations, such as Lepidopteran resistance (Snow et al. 2003).

The high variability within *H. annuus* allowed Russian breeders to develop a wide range of agricultural varieties starting from a few plants introduced to Europe for ornamental purposes. This diversity would comprise a high degree of *endoferality*, which would result in volunteer plants able to interfere with the following crops and *exoferality* to give rise to stable populations through crosses with genuine wild plants (Reagon and Snow 2006). There are large naturalized populations of wild or wild-related *H. annuus* in Spain and France (Müller et al. 2006), Serbia (Stankovic-Kalezic et al. 2007), Italy and other European countries. They most likely originated in seed contaminants (wild or wild-crop seed) imported from the US

Table 9-5 Wild species that produced viable progeny in crosses with cultivated *Helianthus annuus* as male parent (adopted from Jan 1997).

Female parent	Chromosome number (n)	Hybrid[1]	Distribution countries (native or naturalized)
Section Helianthus (annuals)			
H. annuus	17	A, N	US, Argentina, Australia,Spain, France, Italy, Serbia
H. niveus	17	A	US
H. debilis	17	A, N	US, Mozambique
H. praecox	17	A, N	US
H. petiolaris	17	A, N	US, Argentina
H. neglectus	17	A	US
H. bolanderi	17	A, N	US
H. paradoxus	17	A, N	US
H. argophyllus	17	A, N	US, Mozambique
Section Atrorubens (perennials)			
Series Corona-solis			
H. hirsutus	34	A	US
H. decapetalus	17, 34	A	US
H. strumosus	34, 51	A	US
H. tuberosus	51	A, N	US, France, Germany
H. giganteus	17	A	US
H. grosseserratus	17	A	US
H. maximiliani	17	A	US
Series Atrorubentes			
H. rigidus	17	A	US

[1]Artifcial (A) or natural (N) hybrids obtained without embryo culture.

(Bervillé et al. 2005; Muller et al. 2009). In Australia, established *H. annuus* was introduced with ornamental and forage purposes before the 20th century (Dry and Burdon 1986). The ability to develop feral populations also extends to other annual species; *H. argophyllus* and *H. debilis* have naturalized with a high interspecific hybridization rate in Mozambique (Quagliaro et al. 2001; Vischi et al. 2004; Ribeiro et al. 2005) and the perennial *H. tuberosus* has been established in France (Faure et al 2002a) and Germany (Kowarik 2005).

In 2000, the high number of applications for experimental trials and release of transgenic sunflower varieties in Argentina motivated the National Committee for Agricultural Biotechnology (CONABIA) to promote a detailed exploration for wild annual *Helianthus* populations, addressed to evaluate likely wild-crop gene flow. Field exploration trips revealed extensive naturalized populations of *H. petiolaris* and *H. annuus* in central Argentina (Poverene et al. 2002) giving evidence of crop-wild gene flow (Cantamutto et al. 2003).

After the dissemination of these findings, the number of applications for sunflower GM varieties release submitted to CONABIA drastically fell to zero by 2005. Following the acknowledgement of existing naturalized wild populations (Fonseca et al. 2004), the release of transgenic sunflower varieties was presumed as improbable by the Argentine Sunflower Association (ASAGIR) (Ingaramo 2006). A similar situation has taken place since 2004 in the US, where the National Sunflower Association (NSA) began to use the non-transgenic nature of sunflower crop to promote its consumption. On the other hand, there is no clear position of the International Sunflower Association on this subject (*www.isa.cetiom.fr*).

9.6 Prospective View on Transgenic Sunflower Release

Most of the available transgenes and modulators that have been engineered in plants could be expressed in sunflower. However, the traits that are being studied in sunflower for environmental release are limited and mainly consist of insect resistance, herbicide tolerance, and special compound synthesis. Leaving aside product marketing considerations, sunflower crop management would greatly benefit from the introduction of GM varieties (Cantamutto and Poverene 2007).

Botanical files (Conner et al. 2003) would indicate a high ecological risk for GM sunflower release because of the difficulty to keep transgenes restricted within the crop. Being an outcrossing crop, complete isolation demands distances of 5 Km (Anfinrud 1997) because pollen are carried by insects. Likewise, its seeds disperse along wide distances by trucks and machinery, forming ruderal populations also vulnerable to gene flow (Reagon and Snow 2006; Ureta et al. 2008). As well as the drawback posed by volunteer plants for following crops (Robinson 1978), sunflower can develop ferality through sexually compatible crosses with five invading relatives (Table 9.5), being perennial with tubers that can disperse to long distances moving down along river banks (Berville et al. 2005).

Taking for granted food safety assessment by official regulatory systems (Jaffe 2004) and the absence of unpredictable effects from transgene insertion (Clark 2006), it is imperative to predict the environmental impact if transgenes reach wild or feral relatives. Hybridization does not *per se* ensure introgression; to introgress a transgene must have some ecological implication, for example it may suppress a controlling element (Hails and Morley 2005) or enhance fecundity in the receptive population (Lee and Natesan 2006). However, introgression could result in neutral or null depending on natural selection pressure (Chapman and Burke 2006).

Two strategies designed to minimize environmental risks proposed by Gressel and Al-Ahmad (2005) have been suggested for sunflower. Transgenes could be contained in male-sterile varieties, such as those studied for latex production (Cornish et al. 2007) or inserted in chloroplasts. However, those strategies would not ensure a complete barrier for transgene containment. As in other species, a small part of sunflower paternal cytoplasm carrying cp-DNA could be transmitted to the progeny (Haygood et al. 2004) while volunteer plants would obviously carry them through maternal inheritance.

Bervillé et al. (2005) proposed to mitigate gene flow by using transgenic varieties carrying chromosome translocations and inversions. For example, *Helianthus laciniatus* differs from sunflower by at least eight reciprocal translocations and one inversion, which render gene flow almost impossible (Jan and Seiler 2007). Hybrids of wild or volunteer plants pollinated by cultivated biotypes carrying these chromosome rearrangements would lack a normal meiotic pairing having a very low fertility. Although these varieties would show a high fecundity under normal crop conditions, chromosome segments from wild species would carry agronomically undesirable traits, as it occurred with the first generation of IMI-tolerant sunflower varieties.

An alternative would consist in transgene linkage to traits of low persistence in the wild. Snow et al. (1998) reported a low fecundity in domesticated-like genotypes, whereas wild types predominated following introgression of crop genes within wild populations. In order to mitigate gene flow, transgenes should be linked to domesticated traits, such as no branching. One-headed plants would produce large seeds easily detected by predators (Alexander et al. 2001), lessening in this way the soil seed bank, which constitutes the main component of population dynamics (Claessen et al. 2005).

In the present state of art, this option would be difficult to realize given the low probability of a transgene being inserted in a specified locus within the genome. Moreover, as the genus has one of the highest recombinational rates among plants (Burke et al. 2004) this strategy would not be reliable enough and linkage should be extremely tight not to be broken.

A fecundity increase due to Bt transgene acquisition could be expected in wild plants (Snow et al. 2003) however other events would have the opposite or neutral effects. Gene flow-acquired herbicide tolerance could increase weedy sunflower interference, both of volunteers and wild-crop hybrids, but even in the worst scenario this trait would not entail the creation of "superweeds". These variants would be controlled through herbicide rotation, a technique traditionally recommended to avoid mutants tolerant to the chemical groups utilized (*www.weedscience.org*).

If environmental constraints are to be overcome through regulation flexibility or by obtaining varieties harboring containment or mitigation mechanisms, transgenic sunflower diffusion would be strongly conditioned by consumers' attitude. At present one can expect a complete acceptance in the increasing biofuels market (Table 9-6). However this competition seems to leave sunflower behind because other suitable crops like soybean and rapeseed already have available transgenic varieties.

Although sunflower has a high oil concentration, agro-ecological adaptation would not be an advantage. Rapeseed has diffused in cold-temperate regions (like Canada, Australia, and the US), soybean predominates in warm-temperate ones (like Brazil, Paraguay, and Bolivia) while in tropical regions (Malaysia, Indonesia) edible oil is obtained from palm (*Elaeis* spp.) whose oil production is five-fold compared to sunflower.

In high quality edible oil markets sunflower presents advantages that make it a very competitive crop and should be used to increase its value. Mid-oleic varieties obtained from natural mutation (NuSun) present rates of saturated, mono- and poly-unsaturated fatty acids close to those recommended by WHO (FAO 1995), making sunflower oil superior to olive, in which mono-unsaturated oleic acid predominates. This makes sunflower oil very healthy for cardiovascular care (Chapter 6). Sunflower oil is also rich in tocopherol (vitamin E) with antioxidant effects (Fernández Martínez et al. 2004). Would these existing markets accept edible oil coming from a transgenic crop?

The future of transgenic sunflower will be defined by a sharp balance between its industrial use and changes in consumer perception. Environmental risk mainly related to difficult-to-control novel sunflower feral forms can be diminished, but not made eliminated. Nevertheless, risk does not involve merely transgenic varieties, but extends to every new germplasm obtained through classical breeding. Consumer perception would change

Table 9-6 Expected acceptance of sunflower transgenic varieties under present market perception (based on Fernández Martínez et al. 2004; Fonseca et al. 2004 and Vannozzi 2006).

Destination	Attributes	GM acceptance	Competitor crop	Available GM varieties
Biodiesel	high oleic acid	total	palm	no
			canola	yes
			soybean	yes
Bio-lubricants	low linoleic acid antioxidants	total	flax	no
			canola	yes
Edible oil Fried products	saturated fatty acids	parcial	canola	yes
Edible oil Salads	flavor, unsaturated fatty acids	low	olive	no
			corn	yes
Confectionary	big and healthy achenes	none	peanut	no
			pistachio	no

dramatically if transgenic varieties meant an outstanding improvement in the quality of life, including environment and health. Up to that time, advances in other new transgenic crop development will postpone the usage of sunflower transgenic varieties.

References

AGBIOS (2007) GM Database: *http: //www.agbios.com/dbase.php?action=Submit& evidx*: (Accessed on September 20, 2009).

Alexander HM, Cummings CL, Kahn L, Snow AA (2001) Seed size variation and predation of seeds produced by wild and crop-wild sunflowers. Am J Bot 88: 623–627.

Alibert G, Aslane-Chanabe C, Burrus M (1994) Sunflower tissue and cell-cultures and their use in biotechnology. Plant Physiol Biochem 32: 31–44.

Alibert B, Lucas O, Le Gall V, Kallerhoff J, Alibert G (1999) Pectolytic enzyme treatment of sunflower explants prior to wounding and cocultivation with *Agrobacterium tumefaciens*, enhances efficiency of transient beta-glucuronidase expression. Physiol Plant 106: 232–237

Alonso L, Rodríguez Ojeda M, Fernández Escobar J, Lopez Ruiz Calero G (1998) Chemical control of broomrape (*Orobanche cernua* Loefl.) in sunflower (*Helianthus annuus* L.) resistant to imazethapyr herbicide. Helia 21: 45–53.

Anfinrud MN (1997) Planting hybrid seed production and seed quality evaluation. In: AA Schneiter (ed) Sunflower Technology and Production. Am Soc Agron. Madison, Wisconsin, USA, pp 697–708.

Azadi P, Moieni A, Ahmadi MR (2002) Shoot organogenesis from cotyledons of sunflower. Helia 25: 19–26.

Baker C, Muñoz-Fernandez N, Carter C (1999) Improved shoot development and rooting from mature cotyledons of sunflower. Plant Cell Tiss Org Cult 58: 39–49.

Baldini M, Turi M, Vischi M, Vannozzi G, Olivieri A (2002) Evaluation of genetic variability for *Sclerotinia sclerotiorum* (Lib.) de Bary resistance in sunflower and utilization of associated molecular markers. Helia 25: 177–189.

Baumgartner J, Al Khatib K, Currie R (1999) Common sunflower resistance to imazethapyr and chlorimuron in northeast Kansas. Proc Western Soc Weed Sci 52: 12–15.

Bazzalo ME, Bridges I, Galella T, Grondona M, Leon A, Scott A, Bidney D, Cole G, D'Hautefeuille JL, Lu G, Mancl M, Scelonge C, Soper J, Sosa Dominguez G, Wang L (2000) *Sclerotinia* head rot resistance conferred by wheat oxalate oxidase gene in transgenic sunflower. In: Proc 15th Int Sunflower Conf, June 12–15, Toulouse, France, pp K60–K65.

BCBA Bolsa de Cereales de Buenos Aires (2009) Información Histórica. Series Históricas Anuales en *http://www.bolcereales.com.ar/precios.asp?idioma=esp* (Accessed September 15, 2009).

Becelaere G van, Miller J (2004 a) Combining ability for resistance to Sclerotinia head rot in sunflower. Crop Sci 44: 1542–1545.

Becelaere G van, Miller JF (2004 b) Methods of inoculation of sunflower heads with *Sclerotinia sclerotiorum*. Helia 27: 137–142.

Bervillé A, Muller MH, Poinso B, Serieys H (2005) Crop ferality and volunteerism: A threat to food security in the transgenic Era? In: J Gressel (ed) Ferality: Risks of Gene Flow between Sunflower and other *Helianthus* Species. CRC Press, Boca Raton, Florida, USA, pp 209–230.

Bidney D, Scelonge C, Martich J, Burrus M, Sims L, Hufmann G (1992) Microprojectile bombardment of plant tissues increases transformation frequency by *Agrobacterium tumefaciens*. Plant Mol Biol 18: 301–313.

Binsfeld PC, Wingender R, Schnabl H (2000) Characterization and molecular analysis of transgenic plants obtained by microprotoplast fusion in sunflower Theor Appl Genet 101: 1250–1258.

Bohorova NE, Cocking EC, Power JB (1986) Isolation, culture and callus regeneration of protoplasts of wild and cultivated *Helianthus* hybrids. Plant Cell Rep 5: 256–258.

Bolandi AR, Branchard M, Alibert G, Genzbitel L, Berville A, Sarrafi A (2000) Combining-ability analysis of somatic embryogenesis from epidermal layers in the sunflower (*Helianthus annuus* L.). Theor Appl Genet 100: 621–624.

Boulter D, Gatehouse A, Hilder V (1989) Use of cowpea trypsin inhibitor (CpTl) to protect plants against insect predation. Biotechnol Adv 7: 489–497.

Bronner R, Jeannin G, Hahne G (1994) Early cellular events during organogenesis and somatic embryogenesis induced on immature zygotic embryos of sunflower (*Helianthus annuus*). Can J Bot 72: 239–248.

Brookes G, Barfoot P (2006) GM Crops: The First Ten Years—Global Socio-Economic and Environmental Impacts. ISAAA Brief No 36, ISAAA, Ithaca, New York, USA

Burke J, Rieseberg L (2003) Fitness effects of transgenic disease resistance in sunflowers. Science 300: 1250.

Burke J, Gardner K, Rieseberg L (2002) The potential for gene flow between cultivated and wild sunflower (*Helianthus annuus*) in the United States. Am J Bot 89: 1550–1552.

Burke JM, Lai Z, Salmaso M, Nakazato T, Tang S, Heesacker A, Knapp S J, Rieseberg LH (2004) Comparative mapping and rapid karyotypic evolution in the genus *Helianthus*. Genetics 167: 449–457.

Burrus M, Chanabe C, Alibert G, Bidney D (1991) Regeneration of fertile plants from protoplasts of sunflower (*Helianthus annuus* L.). Plant Cell Rep 10: 161–166.

Burrus M, Molinier J, Himber C, Hunold R, Bronner R, Rousselin P, Hahne G (1996) *Agrobacterium*-mediated transformation of sunflower (*Helianthus annuus* L.) shoot apices: transformation patterns. Mol Breed 2: 329–338.

Campbell LG, Snow AA, Ridley CE (2006) Weed evolution after crop gene introgression: greater survival and fecundity of hybrids in a new environment. Ecol Lett 9: 1198–1209.

Cantamutto M, Poverene M (2007) Genetically modified sunflower release: Opportunities and risks. Field Crops Res 101: 133–144.

Cantamutto M, Ureta S, Carrera A, Poverene M (2003) Indicios de hibridación entre especies anuales del género *Helianthus* en Argentina. ASAGIR 2003. 2nd Congr Argentino de Girasol, Buenos Aires, Argentina: *http//www.asagir.org.ar*

Carmona D (2001) Plagas emergentes en siembra directa: *http: //www.inta.gov.ar/ balcarce/ info/documentos/agric/sd/plagasem.htm*

Carvalho H, Lima L, Lescure N, Camut S, Salema R, Cullimore J (2000) Differential expresión of the two cytosolic glutamine synthetase genes in various organs of *Medicago truncatula*. Plant Sci 159: 301–312.

Ceriani MF, Hopp HE, Hahne G, Escandón AS (1992) Cotyledons: an explant for routine regeneration of sunflower plants. Plant Cell Physiol 33: 157–164.

Chapman MA, Burke JM (2006a) Polluting gene flow from crops: Radishes gone wild. Heredity 97: 379–380.

Chapman MA, Burke JM (2006b) Letting the gene out of the bottle: the population genetics of genetically modified crops. New Phytol 170: 429–443.

Charlet L, Brewer G (2001) Sunflower Insect Pest Management in North America: *http: //www.ipmworld.umn.edu/ipmchap.htm*

Charlet LD, Brewer GJ, Franzmann BA (1997) Sunflower insects. In: AA Schneiter (ed) Sunflower Technology and Production. Am Soc Agron. Madison, Wisconsin, USA, pp 183–261.

Charlet LD, Aiken RM, Miller JF, Seiler GJ, Grady KA, Knodel JJ (2007) Germplasm evaluation for resistance to insect pests of the sunflower head. 29th Sunflower Res Workshop, Jan 10–11, 2007, Fargo, ND, USA: *http: //www.sunflowernsa.com/research/ research–workshop/documents/Charlet_etal_Germplasm_07.pdf*

Chen Y, Welter S (2003) Confused by domestication: incongruent behavioral responses of the sunflower moth, *Homeoesoma electellum* (Lepidoptera: Pyralidae) and its parasitoid, *Dolichogenidea homoeosomae* (Hymenoptera: Braconidae), towards wild and domesticated sunflowers. Biol Con 28: 180–190.

Chraibi KM, Latche A, Roustan JP, Fallot J (1991) Stimulation of shoot regeneration from cotyledons of *Helianthus annuus* by the ethylene inhibitors, silver and cobalt. Plant Cell Rep 10: 204–207.

Chraibi KM, Castelle JC, Latche A, Roustan JP, Fallot J (1992a) A genotype-independent system of regeneration from cotyledons of sunflower (*Helianthus annuus* L.). The role of ethylene. Plant Sci 86: 215–221.

Chraibi KM, Castelle JC, Latche A, Roustan JP, Fallot J (1992b) Enhancement of shoot regeneration potential by liquid medium culture from mature cotyledons of sunflower (*Helianthus annuus* L.). Plant Cell Rep 10: 617–620.

Claessen D, Gilligan Ch A, Lutman P J W, van den Bosch F (2005) Which traits promote persistence of feral GM crops? Part 1: implications of environmental stochasticity. OIKOS 110: 20–29.

Clark EA (2006) Environmental risk of genetic engineering. Euphytica 148: 47–60.

Collin J (1999) *Orobanche ramosa*–its development and continued cause for concern on rapeseed crops in the Poitou-Charentes region. Phytoma 515: 19–20.

Conner AJ, Glare TR, Nap JP (2003) The release of genetically modified crops into the environment, Part II. Overview of ecological risk assessment. Plant J 33: 19–46.

Cornish K, Scott D (2005) Biochemical regulation of rubber biosynthesis in guayule (*Parthenium argentatum* Gray). Indus Crop Prod 22: 49–58.

Cornish K, Pearson CH, Rath D.J, Dong N, McMahan CM, Whalen M (2007) The potential for sunflower as a rubber producing crop for the United States. Helia 30: 157–166.

Deglene L, Lesignes P, Alibert G, Sarrafi A (1997) Genetic control of organogenesis in cotyledons of sunflower (*Helianthus annuus*). Plant Cell Tiss Org Cult 48: 127–130

Dhaka N, Kothari SL (2002) Phenylacetic acid improves bud elongation and in vitro plant regeneration efficiency in *Helianthus annuus* L. Plant Cell Rep 21: 29–34.

Donaldson P, Anderson T, Lane B, Davidson A, Simmonds DH (2001) Soybean plants expressing an active oligomeric oxalate oxidase from the wheat gf-2.8 (germin) gene are resistant to the oxalate secreting pathogen *Sclerotinia sclerotiorum*. Physiol Mol Plant Pathol 59: 297–307.

Dry PJ, Burdon JJ (1986) Genetic structure of natural populations of wild sunflowers (*Helianthus annuus* L.) in Australia. Aust J Biol Sci 39: 255–270.

Dussle CM, Hahn V, Knapp SJ, Bauer E (2004) PlArg from *Helianthus argophyllus* is unlinked to other known downy mildew resistance genes in sunflower. Theor Appl Genet 109: 1083–1086.

Encheva J, Christov M (2005) Intergeneric hybrids between cultivated sunflower (*Helianthus annuus* L.) and *Verbesina helianthoides* (genus *Verbesina*)—morphological and biochemical aspects. Helia 28: 27–35.

Escandon A, Hahne G (1991) Genotype and composition of culture-medium are factors important in the selection for transformed sunflower (*Helianthus annuus*) callus. Physiol Plant 81: 367–376.

Espinasse A, Lay C (1989) Shoot regeneration of callus derived from globular to torpedo embryos from 59 sunflower genotypes. Crop Sci 29: 201–205.

Everett NP, Robinson KEP, Mascarenhas D (1987) Genetic engineering of sunflower (*Helianthus annuus* L.) Bio/Technology 5: 1201–1204.

Fambrini M, Cionini G, Pugliesi C (1996) Development of somatic embryos from morphogenetic cells of the interspecific hybrid *Helianthus annuus x Helianthus tuberosus*. Plant Sci 114: 205–214.

FAO (2008) Food Outlook. Global Market Analysis. FAO Corporate Document Repository: *ftp://ftp.fao.org/docrep/fao/011/ai474e/ai474e00.pdf*

Faure N, Serieys H, Bervillé A (2002a) Potential gene flow from cultivated sunflower to volunteer, wild *Helianthus* species in Europe. Agri Ecosyst Environ 89: 183–190.

Faure N, Serieys H, Kaan F, Bervillé A (2002b) Partial hybridization in crosses between cultivated sunflower and the perennial *Helianthus mollis*: effect of in vitro culture compared to natural crosses. Plant Cell Rep 20: 943–947.

Fernandez-Cornejo J, Caswell M, Mitchell L, Golan E, Kuchler F (2006) The first decade of genetically engineered crops in the United States. USDA Bull N° 11: *http: // www.ers.usda.gov/publications/eib11/eib11.pdf* (Accessed on September 20, 2009).

Fernández-Martínez J, Melero-Vera J, Muñoz-Ruz J, Ruso J, Domínguez J (2000) Selection of wild and cultivated sunflower for resistance to a new broomrape race that overcomes resistance of the *Or5* gene. Crop Sci 40: 550–555.

Fernández Martínez JM, Velasco L, Pérez-Vich B (2004) Progress in the genetic modification of sunflower oil quality.16th Int Sunflower Conf, Aug. 29–Sept. 2, Fargo ND, USA, pp 1–14.

Ferrario-Méry S, Valadier M, Godefroy N, Miallier D, Hirel B, Foyer Ch, Suzuki A (2002) Diurnal changes in ammonia assimilation in transformed tobacco plants expressing ferredoxin-dependet glutamate synthase mRNA in the antisense orientation. Plant Sci 163: 59–67.

Finner J (1987) Direct somatic embryogenesis and plant regeneration from immature embryos of hybrid sunflower (*Helianthus annuus* L.) on a high sucrose medium. Plant Cell Rep 6: 372–374.

Fitches E, Edwards M, Mee C, Grishin E, Gatehouse A, Edwards J, Gatehouse J (2004) Fusion proteins containing insect-specific toxins as pest control agents: snowdrop lectin delivers fused insecticidal spider venom toxin to insect haemolymph following oral ingestion. J Insect Physiol 50: 61–71.

Flores Berrios E, Gentzbittel L, Alibert G, Griveau Y, Bervillé A, Sarrafi A (1999) Genetic control of in vitro-organogenesis in recombinant inbred lines of sunflower (*Helianthus annuus* L.). Plant Breed 118: 359–361.

Flores Berrios E, Gentzbittel L, Mokrani L, Alibert G, Sarrafi A (2000) Genetic control of early events in protoplast division and regeneration pathways in sunflower. Theor Appl Genet 101: 606–612.

Fonseca EA, López Bilbao M, Luders ML, Nogués JJ, Parsons AT, Regúnaga M, Sturzenegger AC (2004) Estudio sobre el impacto económico de la eventual utilización de eventos transgénicos de girasol en Argentina: *http//www.asagir.org.ar/transgenicos.asp*

Frey KJ (1996) National Plant Breeding Study—Human and Financial Resources Devoted to Plant Breeding Research and Development in the United States in 1994. Spl Rep 98, Iowa State Univ: *http: //www.ers.usda.gov/Data/PlantBreeding/Plant%20Breeding.pdf*

Freyssinet M, Freyssinet G (1988) Plant regeneration from sunflower (*Helianthus annuus* L.) immature embryos. Plant Sci 56: 177–181.

Friedt W (1992) Present state and future prospects of biotechnology in sunflower breeding. Field Crops Res 30: 425–442.

Gelebart P, San L (1987) Obtention de plantlets haploids par culture in vitro d'ovaires non fecondes de tournesol (*Helianthus annuus* L.). Agronomie 7: 81–86.

Godoy M, Castaño F, Re J, Rodriguez R (2005) *Sclerotinia* resistance in sunflower: I. Genotypic variations of hybrids in three environments of Argentina. Euphytica 145: 147–154.

Grayburn WS, Vick BA (1995) Transformation of sunflower (*Helianthus annuus* L) following wounding with glass-beads. Plant Cell Rep 14: 285–289.

Greco B, Tanzarella OA, Carozzo G, Blanco A (1984) Callus induction and shoot regeneration in sunflower (*Helianthus annuus* L.). Plant Sci Lett 36: 73–77.

Gressel J, Al-Ahmad H (2005) Molecular containment and mitigation of genes within crops–Prevention of gene establishment in volunteer offspring and feral strains. In: J Gressel (ed) Crop Ferality and Volunteerism: A Threat to Food Security in the Transgenic Era? CRC Press, Boca Raton, Florida, USA, pp 371–387.

Gressel J, Segel L, Ransom J (1996) Managing the delay of evolution of herbicide resistance in parasitic weeds. Int J Pest Manag 42: 113–129.

Hails RS, Morley K (2005) Genes invading new populations: a risk assessment perspective. Trends Ecol Evol 20: 245–252.

Haygood R, Ives AR, Andow DA (2004) Population genetics of transgene containment. Ecol Lett 7: 213–220.

Heiser CB (1978) Taxonomy of *Helianthus* and origin of domesticated sunflower. In: JF Carter (ed) Sunflower Science and Technology. Am Soc Agron, Madison, Wisconsin, USA, pp 31–53.

Heiser CB, Smith DM, Clevenger SB, Martin WC (1969) The North American sunflowers (*Helianthus*). Mem Torrey Bot Club 22: 1–28.

Henn HJ, Wingender R, Schnabl H (1998) Regeneration of fertile interspecific hybrids from protoplast fusions between *Helianthus annuus* L. and wild *Helianthus* species. Plant Cell Rep 18: 220–224.

Hérouet C, Esdaile D, Mallyon B, Debruyne E, Schulz A, Currier T, Hendrick K, Jan van der Klis R, Rouan D (2005) Safety evaluation of the phosphinotricin acetyltransferase proteins encoded by the pat and bar sequences that confer tolerance to glufosinate-ammonium herbicide in transgenic plants. Regul Toxicol Pharmacol 41: 134–149.

Herrera-Rodríguez M, Maldonado J, Pérez-Vicente R (2004) Light and metabolic regulation of HAS1, HAS1.1 and HAS2, three asparagine synthetase genes in *Helianthus annuus*. Plant Physiol Biochem 42: 511–518.

Hess D (2006) On the Path to Plant Gene Technology. Sci Rep of a Botanist. Cuvillier Verlag, Göttingen, Germany, p 232.

Hewezi T, Perrault A, Alibert G, Kallerhoff J (2002) Dehydrating immature embryo split apices and rehydrating with *Agrobacterium tumefaciens*: A new method for genetically transforming recalcitrant sunflower. Plant Mol Biol Rep 20: 335–345.

Hewezi T, Alibert G, Kallerhoff J (2004) Genetic transformation of sunflower (*Helianthus annuus* L.). In: I Curtis (ed) Transgenic Crops of the World.–*Essential Protocols*. Kluwer Academic Publ, Dordrecht, The Netherlands, pp 435–451.

Hewezi T, Alibert G, Kallerhoff J (2005) Local infiltration of high- and low-molecular-weight RNA from silenced sunflower (*Helianthus annuus* L.) plants triggers post-transcriptional gene silencing in non-silenced plants. Plant Biotechnol J 3: 81–89.

Hewezi T, Mouzeyar S, Thion L, Rickauer M, Alibert G, Nicolas P, Kallerhoff J (2006) Antisense expression of a NBS-LRR sequence in sunflower (*Helianthus annuus* L.) and tobacco (*Nicotiana tabacum* L.): evidence for a dual role in plant development and fungal resistance. Transgen Res 15: 165–180.

Hilder V, Boulter D (1999) Genetic engineering of crop plants for insect resistance—a critical review. Crop Protec 18: 177–191.

Hu X, Bidney DL, Yalpani N, Duvick JP, Crasta O, Folkerts O, Lu GH (2003) Overexpression of a gene encoding hydrogen peroxide-generating oxalate oxidase evokes defense responses in sunflower. Plant Physiol 133: 170–181.

Hunold R, Burrus M, Bronner R, Duret JP, Hahne G (1995) Transient gene expression in sunflower (*Helianthus annuus* L.) following microprojectile bombardment. Plant Sci 105: 95–109.

Ingaramo J (2006) Plan Estratégico ASAGIR 2006–2015. Cuadernillo Informativo N° 11: *http//www.asagir.org.ar*

ISB Information Systems for Biotechnology (2007) Field Test Release Applications in the U.S: *http://www.isb.vt.edu/CFDOCS/fieldtests1.cfm* (Accessed on September 20, 2009).

Jaffe G (2004) Regulating transgenic crops: a comparative analysis of different regulatory processes. Transgen Res 13: 5–19.

James C (2008) ISAAA Brief 39-2008 Global Status of Commercialized Biotech/GM Crops: *http: //www.isaaa.org/resources/publications/briefs/39/executivesummary/ default.html* (Accessed on September 20, 2009).

Jan CC (1997) Cytology and Interspecific Hybridization. In: AA Schneiter (ed) Sunflower Technology and Production. Am Soc of Agron, Madison, Wisconsin,USA, pp 595–670.

Jan CC, Seiler G (2007) Sunflower. In: RJ Singh (ed) Genetic Resources, Chromosome Engineering, and Crop Improvement. Vol 4: Oilseed Crops. CRC Press Boca Raton, Florida, USA, pp 103–165.

Jeannin G, Hahne G (1991) Donor plant growth conditions and regeneration of fertile plants from somatic embryos induced on immature zygotic embryos of sunflower (*Helianthus annuus* L.). Plant Breed 107: 280–287.

Jeannin G, Bronner R, Hahne G (1995) Somatic embryogenesis and organogenesis induced on the immature zygotic embryo of sunflower (*Helianthus annuus* L) cultivated in vitro: role of the sugar. Plant Cell Rep 15: 200–204.

Kallerhoff J, Alibert GF (1996) Sunflower and biotechnology: Current situation and outlook. OCL-Oleagineux Corps Gras Lipides 3: 154–158.

Knittel N, Escandón A, Hahne G (1991) Plant regeneration at high frequency from mature sunflower cotyledons. Plant Sci 73: 219–226.

Knittel N, Gruber V, Hahne G, Lenee P (1994) Transformation of sunflower (*Helianthus annuus* L)—A reliable protocol. Plant Cell Rep 14: 81–86.

Kolkman J, Slabaugh M, Bruniard J, Berry S, Bushman B, Olungu C, Maes N, Abratti G, Zambelli A, Miller J, Leon A, Knapp S (2004) Acetohydroxyacid synthase mutations conferring resistance to imidazolinone or sulfonylurea herbicides in sunflower. Theor Appl Genet 109: 1147–1159.

Koopmann V, Kutschera U (2005) In-vitro regeneration of sunflower plants: effects of a *Metylobacterium* strain on organ development. J Appl Bot Food Qual 79: 59–62

Kowarik I (2005) Urban ornamentals escaped from cultivation. In: J Gressel (ed) Crop Ferality and Volunteerism: A Threat to Food Security in the Transgenic Era? CRC Press, Boca Raton, Florida, USA, pp 97–121.

Krasnyanski S, Menczel L (1993) Somatic embryogenesis and plant regeneration from hypocotyl protoplasts of sunflower (*Helianthus annuus* L.). Plant Cell Rep 12: 260–263.

Labrousse P, Arnaud M C, Griveau Y, Fer A, Thalouarn P (2004) Analysis of resistance criteria of sunflower recombined inbred lines against *Orobanche cumana* Wallr. Crop Protec 23: 407–413.

Lack G (2002) Clinical risk assessment of GM foods. Toxicol Lett 127: 337–340.

Lacombe S, Bervillé A (2000) Problems and goals in studying oil composition variation in sunflower. 15th Int Sunflower Conf, June 12–16, Toulouse, France, pp PI D16–27.

Laparra H, Burrus M, Hunold R, Damm B, Bravoangel Am, Bronner R, Hahne G (1995) Expression of foreign genes in sunflower (*Helianthus annuus* L)—Evaluation of 3 gene-transfer methods. Euphytica 85: 63–74.

Leclercq P (1969) Une sterilite cyplasmique chez le tournesol. Ann Amelior Plant 19: 99–106.

Lee D, Natesan E (2006) Evaluating genetic containment strategies for transgenic plants. Trends Biotechnol 24: 109–114.

Linder CR, Taha I, Seiler GJ, Snow AA, Rieseberg LH (1998) Long-term introgression of crop genes into wild sunflower populations. Theor Appl Genet 96: 339–347.

Lopez-Bellido R J, Lopez-Bellido L, Castillo J E, Lopez-Bellido F J (2003) Nitrogen uptake by sunflower as affected by tillage and soil residual nitrogen in a wheat-sunflower rotation under rainfed Mediterranean conditions. Soil Till Res 72: 43–51.

Lu G (2003) Engineering *Sclerotinia sclerotiorum* resistance in oilseed crops. Afr J Biotechnol 2: 509–516.

Lu G, Scelonge C. Wang L, Norian L, Macnl M, Parson M, Cole G, Yalpani N, Bao Z, Hu X, Heller J, Kulisek E, Schmidt H, Tagliani L, Duvick J, Biney D (1998) Expression of oxalate oxidase in sunflower to combat *Sclerotinia* disease. Proc Int *Sclerotinia* Workshop, Fargo, ND, USA, 9–12 Sept 1998, p 43.

Lucas O, Kallerhoff J, Alibert G (2000) Production of stable transgenic sunflowers (*Helianthus annuus* L.) from wounded immature embryos by particle bombardment and co-cultivation with *Agrobacterium tumefaciens*. Mol Breed 6: 479–487.

Lupi MC, Bennici A, Locci F, Gennai D (1987) Plantlet formation from callus and shoot-tip culture of *Helianthus annuus* (L.). Plant Cell Tiss Org Cult 11: 47–55.

Malone-Schoneberg J, Scelonge CJ, Burrus M, Bidney DL (1994) Stable transformation of sunflower using *Agrobacterium* and split embryonic axis explants. Plant Sci 103: 199–207.

Manavella PA, Arce AL, Dezar CA, Bitton F, Renou JP, Crespi M, Chan RL 2006 Cross-talk between ethylene and drought signalling pathways is mediated by the sunflower Hahb-4 transcription factor. Plant J 48: 125–137.

Marcellino L, Neshich G, Grossi de Sá M, Krebbers E, Gander E (1996) Modified 2S albumins with improved tryptophan content are correctly expressed in transgenic tobacco plants. FEBS Lett 385: 154–158.

Mayor ML, Nestares G, Zorzoli R, Picardi LA (2003) Reduction of hyperhydricity in sunflower tissue culture. Plant Cell Tiss Org Cult 72: 99–103.

Mayor ML, Nestares G, Zorzoli R, Picardi LA (2006) Analysis for combining ability in sunflower organogenesis-related traits. Aust J Agri Sci 57: 1123–1129

Miller JF, Gulya TJ, Vick BA (2006) Registration of two maintainer (HA 451 and HA 452) and three restorer (RHA 453–RHA 455) *Sclerotinia*-tolerant oilseed sunflower germplasms. Crop Sci 46: 2727–2728.

Mohamed S, Binsfeld PC, Cerboncini C, Schnabl H (2003) Regeneration systems at high frequency from high oleic *Helianthus annuus* L. genotypes. J Appl Bot 77: 85–89.

Mohamed S, Boehm R, Schnabl H (2006a) Stable genetic transformation of high oleic *Helianthus annuus* L. genotypes with high efficiency. Plant Sci 171: 546–554.

Mohamed S, Boehm R, Schnabl H (2006b) Particle bombardment as a strategy for the production of transgenic high oleic sunflower (*Helianthus annuus* L.). J Appl Bot Food Qual 80: 171–178.

Molinier J, Thomas C, Brignou M, Hahne G (2002) Transient expression of *ipt* gene enhances regeneration and transformation rates of sunflower shoot apices (*Helianthus annuus* L.). Plant Cell Rep 21: 251–256.

Monsanto Co (2002) Safety assessment of Roundup Ready soybean Event 40-3-2: *http: //www.monsanto.com/monsanto/content/sci_tech/prod_safety/roundup_soybean/pss.pdf*

Müller A, Iser M, Hess D (2001) Stable transformation of sunflower (*Helianthus annuus* L.) using a non-meristematic regeneration protocol and green fluorescent protein as a vital marker. Transgen Res 10: 435–444.

Muller MH, Arlie G, Bervillé A, David J, Delieux F, Fernandez-Martinez JM, Jouffret P, Lecomte V, Reboud X, Rousselle Y, Serieys H, Teissere N, Tsitrone A (2006) Le compartiment spontané du tournesol *Helianthus annuus* en Europe: prospections et premières caractérisations génétiques. Les Actes du BRG 6: 335–353.

Muller MH, Delieux F, Fernandez-Martínez JM, Garric B, Lecomte V, Anglade G, Leflon M, Motard C, Segura R. (2009) Ocurrence, distribution and distinctive morphological traits of weedy *Helianthus annuus* L. populations in Spain and France. Genet Res Crop Evol 55: 869–877.

Nestares G, Zorzoli R, Mroginski L, Picardi L (1998) Cytoplasmic effects on the regeneration ability of sunflower. Plant Breed 117: 188–190.

Nestares G, Zorzoli R, Mroginski L, Picardi L (2002) Heritability of in vitro plant regeneration capacity in sunflower. Plant Breed 121: 366–368.

NSA National Sunflower Association (2009): *http: //www.sunflowernsa.com* (Accessed September 15, 2009).

Obschatko ES, Ganduglia F, Román F (2006) El sector agroalimentario argentino 2000–2005 Instituto Interamericano de Cooperación para la Agricultura (IICA Ed.) p 282.

OECD Organisation for Economic Co-operation and Development (2004) Consensus Document on the Biology of *Helianthus annuus* L. (SUNFLOWER). Sr on Harmonisation of Regul Oversight in Biotechnol No 31.

Paniego N, Heinz R, Fernandez P, Talia P, Nishinakamasu V, Hopp HE (2007) Sunflower. In: C Kole (ed) Genome Mapping and Molecular Breeding in Plants. vol 2: Oilseeds. Springer-Verlag, Berlin, Heidelberg, New York, pp 153–177.

Panković D, Jocic S, Lacok N, Sakac Z, Skoric D (2004) The use of PCR-based markers in the evaluation of resistance to downy mildew in NS-breeding material. Helia 40: 149–158.

Paterson KE, Everett NP (1985) Regeneration of *Helianthus annuus* inbred plants from callus. Plant Sci 42: 126–132.

Pedraza M, Pereyra V, Escande A (2004) Infection courts and length of susceptible period related to sunflower head rot (*Sclerotinia sclerotiorum*) resístance. Helia 27: 171–182.

Pelissier B, Bouchfra O, Pepin R, Freyssinet G (1990) Production of isolated somatic embryos from sunflower thin layers. Plant Cell Rep 9: 47–50.

Poverene MM, Cantamutto MA, Carrera AD, Ureta MS, Salaberry MT, Echeverria MM, Rodriguez RH (2002) El girasol silvestre (*Helianthus* spp.) en la Argentina: Caracterización para la liberación de cultivares transgénicos. Rev Invest Agropec 31: 97–116.

Power CJ (1987) Organogenesis from *Helianthus annuus* inbreds and hybrids from the cotyledons of zygotic embryos. Am J Bot 74: 497–503.

Pozzi G, Lopez M, Cole G, Sosa Dominguez G, Bidney D, Scelonge C, Wang L, Lu G, Müller-Cohn J, Bradfisch G (2000) Bt-mediated insect resistance in sunflower (*Helianthus annuus* L.) In: Proc 15th Int Sunflower Conf, Tome II, Int Sunflower Assoc, Paris, France, pp H46-50.

Prado E, Berville A (1990) Induction of somatic embryo development by liquid culture in sunflower (*Helianthus annuus* L.). Plant Sci 67: 73–82.

Prats E, Bazzalo M, Leon A, Jorrin J (2003) Accumulation of soluble phenolic compounds in sunflower capitula correlates with resistance to *Sclerotinia sclerotiorum*. Euphytica 132: 321–329.

Prats E, Bazzalo M E, Leon A, Jorrin J V (2006) Fungitoxic effect of scopolin and related coumarins on *Sclerotinia sclerotiorum*. A way to overcome sunflower head rot. Euphytica 147: 451–460.

Pugliesi C, Biasini MG, Fambrini M, Baroncelli S (1993) Genetic transformation by *Agrobacterium tumefaciens* in the interspecific hybrid *Helianthus annuus* × *Helianthus tuberosus*. Plant Sci 93: 105–115.

Quagliaro J, Vischi M, Tyrka M, Olivieri AM (2001) Identification of wild and cultivated sunflower for breeding purposes by AFLP markers. J Hered 92: 38–42.

Radonic LM, Zimmermann JM, Zavallo D, Lopez N, Lopez Bilbao M (2006) Rooting in Km selective media as efficient *in vitro* selection method for sunflower genetic transformation. Electr J Biotechnol DOI: 10.2225/vol9-issue 3–19.

Ramputh A, Arnason J, Cass L, Simmonds J (2002) Reduced herbivory of the European corn borer (*Ostrinia nubilalis*) on corn transformed with germin, a wheat oxalate oxidase gene. Plant Sci 162: 431–440.

Rao KS, Rohini VK (1999) *Agrobacterium*-mediated transformation of sunflower (*Helianthus annus* L.): A simple protocol. Ann Bot 83: 347–354.

Reagon M, Snow A (2006) Cultivated *Helianthus annuus* (Asteraceae) volunteers as a genetic "bridge" to weedy sunflower populations in North America. Am J Bot 93: 127–133.

Reichman JR, Watrud LS, Lee EH, Burdick CA, Bollman MA, Storm MJ, King GA, Mallory-Smith C (2006) Establishment of transgenic herbicide-resistant creeping bentgrass (*Agrostis stolonifera* L.) in nonagronomic habitats. Mol Ecol 15: 4243–4255.

Rieseberg L, Kim M, Seiler G (1999) Introgression between cultivated sunflower and a sympatric wild relative *Helianthus petiolaris* (Asteraceae). Int J Plant Sci 160: 102–108.

Ribeiro A, Bessa A, Faria T (2005) Evaluation of wild *Helianthus* species in Mozambique. Eur Coop Res Network on Sunflower, FAO Working Group, Progr Rep 2001–2004, pp 46–48.

Robinson RG (1978) Production and culture. In: JF Carter (ed) Sunflower Science and Technology. Am Soc Agron, Madison, Wisconsin, USA, pp 89–143.

Rogers CE (1992) Insect pest and strategies for their management in cultivated sunflower. Field Crop Res 30: 301–332.

Roldan Serrano A, Luna del Castillo J, Jorrin Novo J, Fernandez Ocana A, Gomez Rodriguez MV (2007) Chitinase and peroxidase activities in sunflower hypocotyls: effects of BTH and inoculation with *Plasmopara halstedii*. Biol Plant 51: 149–152.

Rousselin P, Molinier J, Himber C, Schontz D, Prieto-Dapena P, Jordano J, Martini N, Weber S, Horn R, Ganssmann M, Grison R, Pagniez M, Toppan A, Friedt W, Hahne G (2002) Modification of sunflower oil quality by seed-specific expression of a heterologous Delta 9-stearoyl-(acyl carrier protein) desaturase gene. Plant Breed 121: 108–116.

Sala CA, Bulos M, Echarte AM (2008) Genetic analysis of an induced mutation conferring imidazolinone resistance in sunflower. Crop Sci 48: 1817–1822.

Sarrafi A, Bolandi AR,Serieys H, Berville A, Alibert G (1996) Analysis of cotyledon culture to measure genetic variability for organogenesis parameters in sunflower (*Helianthus annuus* L.). Plant Sci 121: 213–219.

Sawahel W, Hagran A (2006) Generation of white mold disease-resistant sunflower plants expressing human lysozyme gene. Biol Plant 50: 683–687.

Schnabl H, Binsfeld P, Cerboncini C, Dresen B, Peisker H, Wingender R, Henn A (2002) Biotechnological methods applied to produce *Sclerotinia sclerotiorum* resistant sunflower. Helia 25: 191–197.

Schrammeijer B, Sijmons PC, van den Elzen JM, Hoekema A (1990) Meristem transformation of sunflower via *Agrobacterium*. Plant Cell Rep 9: 55–60.

Scelonge C, Wang L, Bidney D, Lu G, Hastings C, Cole G, Mancl M, D'Hautefeuille JL, Sosa Dominguez G, Coughlan S (2000) Transgenic *Sclerotinia* resistance in sunflower (*Helianthus annuus* L.). In: Proc 15th Int Sunflower Conf, Tome II, Int Sunflower Assoc, Paris, France, pp K66–71.

Skoric D (1992) Achievements and future directions of sunflower breeding. Field Crops Res 30: 231–270.

Snow A, Moran Palma P, Rieseberg L, Wszlaki A, Seiler G. (1998) Fecundity, phenology, and seed dormancy of F1 wild-crops hybrids in sunflower (*Helianthus annuus*, Asteraceae). Am J Bot 85: 794–801.

Snow A, Pilson D, Rieseberg L, Paulsen M, Pleskac N, Reagon M, Wolf D, Selbo M (2003) A Bt transgene reduces herbivory and enhances fecundity in wild sunflowers. Ecol Appl 13: 279–286.

Sossey-Alaoui K, Serieys H, Tersac M, Lambert P, Schilling E, Griveau Y, Kaan F, Bervillé A (1998) Evidence for several genomes in *Helianthus* Theor Appl Genet 97: 422–430.

Stankovic-Kalezic R, Kojic M, Vrbnicanin S, Radivojevic Lj (2007) *Helianthus annuus*—A new important element of the non-arable and arable flora in Serbia's region of Southern Banat. Helia 30: 37–42.

Sujatha M, Lakshminarayana M (2007) Resistance to *Spodoptera litura* (Fabr.) in *Helianthus* species and backcross derived inbred lines from crosses involving diploid species. Euphytica 155: 205–213.

Sujatha M, Prabakaran AJ (2001) High frequency embryogenesis in immature zygotic embryos of sunflower. Plant Cell Tiss Org Cult 65: 23–29.

Tan S, Evans RR, Dahmer ML, Singh BK, Shaner DL (2005) Imidazolinone-tolerant crops: history, current status and future Source. Pest Manag Sci 61 (3): 246–257.

Taski-Ajdukovic K, Vasic D, Nagl N (2006) Regeneration of interspecific somatic hybrids between *Helianthus annuus* L. and *Helianthus maximiliani* (Schrader) via protoplast electrofusion. Plant Cell Rep 25: 698–704.

Thengane SR, Joshi MS, Khuspe SS, Mascarenhas AF (1994) Anther culture in *Helianthus annuus* L. Influence of genotype and culture conditions on embryo induction and plant regeneration. Plant Cell Rep 13: 222–226.

Trigo E, Cap E (2006) Ten years of Genetically Modified Crops in Argentine Agriculture: *http: //www.inta.gov.ar/ies/docs/otrosdoc/Ten_Years_GM_Crops_Argentine_Agriculture.pdf*

Ureta MS, Carrera AD, Cantamutto MA, Poverene MM (2008) Gene flow among wild and cultivated sunflower, *Helianthus annuus* in Argentina. Agri, Ecosyst Environ 123: 343–349.

USDA Foreign Agricultural Service (2009a) Grain: World Market and Trade. Circular Series FG 09-09: http://www.fas.usda.gov/grain/circular/2009/09-09/grainfull0909.pdf (Accessed Sept 14, 2009).

USDA Foreign Agricultural Service (2009b) Oilseeds: World Market and Trade. Circular Series FOP 9-09: http://www.fas.usda.gov/psdonline/circulars/oilseeds.pdf (Accessed Sept. 14, 2009).

Valkov V, Bachvarova R, Slavov S, Atanassova S, Atanassov A (1998) Genetic transformation of tobacco for resistance to BastaReg. Bulg J Agri Sci 4: 1–7

Vannozzi G P (2006) The perspectives of use of high oleic sunflower for oleochemistry and energy raws. Helia 29: 1–24.

Vasic D, Pajevic S, Saric M, Vasiljevic L, Skoric D (2001) Concentration of mineral elements in callus tissue culture of some sunflower inbred lines. J Plant Nutr 24: 1987–1994.

Vasic D, Skoric D, Taski K, Stosic L (2002) Use of oxalic acid for screening intact sunflower plants for resistance to *Sclerotinia* in vitro. Helia 25: 145–152.

Veatch M, Ray D, Mau C, Cornish K (2005) Growth, rubber, and resin evaluation of two-year-old transgenic guayule. Indus Crops Prod 22: 65–74.

Velasco L, Perez-Vich B, Jan CC and Fernandez-Martinez JM (2007) Inheritance of resistance to broomrape (*Orobanche cumana* Wallr.) race F in a sunflower line derived from wild sunflower species. Plant Breed 126: 67–71.

Ves Lozada J, Teppaz E, Manetti P (2007) Manejo de plagas animales: estado actual y novedades. Cuarto Congreso Argentino de Girasol: *http: //asagir.org.ar/4to_congreso*

Vischi M, Di Bernardo N, Scotti I, Della Casa S, Seiler G, Olivieri A (2004) Comparison of populations of *Helianthus argophyllus* and *H. debilis* ssp. *cucumerifolius* and their hybrids from the African Coast of the Indian Ocean and the USA using molecular markers. Helia 27: 123–132.

Vranceanu AV (1977) El Girasol. Ediciones Mundi-Prensa, Madrid, Spain, 379 p.

Vuong T, Hoffman D, Diers B, Miller J, Steadman J, Hartman G (2004) Evaluation of soybean, dry bean, and sunflower for resistance to *Sclerotinia sclerotiorum*. Crop Sci 44: 777–783.

Watanabe M, Shinmachi F, Noguchi A, Hasegawa I 2005 Introduction of yeast metallothionein gene (CUP1) into plant and evaluation of heavy metal tolerance of transgenic plant at the callus stage. Soil Sci Plant Nutr 51: 129–133.

Weber S, Friedt W, Landes N, Molinier J, Himber C, Rousselin P, Hahne G, Horn R (2003) Improved *Agrobacterium*-mediated transformation of sunflower (*Helianthus annuus* L.): assessment of macerating enzymes and sonication. Plant Cell Rep 21: 475–482.

Whitton J, Wolf D, Arias D, Snow A, Rieseberg (1997) The persistence of cultivar alleles in wild populations of sunflowers five generations after hybridization. Theor Appl Genet 95: 33–40.

Wingender R, Henn HJ, Barth S, Voeste D, Machlab H, Schnabl H (1996) A regeneration protocol for sunflower (*Helianthus annuus* L.) protoplasts. Plant Cell Rep 15: 742–745.

Witrzens B, Scowcrofi WR, Downes RW, Larkin PJ (1988) Tissue culture and plant regeneration from sunflower (*Helianthus annuus*) and interspecific hybrids (*H. tuberosus* × *H. annuus*). Plant Cell Tiss Org Cult 13: 61–76.

10

Future Prospects

Brady A. Vick[1]* and *Jinguo Hu*[2]

ABSTRACT

Cultivated sunflower has undergone significant improvement in yield and other agronomic traits during the past half century. However, non-GMO sunflower faces major challenges from insect and herbicide resistant GMO crops such as maize and soybean for a place in the producer's crop rotation plan. In several sunflower producing countries, sunflower production has shifted in recent years to marginal production regions in response to competition from soybean and maize. Despite displacement to less fertile growing areas, yields have steadily increased as a result of improved hybrids. Yet the potential for increased sunflower cultivation world-wide remains optimistic. Many countries have identified prospective new regions for profitable sunflower production. New advances in sunflower genomics are expected to have enormous impact on the genetic improvement of the sunflower crop. Likewise, expanded uses of sunflower, such as biofuels, biolubricants, or for plant production of pharmaceuticals, are expected to encourage increased sunflower production.

Keywords: sunflower; yield; production; expansion; genomics; uses

[1]USDA-Agricultural Research Service, Northern Crop Science Laboratory, Fargo, ND 58105, USA.
[2]USDA-Agricultural Research Service, Western Regional Plant Introduction Station, 59 Johnson Hall, Washington State University, Pullman, WA 99164, USA.
*Corresponding author: *brady.vick@ars.usda.gov*

10.1 The Past

If we look back to the past to foretell the future of cultivated sunflower (*Helianthus annuus* L.), we can predict that a series of incremental, subtle improvements will occur, punctuated by periodic notable technological advances. The history of sunflower advancement as an important oilseed crop has followed this pattern as sunflower evolved into one of the leading oilseed crops worldwide. The domestication of wild *Helianthus annuus* by Native Americans, probably in east-central North America over 4,300 years ago (Harter et al. 2004), can be considered as the first major technological advance of sunflower as an agricultural crop. In contrast to multibranched wild sunflower with many small heads containing small, easily-shattered seeds, domesticated sunflower was characterized by an unbranched stem with a single, large head holding relatively large achenes that were retained until harvest.

Following its introduction in western Europe in the 16th century by early New World explorers, the domesticated sunflower spread eastward as it became increasingly popular as a garden ornamental. When sunflower reached Russia, its value as a good source of cooking oil was recognized. By the 1850s many sunflower crushing mills were in operation in central European Russia. The oil content of sunflower seed at the time was about 25%, but starting in the 1920s Russian sunflower breeders were successful in another significant technological advance: increasing the oil content to over 40%. The most successful of these research programs was led by V. S. Pustovoit at Krasnodar near the Black Sea. By 1965, Pustovoit was testing open-pollinated lines with more than 50% oil while maintaining or increasing total yield (Heiser 1976).

A third important advance came with the discovery of cytoplasmic male sterility in sunflower by Leclercq (1969) in France and the corresponding fertility restorer gene by Kinman (1970) in the US. Together, these two discoveries made it possible to produce hybrid sunflower that resulted in enhanced yield. By the year 2007, oilseed sunflower production in Argentina, Ukraine, Europe, US, India, and Turkey was almost exclusively from hybrid sunflower (Bjerke, Mulpuri, Kaya, pers. comm.). In Russia, about 60% of total sunflower cultivation was still with open-pollinated cultivars in 2007. However, regional differences in production practices are evident in Russia, with the sunflower growing regions around Krasnodar, Stavropol, and Rostov, utilizing hybrid oilseed sunflower for about 70% of production (Bochkaryov, pers. comm.). Almost all oilseed-type sunflower production in China has been converted to hybrids. However, China still adheres to open-pollinated cultivars for about 85% of its confectionery sunflower, which accounts for more than 60% of total sunflower production (Liu, pers. comm.).

Finally, the creation of a high-oleic acid sunflower mutant (Soldatov 1976) was another significant advance that provided a new tool for breeders to create variation in the unsaturated fatty acid composition of sunflower oil. This new genetic resource has increased the versatility of sunflower oil for both food and industrial applications.

The quantum leaps in technological advancements in sunflower breeding during the past half century are less publicized, yet are highly significant with regard to the progress in sunflower improvement. Breeding for disease resistance became a priority, and resistance genes for several serious diseases were identified in the cultivated germplasm and incorporated into hybrids. In addition, the wild sunflower species were proven to be significant sources of important genes for herbicide tolerance and for resistance to evolving races of fungal and parasitic diseases.

It is reasonable to expect similar progress in sunflower improvement in the next decades. The future of the cultivated oilseed sunflower crop will be shaped by many fluid forces, including market dynamics, germplasm enhancement, genomics, new uses of sunflower oil, new products from the sunflower plant, and environmental concerns. The following sections explore potential changes, including opportunities and obstacles, in the continuing evolution of sunflower production and utilization in the approaching decade.

10.2 Future Potential to Increase Productivity

During the past decade, sunflower production in several countries has migrated from regions of adequate rainfall and fertile soil to less productive areas where water and fertility restrict yield. In China, there was a recent shift from the major production areas of the northeastern provinces to the northern and western semiarid and arid provinces of Inner Mongolia and Xinjiang. Argentina also saw a decline in sunflower production area during the past decade, from more than 4 million ha in 1998 to less than 2 million ha in 2005 (Vasquez and de Romano 2006). Favorable profits for maize and soybean, coupled with ease of cultivation and pest protection of these two GMO crops, helped to displace sunflower to regions of inferior soils. Although hybrids improved during this transition, the end result of the shift in production areas was static yields.

The situation in the US parallels China and Argentina, in which the current areas of sunflower production are removed from the original regions where open-pollinated sunflower was introduced to farmers in the 1960s. However, there has been an overall trend towards increased yields in the US despite the shift in production regions. Hybrid sunflower was introduced in the US in the early 1970s, with the potential to increase yield through heterotic effects of genetically distant parental lines. Figure 10-1 shows the

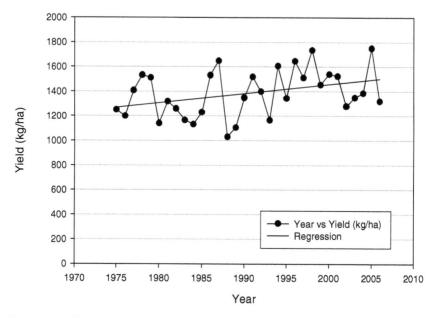

Figure 10-1 US oilseed sunflower yields from 1975 to 2006 (USDA, National Agricultural Statistics Service 2009).

average annual oilseed sunflower yields in the US from 1975 through 2006. Although environmental conditions varied considerably from year to year and affected crop production, a regression line shows that there was a gradual annual increase in yield of about 7.5 kg ha^{-1}. This translates into an increase of about 240 kg ha^{-1} over the 32-year reporting period. The trend to increase yield can be expected to continue as sunflower heterotic groups become better defined and the effects of their interactions are understood and exploited.

The US national annual yield increase occurred despite a gradual westward shift in production from fertile soils with adequate rainfall to semiarid regions with less productive soil. Diseases and insect problems were among the reasons for the shift in the production areas. However, in a situation similar to Argentina, competition for production area by maize and soybeans, which many producers regarded as more profitable and easier to grow, also played a role in dislocating sunflower to the western, drier areas of the Great Plains where sunflower with its deep taproot could thrive better than maize or soybean. Coupled with the displacement of sunflower growing areas in the US in the past two decades was a concomitant decline in total production due to decreased land area planted to sunflower. This was largely due to increased competition from canola, which has established itself across the northern tier of the US, and from new maize and soybean cultivars that are adapted to northern environments.

Other factors in addition to displacement to less fertile regions also contributed to static or intermittent decreased yields. Small, but noticeable yield declines were occasionally observed during the hybrid sunflower era when significant new traits were introduced into hybrids. When NuSun® hybrids, with mid-oleic fatty acid composition, were first introduced into the US early in the first decade of the millennium, a yield "drag" was perceived by some producers. This was quickly overcome in the subsequent years by commercial breeders such that current NuSun® hybrids outperform most traditional high-linoleic hybrids. A similar transitional period is currently in progress with herbicide resistant sunflower hybrids. In the future, parallel transient decreases in yield followed by rapid recovery can be expected when major nonyield genetic improvements are made to sunflower. Sunflower that has been genetically modified (GMO) to incorporate genes from other species for useful new agronomic traits would, if introduced into commercial cultivation, likely follow a similar path.

Irrigation, as expected, increases yields dramatically. Company-sponsored sunflower yield contests have not typically been offered in the US, but anecdotal evidence from producers suggests that yields of over 4,500 kg ha^{-1} are possible for irrigated oilseed sunflower hybrids under optimum fertilization. Such a yield is three-fold higher than the average US yield in recent years. Market reality suggests, however, that large-scale irrigation of sunflower is unlikely in the future. Producers with an irrigation option are more likely to put other lucrative crops such as maize under irrigation, and to utilize sunflower under dryland conditions due to its superior ability to capture available moisture. Thus, in many countries, if formerly productive regions of sunflower cultivation were to re-establish themselves as major growing areas, the average annual yield of sunflower would increase dramatically. However, this is not likely in the near future due to increased attention to maize and soybean production worldwide.

10.3 Impact of Advances in Sunflower Genomics

The past decade has been marked with milestones in sunflower genomics research that included the construction of linkage maps using various molecular markers, such as RFLP, RAPD, AFLP, SSR, TRAP, and InDel (insertion/deletion), the establishment of genetic transformation methods, and the sequencing of thousands of ESTs (expressed sequence tags) of the cultivated sunflower and its related species. These milestones have an enormous impact on addressing problems in both fundamental biology and applied genetic improvement of this crop.

On one hand, the massive sequence data will offer access for developing DNA markers for genes controlling economically important traits. These markers will be useful for rapidly moving genes among elite lines to enhance the target traits via marker-assisted breeding. Marker-assisted selection evolved from the idea of indirect selection, which has been a common practice in both plant and animal breeding. For two highly correlated traits, selection imposed on trait A will result in a genetic gain for trait B. Similarly, if a visible morphological trait is affecting or associated with a desirable economical trait like high yield, this visible morphological trait will be useful as a marker for selecting to increase yield. However, for a given species, the number of visible and stable morphological markers is limited. In sunflower the anthocyanin pigment gene (T), which was tightly linked to a nuclear male sterile gene (Ms_{10}), was proposed by Leclercq (1966) to be used as a marker for utilizing genetic male sterility for hybrid production. When isozyme markers were used to reveal genetic polymorphism in plants, Tanksley and Rick (1980) applied the isozyme markers to tomato breeding and proposed to use them in other crops.

With the advent of DNA-based markers, marker-assisted selection became a reality for many important crop species. It has proven to be a powerful tool for incorporating target genes into a recipient line at an accelerated pace. Following the same trend of other crops, sunflower researchers have applied marker-assisted selection to improve this crop. Marker-assisted selection is the primary level for application of molecular markers to breeding because only one or a few single-gene controlled traits are targeted and low throughput techniques are used. The secondary level is marker-assisted breeding in which a large number of qualitative and quantitative traits are targeted and a high throughput genotyping facility is required. The highest level for application of molecular markers to breeding will be breeding by design, a fairly new concept which aims to optimize all allelic variation for all genes governing traits of economic importance. The achievement of such optimization will depend on a combination of precise genetic mapping, high-resolution chromosome haplotyping and extensive phenotyping (Peleman and van der Voort 2003). The practice of breeding by design in sunflower waits for more available genetic information and genomics tools to be developed in the near future.

On the other hand, the sequence data will enable scientists to determine the genetic and molecular bases of agriculturally important genes through massively parallel genetic analysis on microarray chips. Once a gene conditioning a given trait is identified, it is possible and relatively easy to replace, suppress, or otherwise modify that gene to achieve an ideal phenotype through genetic engineering.

Up to now, there has not been commercial production of transgenic sunflower. This has been due to two major concerns. The first relates to the public acceptance of transgenic sunflower because sunflower products are primarily for human consumption—sunflower oil is a premium healthy vegetable oil for home cooking and food processing, and confection sunflower is a snack food. The second is the concern about genetic pollution (Rifkin 1998) which is currently defined as the "uncontrolled spread of genetic information (frequently referring to transgenes) into the genomes of organisms in which such genes are not present in nature" (Zaid et al. 2001). Sunflower is one of the few crops that originated and was domesticated in North America and can hybridize freely with its abundant wild relatives. As transgenic crops have become more widespread, these concerns have become less worrisome. The real challenge for commercialization of transgenic sunflower is the lack of a superior transgenic line which overcomes the limiting factors for sunflower production. The continued in-depth investigation into sunflower genomics will yield unprecedented insights into basic mechanisms of sunflower development, tolerance to biotic and abiotic stresses, and other metabolic processes related to productivity. The translation of these new insights into applications is anticipated to result in superior sunflowers.

The advances in sunflower genomics have an enormous impact on the genetic improvement of the sunflower crop. Together with the conserved diverse germplasm collection, the available and yet-to-come genomics tools will enable sunflower researchers to effectively and efficiently shape sunflower as a competitive agricultural crop to meet the needs of mankind.

10.4 Future Potential to Expand into Other Geographical Areas

Many developing countries have the potential to increase sunflower production. Considering Pakistan as one example, one observes a country where consumers prefer sunflower oil for cooking, while 70% of the country's total vegetable oil is imported. Yet despite Pakistan's inability to meet its domestic demand for sunflower and other vegetable oils, the sunflower production area in 2002–03 decreased by 25% from its peak in 1998–99. In a report that recognized the potential of sunflower oil to contribute to the increased domestic production of vegetable oil (Shah et al. 2005), proposals were made to increase production by: 1) expanding sunflower oilseed cultivation into regions which previously did not grow sunflower, 2) providing government support prices, and 3) enhancing research to target a three-fold yield increase by adopting improved cultural practices and developing higher yielding varieties and hybrids. Progress toward these goals was apparently made, as national agricultural statistics indicate that the area of production had increased from 264,000 ha in 2004–05 to 397,000 ha in 2007–08 (Pakistan Federal Bureau of Statistics 2008). Several regions of south central and coastal Pakistan have been identified as potential growth areas for sunflower production.

Improvement in agronomic practices might also contribute significantly to increased production in some countries where sunflower is currently grown. Beg et al. (2007) report that introduction of high yielding sunflower varieties in Iran, along with standardization of sunflower production practices such as optimal plant populations and row spacing for different climatic conditions, has the potential to enhance productivity in that region significantly. However, under dryland conditions, some open-pollinated varieties were as productive as hybrids and may be more attractive to induce farmers to increase sunflower production due to the advantage of both cost and the ability to save seed for the next growing season (Pourdad and Beg 2008).

Potential new areas for sunflower production have also been explored in the Mediterranean countries. Ismailia Governorate in northeastern Egypt, where sandy soil regions have been reclaimed, has been identified as a promising area for sunflower crops, provided high yielding hybrids are used along with proper fertilization and adequate irrigation regimes (Abdel-Wahab et al. 2005). In Italy, sunflower has traditionally been cultivated in the rainfed regions of central Italy. However, interest in sunflower as a biofuel crop has stimulated research, which showed sunflower can be a viable crop under the typical environmental conditions of southern Italy (Flagella et al. 2006).

In the US, the two predominant oilseed sunflower production regions are the northern Great Plains and the western central Great Plains. A subtle production shift to the west within the northern Great Plains area has occurred since the introduction of hybrid sunflower in the mid-1970s. Among the reasons for the relocation was a predisposition for disease, especially Sclerotinia, in the eastern Great Plains areas with higher rainfall. Insects, particularly the sunflower midge (*Contarinia schultzi* Gagné), which is consistently destructive in certain growing areas, were also a factor in the decline in sunflower production in this region. The recent release of sunflower germplasm with improved Sclerotinia head and stalk rot tolerance should lead to the development of more resistant commercial hybrids and improve the potential for sunflower to move back into previously productive growing areas. Germplasm with improved tolerance to the sunflower midge has also been released, and experimental lines have been identified with increased tolerance to other US insects. These new germplasms will make it possible for a return of production to areas most susceptible to insect damage. Migrating blackbirds continue to be a serious problem in the heart of migratory pathways, and chemical, biological, and behavioral control methods have met with only limited success.

Many areas of the US are favorable for sunflower production, but economics and point-of-delivery availability will continue to determine where sunflower is grown. Zheljazkov et al. (2008) have demonstrated that most areas of Mississippi are potential growing areas for producing and extracting sunflower oil for local biodiesel production. Other states in the southeastern US have explored the possibility of double cropping sunflower with other crops, for example wheat in Kentucky and cotton in Georgia, with a view to selling the seed for oil processing or for using the oil as a biodiesel fuel. Thus, several new regions of sunflower production in the US are possible, and expansion into those areas will be driven by the market and the related need for new, renewable energy sources.

Similar opportunities for sunflower expansion into new areas undoubtedly exist in many countries. We can expect that many of the new areas of production will be marginally fertile lands in hot, arid climates, where sunflower can outcompete many other crops to produce a profitable yield.

10.5 New uses

10.5.1 Biofuels and Lubricants

Sunflower has been used as a source for biofuel by two quite different systems. It is possible to use whole sunflower plants for biomass production and conversion in biogas reactors to methane gas. Hahn et al. (2006) tested different sunflower types, and estimated that 6,000 $m^3\,ha^{-1}$ of methane gas is a realistic breeding goal for biogas production from sunflower. In Germany, district biogas reactors have been built to convert biomass to methane as an energy source. Whether high biomass sunflowers can be grown economically and can compete with other high biomass sources for commercial methane production will be determined by market economics.

Sunflower oil, along with most other vegetable oils, has been recognized as an excellent biodiesel fuel for several decades. Early utilization of vegetable oils for biodiesel fuel was with unmodified oil. Vegetable oils are now transesterified to release the fatty acids as methyl or ethyl esters for use as biodiesel fuels. High-oleic sunflower oil, whether directly or as feedstock for synthetic esters, has great potential both as a biodiesel fuel and as a lubricant due to its superior wear-reducing properties. In the EU, where regulatory rules favor environmentally friendly applications of industrial oils, natural oils like sunflower are in a good position to compete with petroleum-based products when overall costs are compared (Vannozzi 2006). Reduced consumption, high quality, low maintenance, and biodegradability decrease the long-term costs associated with disposal of lubricants to the environment.

High-oleic sunflower oil is in a particularly good position for use as a biofuel because of its higher stability than traditional, high-linoleic sunflower oil. In Europe, rules have been proposed to require a minimum percentage of biofuels in transportation fuels, perhaps as much as 8% by the year 2020. Oilseed rape (canola) and high-oleic sunflower are the two crops most likely to serve as sources of biodiesel in Europe. Major sunflower exporting countries, such as Argentina, are moving to increase production of high-oleic sunflower oil for the European market.

In the US, it is unlikely that sunflower oil will be used on a large commercial scale for biodiesel production in the near future. Commercial biodiesel production plants in the US currently use soybean and canola as the primary vegetable oil feedstock (National Biodiesel Board 2009), due to the large amounts grown in North America. Sunflower oil, which sells at a premium and is in high demand as a food oil by large snack food companies, will likely continue to have its major role in this domestic market for the foreseeable future. As soybean and canola oils move into the biodiesel market, sunflower oil has the potential to gain markets formerly held by displaced soybean and canola oils. The role of sunflower oil as a biodiesel fuel in the US will be largely limited to on-farm production or to regional cooperatives that have the ability to convert the oil to esters and deliver the fuel to nearby customers.

10.5.2 Bio-pharming by Sunflower

Bio-pharming is the production of pharmaceutical products by introduction into a host plant of a foreign gene that codes for the drug product. Sunflower is a candidate host for producing pharmaceuticals, and compares favorably with maize, canola, soybeans, and alfalfa for cost of production (Kusnadi et al. 1997). Many private companies and institutions in countries around the world have initiated development of pharmaceutical crops, but as of 2007 only a handful of proteins had been commercialized, all in the US. All were produced in transgenic maize, tobacco, or rice (Mewett et al. 2007). In Canadian field trials, safflower, tobacco, and canola are the most common crops for biopharming trials, and in Europe (primarily France) tobacco and maize have been the crops of choice (Bauer 2006). To date, sunflower has not been chosen as a primary bio-pharming crop in any country.

Sunflower, however, has potential for bio-pharming, but it suffers from several key environmental drawbacks, especially in the US. Transgenic annual sunflower (GMO) pollen will cross with several annual wild sunflower species, which are native to the US. This is of concern to many environmentalists because of unknown consequences to the population dynamics of wild sunflower species, or to the risks to wildlife or humans if the seed is inadvertently introduced into the food chain. In addition, sunflower seeds can remain dormant for several years and then germinate and grow years later. Dormant transgenic sunflower seeds would pose a control problem if transgenic pollen is dispersed near populations of nontransgenic cultivated or wild sunflower over a period of several years.

Bio-pharming of sunflower for drugs or medicines that are expensive to synthesize and produce is a potentially lucrative enterprise. It is not likely to occur in the US except under very rigid control situations. However, it might be feasible and profitable in other countries or regions where wild sunflower populations do not exist. Perhaps we can expect a new "quantum leap" in sunflower production with the development of a technological tool that allows the introduction of an agronomically favorable foreign gene that is not expressed in pollen, which would minimize the possibility of outcrossing with wild species to an acceptable risk. One potential technique would be the introduction of foreign genes into sunflower chloroplast DNA rather than nuclear DNA, as has been demonstrated in tobacco (Ruf et al. 2007; Svab and Maliga 2007). Each cellular chloroplast could act as a factory for production of a pharmaceutical or nutriceutical. Because plant cells possess many chloroplasts compared to only one nucleus, the efficiency of product formation could be substantial over nuclear-transformed plants. Of importance is the presence of foreign genes in the chloroplast rather than the nucleus which would significantly reduce their integration into pollen and minimize the risk of outcrossing with unintended species to a tolerable level, thus allowing for commercial outdoor bio-pharming of valuable proteins in North America with sunflower as the host plant.

10.5.3 Bioremediation

Sunflowers have the potential to assist in bioremediation of contaminated soils. Two approaches have been reported in which sunflowers play an important role: 1) utilization of sunflower oil to extract harmful chemicals, and 2) planting of sunflower (phytoremediation) on soils contaminated with heavy metals.

Gong et al. (2006) have reported that sunflower oil has a great capacity to remove PAHs (polycyclic aromatic hydrocarbons) from contaminated soils. Contaminated soils are transferred to a column and sunflower oil allowed to percolate down through the soil. More than 90% of PAHs were removed, with the treated soil retaining only 4–5% of readily biodegradable sunflower oil.

The ability of sunflower to accumulate heavy metal ions such as cadmium, chromium, zinc, mercury, and lead from the soil is well-known. Li et al. (1997) described the genetic variation in cadmium accumulation by sunflower genotypes, and Van der Lelie et al. (2001) have reported the use of sunflower as a suitable crop for phytoremediation of soils polluted by zinc.

However, the rate of phytoextraction of zinc by sunflower from soils very heavily contaminated by a toxic spill at a pyrite mine was very low and was not a feasible option for zinc removal, although the sunflower plants were effective in stabilizing the contaminated soil (Madejón et al. 2003).

The symbiotic relationship between sunflower and microbes can be exploited to enhance heavy metal ion accumulation by the plant, and is environmentally more favorable than using metal ion chelators, which run the risk of leaching chelated metals into groundwater. Uptake of chromium ions by sunflower roots was enhanced by inoculation of the roots with mycorrhizal fungi (Davies et al. 2001). Similarly, inoculation of sunflower roots with the rhizobacterium *Pseudomonas putida* 06909, which had been engineered to express a synthetic metal-binding peptide, increased cadmium binding by 40% (Wu et al. 2006). Thus, for moderately contaminated soils, the symbiotic combination of sunflower and microbes may provide a practical option for phytoremediation.

10.6 Outlook

In many parts of the world, especially Europe, South Asia, and Central and South America, sunflower oil is well established as a preferred vegetable oil for household cooking. The recent introduction of mid-oleic sunflower into the North American market has driven demand for sunflower oil to such an extent that producers can barely meet the demand. Mid-oleic sunflower oil is currently considered the ideal frying oil for snack foods by the US snack food industry, due to its favorable composition of approximately 65% oleic acid (high frying stability and shelf life), about 22% linoleic acid (precursor to desirable flavor components), and low saturated fatty acid content. The resulting shortage of mid-oleic sunflower oil has made bottled sunflower oil unavailable in supermarkets to most US consumers in recent years.

Sunflower oil as a premium oil will be challenged in future years as soybean breeders make advances to reduce or eliminate the linolenic acid content of soybean oil and to achieve a mid-oleic fatty acid composition similar to sunflower, while continuing to maintain yield and other important agronomic traits. Canola oil, with its low saturated fatty acid composition, will likely persist as a strong competitor to sunflower oil. Despite the continued competition from other vegetable oils, sunflower oil will continue to maintain an image advantage as a heart-healthy vegetable oil over other oils simply because of the visual impact of a beautiful, radiant and showy, bright yellow flower that appealed first to Native Americans, and has now been adopted and admired by the rest of the world for over five centuries.

References

Abdel-Wahab AM, Rhoden EE, Bonsi CK, Elashry MA, Megahed ShE, Baumy TY, El-Said MA (2005) Productivity of some sunflower hybrids grown on newly reclaimed sandy soils, as affected by irrigation regime and fertilization. Helia 28: 167–178.

Bauer A (2006) Pharma crops: state of field trials worldwide. Umweltinstitut München e.V.—Munich Environmental Institute. *http://umweltinstitut.org/download/ field_trials_engl_september06_01-2.pdf* (accessed September 8, 2009)

Beg A, Pourdad SS, Alipour S (2007) Row and plant spacing effects on agronomic performance of sunflower in warm and semi-cold areas of Iran. Helia 30(47): 99–104.

Davies FT, Puryear JD, Newton RJ, Egilla JN, Saraiva Grossi JA (2001) Mycorrhizal fungi enhance accumulation and tolerance of chromium in sunflower (*Helianthus annuus*). J Plant Physiol 158: 777–786.

Flagella Z, De Caterina R, Monteleone M, Giuzio L, Pompa M, Tarantino E, Rotunno T (2006) Potentials for sunflower cultivation for fuel production in Southern Italy. Helia 29: 81–88.

Gong Z, Wilke B-M, Kassem A, Li P, Zhou Q (2006) Removal of polycyclic aromatic hydrocarbons from manufactured gas plant-contaminated soils using sunflower oil: Laboratory column experiments. Chemosphere 62: 780–787.

Hahn V, Oechsner H, Ganssmann M (2006) Sunflower for biogas production. In: Proc 7th Eur Conf on Sunflower Biotechnol, Sept 3–6, 2006, Gengenbach, Germany, p 21.

Harter AV, Gardner KA, Falush D, Lentz DL, Bye RA, Rieseberg LH (2004) Origin of extant domesticated sunflowers in eastern North America. Nature 430: 201–205

Heiser CB Jr (1976) The Sunflower. Univ of Oklahoma Press, Norman, OK, USA, pp 29–46.

Kinman ML (1970) New developments in the USDA and state experiment station sunflower breeding programs. In: Proc 4th Int Sunflower Conf, June 23–25, 1970, Memphis, TN, USA, pp 181–183.

Kusnadi AR, Nikolov ZL, Howard JA (1997) Production of recombinant proteins in transgenic plants: practical considerations. Biotechnol Bioeng 56: 473–484.

Leclercq P (1966) Une stérilité male utilisable pour la production d'hybrides simples de tournesol. (In French.) Ann Amel Plant 16: 135–144.

Leclercq P (1969) Une stérilité mâle cytoplasmique chez le tournesol. Ann Amel Plant 19: 99–106.

Li Y-M, Chaney RL, Schneiter AA, Miller JF, Elias EM, Hammond JJ (1997) Screening for low grain cadmium phenotypes in sunflower, durum wheat and flax. Euphytica 94: 23–30.

Madejón P, Murillo JM, Marañón T, Cabrera F, Soriano MA (2003) Trace element and nutrient accumulation in sunflower plants two years alter the Aznalcóllar mine spill. Sci Total Environ 307: 239–257.

Mewett O, Johnson H, Holtzapffel R (2007) Plant molecular farming in Australia and overseas. Canberra, ACT, Australia: Bureau of Rural Sciences. *http://adl.brs.gov.au/ brsShop/html/brs_prod_90000003645.html* (accessed September 8, 2009).

National Biodiesel Board (2007) Commercial biodiesel production plants (as of September 8, 2009). *http://www.biodiesel.org* (accessed September 8, 2009).

Pakistan Federal Bureau of Statistics (2008) *http://www.statpak.gov.pk/depts/fbs/statistics/ agriculture_statistics/agriculture_statistics.html* (accessed September 8, 2009).

Peleman JD, van der Voort JR (2003) Breeding by design. Trends Plant Sci 8: 330–334.

Pourdad SS, Beg A (2008) Sunflower production: hybrids versus open pollinated varieties on dry land. Helia 31(48): 155–160.

Rifkin J (1998) The Biotech Century: Harnessing the Gene and Remaking the World. JP Tarcher/Putnam, New York, USA.

Ruf S, Karcher D, Bock R (2007) Determining the transgene containment level provided by chloroplast transformation. Proc Natl Acad Sci USA 104: 6998–7002.

Shah NA, Shah H, Akmal N (2005) Sunflower area and production variability in Pakistan: oppotunities and constraints. Helia 28: 165–178.

Soldatov KI (1976) Chemical mutagenesis in sunflower breeding. In: Proc 7th Int Sunflower Conf, June 27–July 3, 1976, Krasnodar, Russia, pp 352–357.

Svab Z, Maliga P (2007) Exceptional transmission of plastids and mitochondria from the transplastomic pollen parent and its impact on transgene containment. Proc Natl Acad Sci USA 104: 7003–7008.

Tanksley SD, Rick CM (1980) Isozymic gene linkage map of the tomato: Applications in genetics and breeding. Theor Appl Genet 57: 161–170.

US Department of Agriculture, National Agricultural Statistics Service (2007): *http:// www.nass.usda.gov/* (accessed September 8, 2009).

Van der Lelie D, Schwitzguébel JP, Glass DJ, Vangronsveld J, Baker A (2001) Assessing phytoremediation's progress in the United States and Europe. Environ Sci Technol 35: 446A–452A.

Vannozzi GP (2006) The perspectives of use of high oleic sunflower for oleochemistry and energy raws. Helia 29: 1–24.

Vasquez A, de Romano A (2006) Sunflower crop in Argentina to date. Helia 29: 159–164.

Wu CH, Wood TK, Mulchandani A, Chen W (2006) Engineering plant-microbe symbiosis for rhizoremediation of heavy metals. Appl Environ Microbiol 72: 1129–1134.

Zaid A, Hughes HG, Porceddu E, Nicholas F (2001) Glossary of Biotechnology for Food and Agriculture—A Revised and Augmented Edition of the Glossary of Biotechnology and Genetic Engineering. A FAO Research and Technology Paper ISSN 1020–0541. FAO of the United Nations. ISBN 92-5-104683-2.

Zheljazkov VD, Vick BA, Ebelhar MW, Buehring N, Baldwin BS, Astatkie T, Miller JF (2008) Yield, oil content, and composition of sunflower grown at multiple locations in Mississippi. Agron J 100: 635–642.

Subject Index

Professor Loren Rieseberg

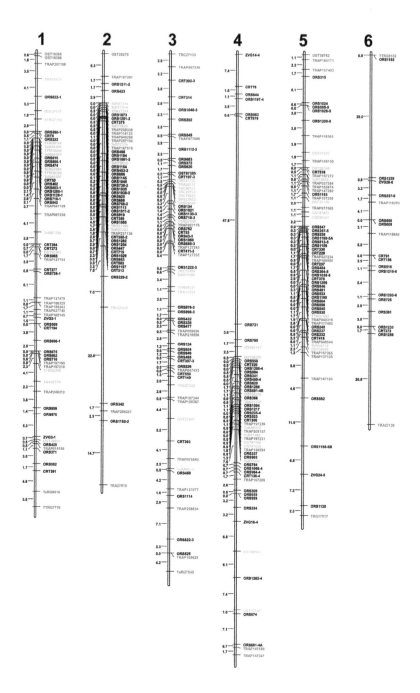

Figure 3-1

Figure 3-1 contd....

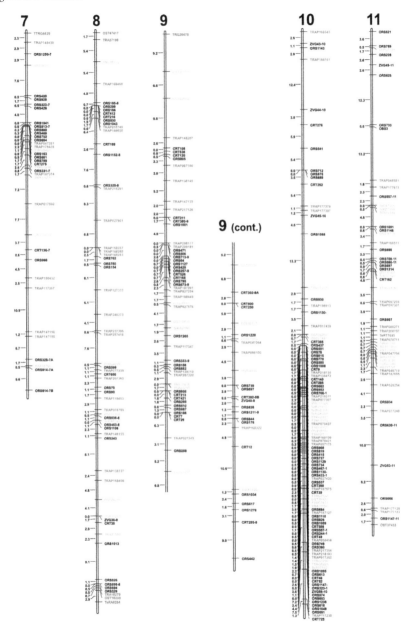

Figure 3-1 contd...

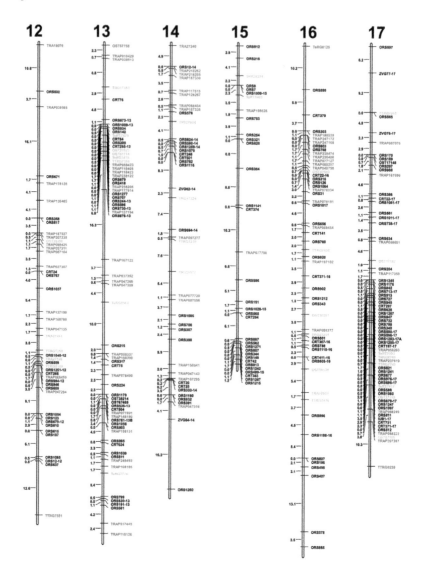

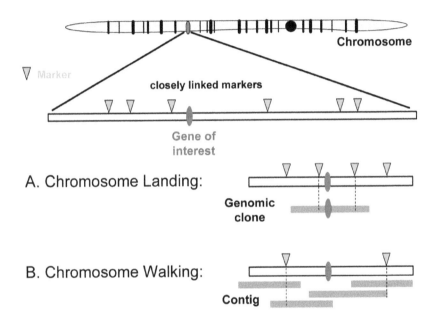

Figure 6-1 Map-based cloning approach. A, Chromosome Landing; B, Chromosome Walking.

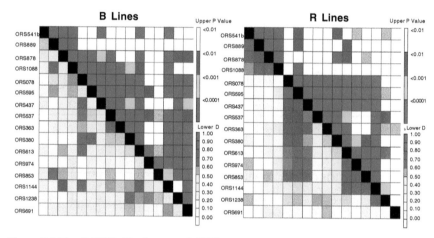

Figure 7-1 Two TASSEL (Bradbury et al. 2007) outputs, which were produced by analyzing the male and female lines separately from a total panel of 192 elite inbreds, genotyped with 16 public SSR markers on LG 10. Several regions of high LD, between tightly-linked SSR markers (as measured by the D'statistic—*shown below the black diagonal line*), can clearly be seen in the R lines (*shown in the right-hand panel*), but these regions are completely absent from the B lines (*shown in the left-hand panel*). The recessive branching gene (*b1*) is located in between ORS1088 and ORS437 and selection for this trait in elite R lines has resulted in a reduction of the allelic diversity at linked SSR loci (Combes and Berry, unpub data).

Milton Keynes UK
Ingram Content Group UK Ltd.
UKHW022045141024
449569UK00022B/811